Trogons and Quetzals of the World

Also by Paul A. Johnsgard

The Pheasants of the World: Biology and Natural History (2d ed. 1999;
  1st ed. 1986)
Earth, Water, and Sky: Stories and Sketches by a Naturalist (1999)
Baby Bird Portraits by George Miksch Sutton: Watercolors in the Field Museum
  (1998)
The Avian Brood Parasites: Deception at the Nest (1997)
The Hummingbirds of North America (2d ed. 1997; 1st ed. 1983)
Ruddy Ducks and Other Stifftails: Their Behavior and Biology
  (with M. Carbonell) (1996)
This Fragile Land: A Natural History of the Nebraska Sandhills (1995)
Arena Birds: Sexual Selection and Behavior (1994)
Cormorants, Darters, and Pelicans of the World (1993)
Ducks in the Wild: Conserving Waterfowl and Their Habitats (1992)
Bustards, Hemipodes, and Sandgrouse: Birds of Dry Places (1992)
Crane Music: The North American Cranes (1991)
Hawks, Eagles, and Falcons of North America: Biology and Natural History
  (1990)
Waterfowl of North America: The Complete Ducks, Geese, and Swans
  (with Robin Hill and S. D. Ripley) (1989)
North American Owls: Biology and Natural History (1988)
The Quails, Partridges, and Francolins of the World (1988)
Diving Birds of North America (1987)
Birds of the Rocky Mountains (1986)
Prairie Children, Mountain Dreams (1985)
The Platte: Channels in Time (1984)
The Cranes of the World (1983)
The Grouse of the World (1983)
Dragons and Unicorns: A Natural History (with Karin Johnsgard) (1982)
Teton Wildlife: Observations by a Naturalist (1982)
Those of the Grey Wind: The Sandhill Cranes (1981)
The Plovers, Sandpipers, and Snipes of the World (1981)
A Guide to North American Waterfowl (1979)
Birds of the Great Plains: Breeding Species and Their Distribution (1979)
Ducks, Geese, and Swans of the World (1978)
The Bird Decoy: An American Art Form (editor) (1976)
Waterfowl of North America (1975)
American Game Birds of Upland and Shoreline (1975)
Song of the North Wind: A Story of the Snow Goose (1974)
Grouse and Quails of North America (1973)
Animal Behavior (rev. ed. 1972; 1st ed. 1967)
Waterfowl: Their Biology and Natural History (1968)
Handbook of Waterfowl Behavior (1965)

TROGONS
AND
QUETZALS
OF THE WORLD

PAUL A. JOHNSGARD

SMITHSONIAN INSTITUTION PRESS
Washington and London

Paintings by Dana Gardner, Daniel Lane,
James D. McClelland, John O'Neill, and David Reiser.
Reproductions of hand-colored lithographs by John Gould

Copy editor and typesetter: Princeton Editorial Associates
Production editor: Ruth Spiegel
Designer: Janice Wheeler

The unidentified drawings are as follows: *p. iii,* golden-headed
quetzal; *p. 2,* resplendent quetzal; *p. 25,* red-naped trogon; *p. 28,*
narina trogon; *p. 44,* resplendent quetzal; *p. 56,* resplendent quetzals
(male and female); *p. 67,* golden-headed quetzal. All are by the
author, and all except those on p. 56 are males.

Library of Congress Cataloging-in-Publication Data

Johnsgard, Paul A.
    Trogons and quetzals of the world / Paul A. Johnsgard.
        p.   cm.
    Includes bibliographical references (p.   ).
    ISBN 1-56098-388-4 (alk. paper)
    1. Trogons.   2. Quetzals.   I. Title.
QL696.T7 J65 2000
598.7'3–dc21   99-047201

British Library Cataloguing-in-Publication Data available

Manufactured in the United States of America

15   03

⊗ The paper used in this publication meets the minimum
requirements of the American National Standard for Information
Sciences–Permanence of Paper for Printed Library Materials
ANSI Z39.48-1984.

TO ALEXANDER F. SKUTCH IN RECOGNITION
OF HIS LIFETIME OF STUDIES OF NEOTROPICAL BIRDS

CONTENTS

PREFACE

More than a century has passed since the last and only comprehensive summary of the trogons and quetzals of the family Trogonidae was published, despite the fact that one species of the family, the resplendent quetzal, has often been nominated as the most beautiful bird in the world and is one of the few bird species to have assumed god-like status, in the form of Quetzalcoatl, the mythic feathered god of the Nahautl-speaking natives of Mexico and Central America. Added to these aesthetic attractions is the fact that the trogonids are among the most typical and charismatic and yet most reclusive of the avifauna of the tropical forests of both the New World and the Old World, and their collective range perhaps defines the limits of the world's tropical and subtropical forests better than that of any other bird family. These birds also provide living barometers of the state of our tropical forests. In Central and South America, Africa, and Southeast Asia these ancient and incredibly diverse forests are rapidly disappearing before our eyes, and with them we are also losing many wonderful birds and uncountable numbers of other, less conspicuous animal and plant species. Deforestation and recent forest fires in Indonesia, Malaysia, and Central America have ravaged many prime trogonid habitats.

The rapid disappearance of habitat makes ours certainly the last generation of bird lovers and ornithologists who could, theoretically, observe and study all the world's trogonids. Yet, little to nothing is known of the biology of many of these wonderful species. This distressing situation convinced me that I should try to summarize the currently available information on the trogons and quetzals, if for no other reason than to provide a sad memorial to a rapidly disappearing avian family before opportunities to acquire any new information are forever gone. I am saddened by the number of times I had to enter "No information available" in the species accounts: for nearly one-quarter of all trogon species neither nest sites nor eggs have been discovered, to say nothing of other details of breeding biology, such as territoriality, ecology, mating systems, social behavior, and reproductive biology.

I have long believed that one of the most attractive of ornithologist John Gould's many bird family monographs is that on the family Trogonidae. This was his second family monograph, following very shortly after the first edition of his memorable work on the toucans (1833–1835). The first edition of his trogonid monograph appeared in multiple fascicles between 1835 and 1838, and in it he recognized some 34 species (one species called the giant trogon, *Trogon gigas*, later proved to be invalid). These species were illustrated in 36 hand-colored lithographic plates (executed by John Gould and his wife Elizabeth).

About two decades later (1858–1875) Gould produced a second edition of his trogonid monograph, in which he recognized 45 species. This edition contained 46 lithographic plates completely redrawn by Gould with the assistance of H. C. Richter and William Hart. Modern ornithologists have reduced Gould's 45 species to 34, with the remainder now considered to represent either geographic races of previously discovered species or invalid forms. Five additional species have been described since the second edition of Gould's great monograph, including 2 from Asia, 2 from Africa, and 1 from South America, for a total of 39 currently accepted species.

Like his other famous monographs, the two editions of Gould's trogonid monograph were published as highly limited printings, and it is unlikely that more than 250 bound copies of either were ever produced. In the 135 years since their publication, many of these copies have been accidentally damaged or destroyed or even disassembled to provide individual plates for resale and framing as fine art prints. As a result, intact copies of this monograph currently sell for at least $20,000. Only partial and reduced-scale reproductions of these sets of plates have been published, with rewritten texts (Lambourne 1992; Rutgers 1969, 1972).

Luckily, some museums and libraries have maintained intact sets of the Gould monographs, and I was lucky enough to have access to two such sets, namely, those at the zoology library of the Field Museum of Natural History in Chicago and at the Kenneth Spencer Research Library at the University of Kansas, Lawrence. It was while I was examining some Gould monographs in the Field Museum in conjunction with a hummingbird study that I decided that a new book on the trogons and quetzals might be a desirable and challenging project. Gould's trogon monograph, unlike several of his others, had not been reproduced; and no competing works had been published, this situation providing an additional reason for revisiting it. Examination of all the plates in both editions convinced me that those in the second edition were of uniformly high quality and could not easily be improved upon. Furthermore, for 11 of the currently recognized species of Trogonidae Gould had included more than one plate, giving me a selection from which to choose. I chose 35 of Gould's plates, one of each available species. Reproduction of these plates was greatly aided by Mr. Benjamin Williams of the Field Museum of Natural History and Mr. James Helyar of the Kenneth Spencer Research Library. (I should mention that soft-part colors shown in the Gould plates were based on prepared museum specimens and are often unreliable. Thus, where the text descriptions of soft-part colors differ from the plates, the former should be relied upon.)

In order to illustrate all 39 of the currently recognized trogon species in color, I was forced either to find other illustrations for reproduction or to commission entirely new plates. James D. McClelland painted one of the African species (the bare-cheeked trogon) using specimens kindly loaned me by the Field Museum of Natural History. The other African trogon (the bar-tailed trogon) and the single American species not illustrated by Gould (the white-eyed trogon) were splendidly painted for me by Daniel Lane of Louisiana State University. For one rare Asian trogon (Whitehead's trogon) I reproduced a color lithograph associated with the species' original description, as specimens were not readily available to me. For the other (Ward's trogon), David Reiser produced a fine oil painting in short order. Dana Gardner kindly allowed me to reproduce his fine painting of Diard's trogon based on studies from life, thus replacing a relatively poor Gould lithograph of this species. Dr. John O'Neill graciously allowed me to use a marvelous and evocative painting of the resplendent quetzal in a misty cloud forest, the painting supplementing the spectacular Gould lithograph of this species. Finally, Dr. Alexander Skutch kindly contributed several monochrome photographs of trogons and their Central American habitats. Lee Schoen of the Houston Zoological Gardens and John Sullivan loaned me transparencies of trogons for possible use in the book.

Writing this book has required the use of museum specimens in the many cases where little or no field information was available for a species. Thus, I have relied greatly on repeated access to the collections of the Field Museum of Natural History, through the kindness of Dr. David Willard, and the museum's associated library holdings, through the assistance of special collections librarian Benjamin Williams. Specimens for paintings were also borrowed from the field museum collections (abbreviated FMNH in text). Specimen records from the National Museum of Natural History in Washington, D.C., were obtained for me by Dr. Gary Graves and Craig Ludwig; and specimen records and egg measurements from the Museum of Vertebrate Zoology, Berkeley, California, came to me with the help of Drs. Ned Johnson and Carla Cicero. I also had access to some specimen records from the Natural History Museum of Louisiana State University (abbreviated LSU in text), thanks to Drs. John O'Neill, J. Van Remsen, and Steve Cardiff. Egg data from the files of the Western Foundation of Vertebrate Zoology were kindly extracted by Freda Kinoshita, and Rene Corado provided supplemental egg measurements. Information on Asian trogons was kindly offered by Dr. G.W.H. Davison and Dr. David Wells, the latter providing me with early access to materials from his then unpublished book on the birds of the Malay Peninsula.

Dr. Josef Kren helped me greatly with software and associated computer graphics.

The decorative drawings and distribution maps are my own. Distribution maps for the trogonids were generally rather difficult to construct, owing to the limited information available for most South American and Asian species. Maps provided by Howell and Webb (1995) for Mexico and northern Central America proved helpful for that region, as did the maps by Fry, Keith, and Urban (1988) for the African species. Trogons of the Thai-Malaysian region have recently been accurately mapped by Wells (1999), but ranges of the forms of the Sunda Islands are still only poorly documented.

I have used the names and spellings of political regions employed by Sibley and Monroe (1990, 1993), whose taxonomic sequence and views on specific and generic limits I have also generally, but not invariably, followed. Like Sibley and Moore and also de Schauensee, who has written the most generally accessible reference on Chinese birds (de Schauensee 1984), I have used the traditional and familiar spellings of Chinese political regions and localities, rather than the most recent and now officially recognized versions.

The help provided by the University of Nebraska librarians and the assistance of the staff of the Wilson Ornithological Society's Joslyn Van Tyne Memorial Library, Ann Arbor, Michigan, are gratefully acknowledged. I also was helped in my literature searches by A. T. Peters of the University of Kansas Natural History Museum, Lawrence, and by Ben Williams of the Field Museum of Natural History, Chicago. Unpublished information on the breeding of the golden-headed quetzal at the Houston Zoological Gardens was sent to me by Trey Todd, one-time supervisor of birds; and the current curator, Lee Schoen, provided personal assistance and information during my visit to the zoo. Photocopies of nesting information on many trogon species from the files of the Western Foundation of Vertebrate Zoology were provided by Dr. Lloyd Kiff; and comments on the white-eyed trogon and orange-breasted trogon were offered by Dr. Gary

Stiles. Sheri Williamson offered similar help for the eared and elegant trogons. Tape recordings of the New World trogons by J. W. Hardy, G. Reynard, and B. B. Coffey greatly assisted me in my attempt to describe the diverse utterances of these highly vocal species. I must also thank Peter Cannell of the Smithsonian Institution Press for encouraging me to take on the task of monographing this beautiful but little-known group of birds. Dr. Linnea Hall, California State University, provided helpful advice on the elegant trogon, critically reviewed the entire manuscript, and offered many constructive comments.

INTRODUCTION

If ever there existed an ultimate symbol of the beauty embodied by birds, the re-splendent quetzal would certainly qualify. Rare, stunningly beautiful, elusive, and limited to the remote and mist-draped cloud forests of Middle America, this bird, with its shimmering golden green and blood-red feathers, is an artist's dream. Most birders, if asked which tropical bird they would like to glimpse in its natural habitat sometime during their lifetime, would probably nominate the resplendent quetzal.

Modern bird lovers are not unique in treasuring the quetzal. To the pre-Colombian aborigines of Middle America the bird's glorious plumes were more highly treas-ured than gold, and only high priests and royalty were allowed to possess and wear them. Ornaments made of the incredibly long, golden-green tail-coverts of the male quetzal adorned the heads of these personages; like the equally coveted and similarly rare jade, these feathers are the color of maize leaves and all the other green vegetation that gives life and beauty to the earth.

Little wonder then that of all the deities of the Nahuatl-speaking tribes, Quetzal-coatl was preeminent. The name of this benevolent god comes from *quetzal* (repre-senting the sky); *co*, the snake (representing earth); and *atl* (denoting water), thus collectively comprising and representing a combination of the earth, the seas, and the heavens. This so-called plumed serpent is often represented in human form as having a headdress of flowers and feathers and holding in one hand a staff of life and in the other a spear with its point representing the morning star, where his heart resides. Perhaps initially based on an actual historic figure, the last hero-king of early Toltec culture, Quetzalcoatl was later gradually transformed into a major deity among the Aztecs of the high plateau of central Mexico. There he gradually metamorphosed into a wind god, presiding over earth, water, and sky, and a peace-loving guardian of arts, crafts, and music. Indeed, in a sixteenth-century Nahuatl manuscript, a poem describes how Quetzalcoatl (in his wind-god persona) stole Music and his attendant musicians from the possessive Sun, in order to bring them and their special treasures back to Earth (Nicholson 1967):

> Bearing them gently lest he should harm their tender melodies,
> the Wind with that tumult of happiness in his arms
> set out on his downward journey, generous and contented.
> Below, Earth raised its dark eyes to heaven and its great face shone,
> and it smiled.
> As the arms of the trees were uplifted,
> there greeted with Wind's wanderers the awakened voice of its people,

> the wings of the quetzal birds,
> the face of the flowers,
> and the cheeks of the fruit.
> When all that flutter of happiness landed on Earth,
> and the Sun's musicians spread to the four quarters,
> then Wind ceased his complaining and sang,
> caressing the valleys, the forests, and seas.
> Thus was Music born on the bosom of Earth.

In a similar way, birds once brought the first real music to our Earth, music that together with their great beauty and grace, we now all enjoy. It would be a much poorer and sadder world if the quetzal and all of its avian relatives were to disappear. Then Earth would become a more silent and fearful place, as it was before we received, at least in these mythic traditions, the gift of Music from Quetzalcoatl.

PART ONE

Comparative Biology

INTRODUCTION AND EVOLUTIONARY RELATIONSHIPS

The trogon and quetzal family (avian family Trogonidae, order Trogoniformes) is most readily characterized by its unique toe arrangement, in which both the first (hind) and second (normally inner front) toes are oriented posteriorly, so that two toes (the third and fourth) are in front and two are behind (see figure 1). This "heterodactyl" arrangement differentiates trogonids from all other birds (such as woodpeckers) with similar x-shaped, or "yoke-toed," toe orientations; woodpeckers have an analogous "zygodactyl" toe arrangement, wherein the first and fourth toes, rather than the first and second, are oriented posteriorly. The purpose of the heterodactyl arrangement is uncertain, but trogons often cling woodpecker-like to the sides of rotted trees as they excavate nesting cavities. The feet are relatively small, and the legs short; walking and hopping are rare. The two anterior toes of most trogonids are often united for as much as half their length (as in the kingfishers and all other members of the avian order Coraciiformes), but the three African trogon species show little or no toe fusion. Presumably trogons use this adaptation for digging or scooping during nest excavation, as do some kingfishers.

Most (25 of 39) of the extant trogonid species are distributed in the New World warm-temperate and tropic zones (collectively between southern Arizona and northern Argentina, plus 2 in the West Indies), but in addition to 3 African forms there are also 11 Asian trogon species (figure 2). These three well-isolated geographical groups are approximately centered around the equatorial forests of the Amazon and Congo Basins and around the tropical forests of Malaysia and Indonesia. Few other families of birds show such clearly pan-tropical distributions, but similar patterns occur in the anhingas (Anhingidae), jacanas (Jacanidae),

Figure 1. External topography of trogons, showing locations of some anatomical and structural terms mentioned in the text.

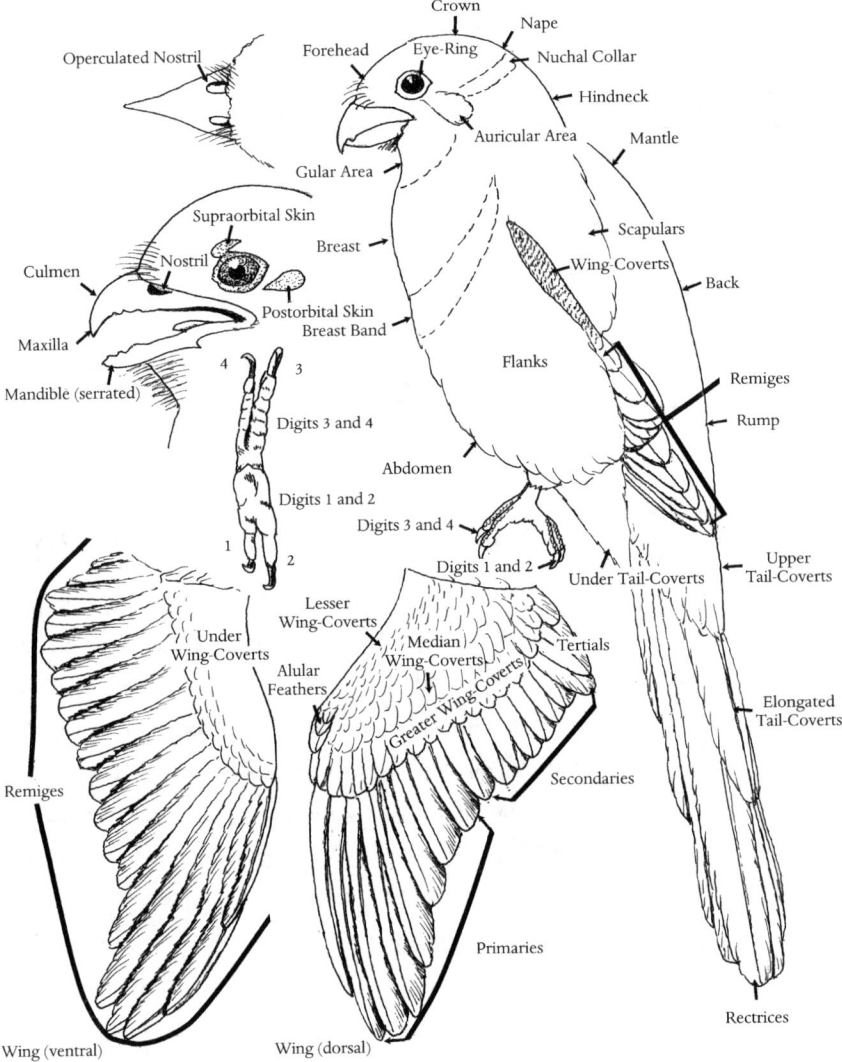

finfoots (Heliornithidae), parrots (Psittacidae), and barbets (Capitonidae). Furthermore, one nonparasitic subfamily (Phaenocophaeinae) of the cuckoo family (Cuculidae) exhibits a nearly identical overall distribution pattern.

The trogonid group may thus be described geographically as predominantly pan-tropical, occurring on almost all the major land masses around the equator. However, the 39 extant species collectively range in elevation from sea level to more than 3,500 m (11,500 ft) and are spread across nearly 35° of latitude on both sides of the equator. Interestingly,

resting metabolic rates of at least the tropically adapted black-throated trogon are quite low (Bennett and Harvey 1987), and this characteristic, if typical of the whole family, may place some limits on the climatic tolerances of the group. Altitudinal-latitudinal relationships are especially evident among the 23 species that occur on the mainland of Central and South America (figures 3–5). In the tropical and subtropical climatic zones, humid to wet evergreen forests at lower montane altitudes seem to hold the largest number of species; but especially toward the northern and southern limits of their range the

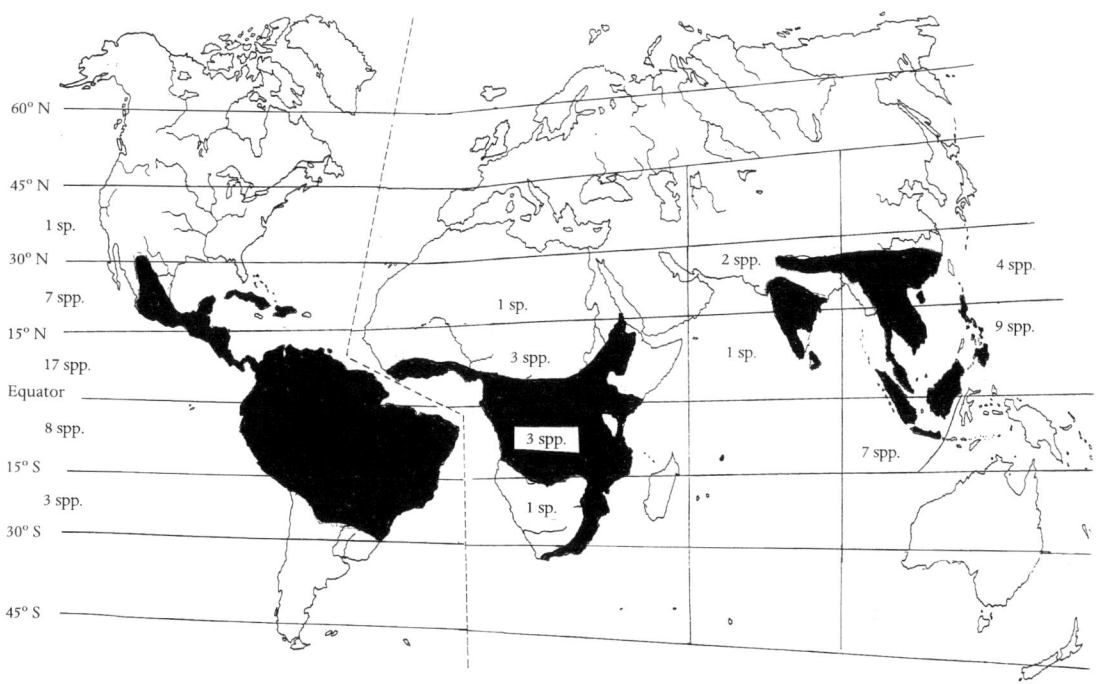

Figure 2. Species diversity of trogons by latitudinal zones in the Old World and New World. Inked areas show approximate collective limits of trogon ranges in each region.

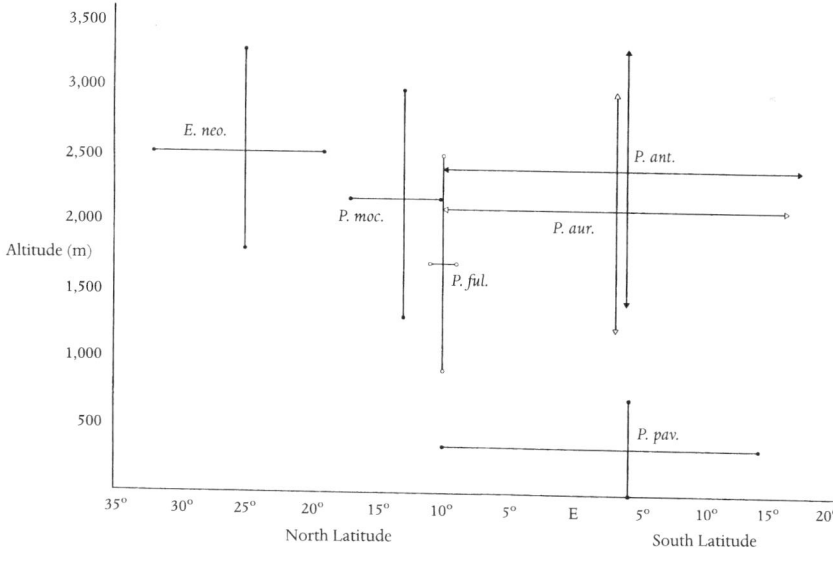

Figure 3. Altitudinal and latitudinal ranges of *Pharomachrus* and *Euptilotis* species in Central and South America. Abbreviations are as follows: *E. neo.* = *Euptilotis neoxenus*, *P. ant.* = *P. antisianus*, *P. aur.* = *P. auriceps*, *P. ful.* = *P. fulgidus*, *P. moc.* = *P. mocinno*, and *P. pav.* = *P. pavoninus*.

birds often extend into savanna, thorn forest, and other xeric woodland habitats. Interestingly, the five species of predominantly fruit-eating quetzals and the closely related eared trogon (genera *Pharomachrus* and *Euptilotis*) are mostly found in temperate high-

altitudes, in montane forests, for example (figure 3). The eight mostly intermediate-size species in the subgenera *Curucujus* and *Trogon* occupy intermediate altitudinal ranges and consume a mixture of fruits and insects (figure 4). Last, the nine relatively

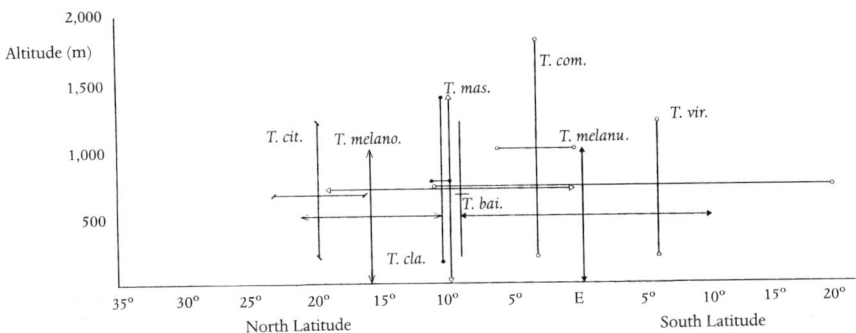

Figure 4. Altitudinal and latitudinal ranges of *Trogon* species (subgenera *Curucujus and Trogon*) in Central and South America. Abbreviations are as follows: *T. bai.* = *T. bairdii*, *T. cit.* = *T. citreolus*, *T. cla.* = *T. clathratus*, *T. com.* = *T. comptus*, *T. mas.* = *T. massena*, *T. melano.* = *T. melanocephalus*, *T. melanu.* = *T. melanurus*, and *T. vir.* = *T. viridis*.

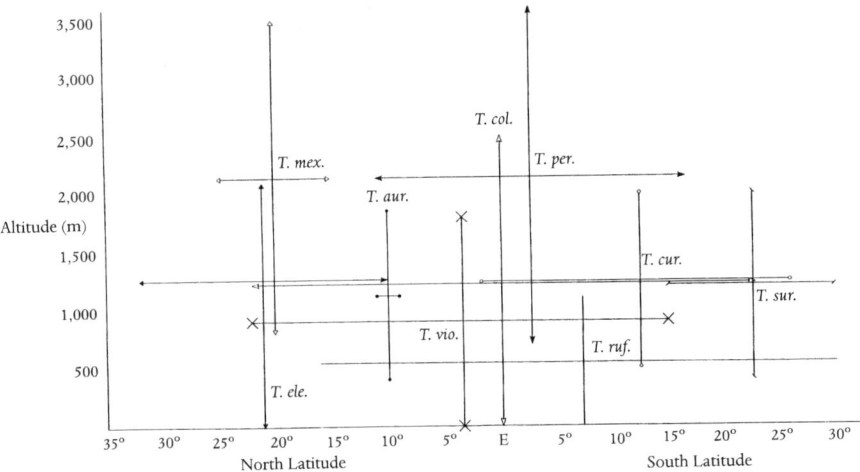

Figure 5. Altitudinal and latitudinal ranges of *Trogon* species (subgenus *Trogonurus*) in Central and South America. Abbreviations are as follows: *T. aur.* = *T. aurantiventris*, *T. col.* = *T. collaris*, *T. cur.* = *T. curucui*, *T. ele.* = *T. elegans*, *T. mex.* = *T. mexicanus*, *T. per.* = *T. personatus*, *T. ruf.* = *T. rufus*, *T. sur.* = *T. surrucura*, and *T. vio.* = *T. violaceus*.

small species in the subgenus *Trogonurus* tend to be low-altitude, tropical forms that eat a high proportion of insects and have proliferated into a considerable number of subspecies (figure 5). Small species might be expected to be the most cold-sensitive of all trogons, if indeed low metabolic rates are typical of the entire group.

Species richness among the trogonids is greatest in Central and South America, with Colombia and Ecuador having particularly high species diversity (table 1). In many ways, this pattern of species diversity parallels that found in the hummingbirds,

although the overall number of hummingbird species is far greater than that of the trogons (Johnsgard 1997). The trogonid family may once have had a much broader geographic range; fossils of the once-presumed ancestral family Archaetrogonidae described from early Cenozoic times (Upper Eocene to early Oligocene) have been found as far away as southern Europe (Brodkorb 1971). However, this fossil group lacked a trogon-like heterodactyl toe arrangement and is now regarded as caprimulgiform rather than trogoniform (Feduccia 1996). The first clear-cut trogonid fossils having heterodactyl

Table 1. Species Diversity of Trogons in the Western Hemisphere, by Country or Region

	Area (1,000 km^2)	Breeding Trogon Species (estimated)	Species Density (no. species/1,000 km^2)
United States	5,972[a]	1	0.0002
Cuba	114	1 (endemic)	0.009
Hispaniola	25	1 (endemic)	0.04
Mexico	1,976	9 (1 endemic)	0.0045
Guerrero	64	4	0.06
Campeche	51	4	0.08
Veracruz	72	6	0.08
Chiapas	74	7	0.09
Oaxaca	94	8	0.09
Belize	23	4	0.17
Guatemala	117	7	0.060
Honduras	153	7	0.046
El Salvador	34	6	0.18
Nicaragua	127	7	0.055
Costa Rica	42	10[b]	0.24
Panama	73	11[b]	0.15
Colombia	1,144	13	0.0114
Ecuador	276	12	0.043
Venezuela	915	10	0.011
Peru	1,253	10	0.0080
Bolivia	1,079	10	0.0093
Brazil	8,549	9	0.0011
Guyana	216	6	0.028
Suriname	140	5	0.036
French Guiana	91	5	0.05
Paraguay	408	3	0.007
Argentina	2,808	3	0.0010

[a]Excluding Alaska and Hawaii.

[b]Costa Rica and Panama share three endemics.

feet (*Protornis* and *Primotrogon*) date to the middle Oligocene (Olson 1975, 1985; Mayr 1999).

On the basis of molecular evidence (gene sequencing of mitochondrial cytochrome-*b* and ribosomal RNA extracts), Espinosa de los Monteros (1997, 1998) suggested that Africa is the ancestral home of the trogonids and that they later invaded Asia and, more recently, the New World. Espinosa de los Monteros (1998) also suggested, on the basis of gene divergence rates, that the trogons probably diverged from the typical coraciiform groups between 25.5 and 46.1 million years ago and that the splitting of the African trogons from the other surviving groups

occurred between 19.7 and 35.6 million years ago. Johansson (1998) has more recently suggested that the New World trogons separated from the Old World forms very early, perhaps during the splitting of Africa and South America in the late Jurassic, whereas the African and Asian trogon groups are phyletically "very close," judging from their gene sequences.

Trogonids are not strong fliers, and a long-distance invasion of the New World from Africa via the distances associated with the present-day Atlantic seems rather unlikely. The invasion of Asia poses less serious problems. Most probably the

Asian invasion occurred as a result of a gradual east-ward expansion out of Africa, an expansion similar to that perhaps experienced by the pheasants, which seemingly also originated in tropical Africa but are now centered in Southeast Asia (Johnsgard 1999). However, this mode of distribution would seem-ingly make the Asian trogons phyletically closer to the African trogons than the African trogons are to the New World trogons, a situation that is contrary to currently held opinions.

The kingfisher family (Alcedinidae) has a present-day distribution somewhat similar to that of the tro-gons, encompassing both the Eastern and Western Hemispheres; but kingfishers are not so clearly re-stricted to the tropical regions as are the trogons. The kingfisher family is generally thought to have originated in Southeast Asia, an area rich in king-fisher species diversity. Curiously, two other coraci-iform families (todies and motmots) are confined to the New World, whereas four other major coraci-iform groups (bee-eaters, rollers, hoopoes, and hornbills) are limited to the Old World.

Espinosa de los Monteros (1997) also believed that the African mousebird group (order Coli-iformes) represents the sister taxon of the trogons (order Trogoniformes) and that the trogons have additional affinities to the parrots (Psittaciformes) and the cuckoos (Cuculiformes). However, both Feduccia (1975) and Maurer and Raikow (1981) found sufficient evidence of coraciiform affinities to place the trogons within the order Coraciiformes. Using evidence from the anatomy of the middle ear bone (the columella), Feduccia erected a new order for this assemblage, which he called Alcediniformes and which included the trogonids (Trogonidae), motmots (Motmotidae), bee-eaters (Meropidae), and kingfishers (Alcedinidae). However, Maurer and Raikow (1981) placed this same assemblage of birds within the traditionally accepted order Coraci-iformes, but in a new subinfraorder Alcedinides, which also included the West Indian todies (family Todidae). Sibley and Alquist (1990) were unable to establish clear phyletic relationships for the trogons with limited blood-DNA samples. Using mitochon-drial DNA, Mindell et al. (1997) had a single species

of trogon (*T. melanurus*) for comparison with other avian orders; their parsimonious phylogram associ-ated the trogon with the owls.

Fry, Keith, and Urban (1988) recently added the family Trogonidae to the traditional order Coraci-iformes, sequentially placing this family immedi-ately before the kingfisher family (Alcedinidae). However, Sibley and Monroe (1990) recognized the trogonids as a separate order (Trogoniformes) and placed them adjacent to the Coraciiformes in a new superorder (Coraciimorphae). A comparison of traits of trogonids with those of mousebirds (Coli-iformes) and kingfisher-like birds (Alcediniodaea; kingfishers, motmots, and bee-eaters) is provided in table 2.

At the intrafamilial level, Espinosa de los Mon-teros (1998) recommended that two subfamilies of trogonids be recognized, with the single African genus *Apaloderma* constituting the subfamily Apalo-derminae. In the second subfamily (Trogoninae), he recommended that the Asian tribe Harpactinini (consisting of the single genus *Harpactes*) precede the New World tribe Trogonini. In this latter group he recognized only three genera: *Priotelus*, *Trogon*, and *Pharomachrus*, in that sequence, and placed the eared trogon within *Pharomachrus*. There seems little question that the eared trogon is closer to the quetzals than to the typical trogons and perhaps might better be called the eared quetzal (Howell and Webb 1995), although this suggestion has not yet been adopted by the American Ornithologists' Union (1998). The eared trogon's quetzal-like traits are most apparent in the iridescent condition of the upper wing-coverts of both sexes, as well as in the tail shape, vocalizations, and pale blue egg color.

The West Indian trogons pose additional taxo-nomic and zoogeographic problems. In their some-what elongated ear-coverts they remind one of the eared trogon, but otherwise they are most similar to various species in the subgenus *Trogonurus*. The in-cised tail feathers of adults of the Cuban species do not appear to warrant generic separation from the Hispaniolan form, but the former species is addi-tionally unusual in that no sexual dichromatism is

Table 2. Comparison of Trogoniformes Traits with Those of Coliiformes and Alcedinidae

	Coliiformes	Alcedinidae	Trogoniformes
Major foods	Fruit, leaves, seeds	Vertebrates, insects	Fruit, insects
Sociality	Gregarious	Nongregarious	Nongregarious
Allopreening	Present	Absent	Absent
Water sucking	Present	Absent	Absent
Clumped roosting	Present	Absent	Absent
Cere	Present	Absent	Absent
Stapes condition[a]	Primitive	Derived	Derived
Plantar tendons[b]	Unfused	Fused	Fused
Palate[b]	Desmognathous	Desmognathous	Schizognathous
Legs and toes	Long	Short	Short
Toe condition	Toes 1 and 4 posterior	Toes not reversible	Toes 1 and 2 posterior
Toe fusion	Absent	All syndactyl	Variably syndactyl
Adult plumage	Hair-like	Coherent	Coherent
Feather tracts	Lacking	Present	Present
Iridescence	Absent	Present in most spp.	Present in most spp.
Feather carotinoids	Absent	Rare	Common
Mating system	Monogamous	Monogamous	Monogamous
Territoriality	Group	Individual in most	Individual
Communal laying	Present in some	Absent	Absent
Cooperative nesting	Present in some	Present in some	Absent
Nest	Open, shallow cup	Excavated cavity	Excavated cavity
Nest substrate	Tree or bush	Earth, wood, termite nest	Wood; termite or wasp nest
Nest lining	Present	Absent	Absent
Egg color	White or spotted	White	White or pale-colored
Egg texture	Rough, thick	Smooth, thin	Smooth, thin
Chicks	Semiprecocial	Altricial	Altricial
Natal down	Sparse or absent	Present in one species	Absent
Tarsi callused	No	Yes	Yes
"Hedgehog" stage[c]	No	Yes	Yes
Nest sanitation	Present	Absent	Absent

[a] See Feduccia 1975.

[b] See Sibley and Alquist 1973.

[c] Nestlings with greatly lengthened pin-feather phase prior to juvenal plumage.

present among adults. It seems most conservative to imagine that the West Indies (Cuba, Hispaniola) were invaded by an ancestral trogon via a relatively short flight from the Central American mainland. Even today that distance is not very great, and it would have been shorter during glacial periods, when ocean levels were lower. Bond (1948) suggested that Neotropical avian groups such as the trogons may have reached the West Indies either via Jamaica (by means of fairly short hops from the Honduran mainland during a period of lowered ocean levels)

or via the Lesser Antilles (by means of island hopping). Faaborg (1985) mapped approximate land boundaries in the Caribbean during Pleistocene times, showing that the water gap between the Yucatan Peninsula and Cuba was about a quarter less than it is at present, and he holds that this pattern represents an easier dispersal route to Cuba than the one via Jamaica or the Lesser Antilles.

This hypothetical invasion pattern in the West Indies perhaps occurred at about the same time that the Greater Sunda Islands of Southeast Asia were part of a wholly exposed Sunda Platform, which would have facilitated an easy trogonid expansion from the Malaysian mainland. The scanty fossil record of trogons provides no assistance in understanding these zoogeographic puzzles.

In this work I have followed the taxonomic sequence suggested by Sibley and Monroe (1990) and most other authorities. However, I believe that the two major modifications proposed by Espinosa de los Monteros (1998), namely, the merger of *Euptilotis* with *Pharomachrus* and the listing of the Harpactinini before rather than after the Trogonini, are

warranted. It seems likely that these changes will eventually be adopted.

ANATOMY AND MORPHOLOGY

The trogons and quetzals may be anatomically characterized as having short, basally broadened bills that are partially concealed by bristly, forward-directed feathers; weak feet; and short, partially feathered tarsi (see figures 6–8). In most trogonid species the cutting edges of the mandible and maxilla are somewhat serrated; the serrations probably

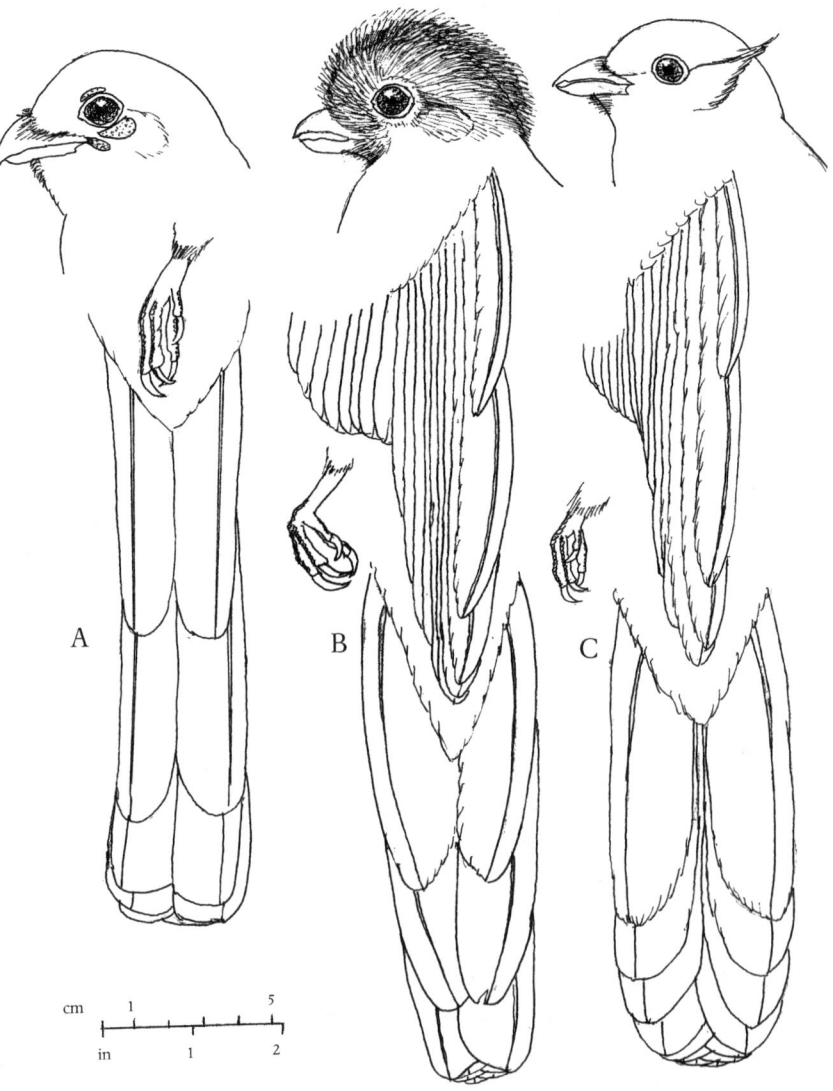

Figure 6. Head, tail, and foot characteristics of the Narina trogon (A), resplendent quetzal (B), and eared trogon (C). Drawn to scale. In part after Ridgway 1911.

Figure 7. Head and tail characteristics of the Cuban trogon (A), Hispaniolan trogon (B), slaty-tailed trogon (C), and black-headed trogon (D). Drawn to scale. In part after Ridgway 1911.

serve as an aid in securely holding live prey or large fruit and, along with the decurved tip of the bill, are undoubtedly useful in cutting food items into smaller pieces. The base of the bill is unusually wide, providing a large gape in relation to bill length (a nearly one-to-one ratio), even in very young nestlings, and allowing for the ingestion of surprisingly large food items.

The muscles of the legs and feet are notably small among trogonids, representing only about 3 percent of the body mass (Hartman 1961; Snow and Snow 1988). Trogons can, however, walk on the ground fairly well, and nestlings of the elegant trogon are surprisingly adept at climbing the vertical inner walls of their nest chamber, even as early as 2 weeks following hatching (Taylor 1994).

The wings of trogons are fairly short and are rounded in outline, and the ten primary flight feathers are strongly decurved, with the sixth or seventh primary the longest and the outermost (tenth) very short. There are eight to ten typical secondaries, which are also rather short; if one counts the even shorter tertials, ten to twelve such feathers are present. In spite of trogons' rather short wings and generally short flights, some species, such as the elegant trogon, maintain surprisingly large territories of about a kilometer in length (Kunzmann, Hall, and Johnson 1998; Taylor 1994). Their flight muscles are large, representing about 20 percent of their total body weight; and their heart is also relatively large (Hartman 1961). Their wings are deeply slotted, and the birds can rise to a stall to snatch a food item at

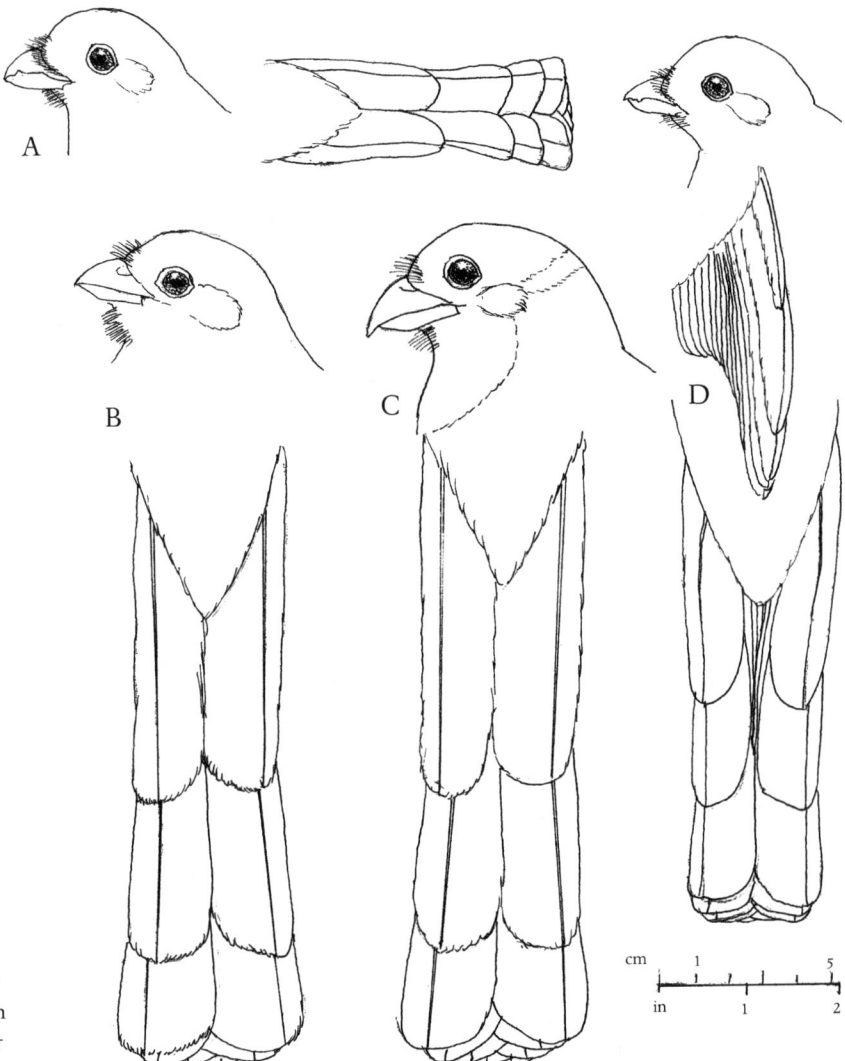

Figure 8. Head and tail characteristics of the violaceous trogon (A), red-headed trogon (B), red-naped trogon (C), and elegant trogon (D). Drawn to scale. In part after Ridgway 1911.

the stalling point while still maintaining flight control (Moermond and Denslow 1985).

The twelve rectrices are relatively long and broad and are often distinctly squared-off (truncated) at their tips, especially in adults. The long tail is believed to be important in braking and steering while taking food items in flight (Moermond and Denslow 1985). The six central rectrices are of similar length and are generally iridescent green to blue above and dark brown or black below. The edges of the outer rectrices are essentially parallel among most of the typical trogons but are more tapered and curved in

the quetzals and the eared trogon. The outer three pairs of rectrices are progressively shorter toward the outside and are often white, white-tipped, or variously freckled, spotted, barred, or otherwise patterned with black and white. Wavy black-and-white (vermiculated, or finely barred) patterning is present on the upper wing-coverts of most trogonid species except for the quetzals, the eared trogon, and the "dark-winged" trogons (subgenus *Trogon*). At least in some species, such as the elegant trogon (and probably others), the tails of yearling birds average about 6 percent longer than those of adults (Kunzmann, Hall, and Johnson 1998). The tails of young birds also tend to be narrower and more rounded at their tips than those of adults and may have patterning that differs from that typical of adults. The occurrence of longer tail feathers in young birds seems unusual, but this phenomenon also occurs in some other avian groups, including hawks.

Adult male New World and African trogons are usually iridescent green, blue, or violet above and are among the most brilliantly arrayed of all birds. The structural basis for this high degree of iridescence bears some microscopic similarities to that of the hummingbirds, shining starlings (*Lamprotornis*), and monal pheasants (*Lophophorus*; Dorst 1951; Durrer and Villiger 1966). Like the feathers of these species, the most-iridescent trogon feathers are modified in such a way that few or no hook-like barbicels are present on the barbules, a modification that tends to destroy the cohesiveness of the feather vanes and make them more airy. The trogons with the least highly modified feathers as to their iridescence are the African and Asian trogons, which have feathers with relatively large, air-filled rods of melanin separated by thick keratin layers. In *Priotelus*, and more especially in *Trogon*, the keratin layer is thinner and the rods are smaller, more numerous, and displaced toward the barbicel surface. These structural differences affect the light-interference effects and the resultant feather colors (Dorst 1951; Durrer and Villiger 1966). In both of these genera, the rod structure allows light to pass into the feather at a 30° angle, which causes the green portion of the visible spectrum to be refracted and reflected back and produces a perceived pure spectral or iridescent color. In the quetzals (as well as the eared trogon) the melanin rods are smaller and more linear than in the other trogons, producing almost continuous and uniformly stacked melanin platelets somewhat similar to those of hummingbirds. The intense golden green of the resplendent quetzal's feathers thus depends on light-interference effects produced by partially air-filled and flattened melanin platelets separated by a keratin matrix and uniformly spaced about 5,400 angstroms apart, a distance that corresponds to the wavelength of

green light (LaBastille, Allen, and Durrell 1972). In the pavonine quetzal the melanin granules are somewhat more rod-like, with little or no keratin matrix separating them. The result is a less intense iridescence.

Most trogons and quetzals are also brightly pigmented with lipochrome carotenoid pigments (e.g., cantaxanthins) of red, pink, orange, or yellow below and are sometimes reddish dorsally as well. Although the iridescent upperparts are seen to best advantage when illuminated by direct overhead sunlight, the red to yellow carotenoid underparts do not require such directed light to maintain their brilliance and thus are effective in the diffuse light of the understory. Asian species (with one exception) usually have non-iridescent adult plumages, and instead red or reddish-brown hues predominate dorsally. The intensities and sometimes the hues of the red to yellow carotenoid pigments are quite variable, and museum specimens are prone to fading under prolonged exposure to light, especially ultraviolet light. Espinosa de los Monteros (1998) concluded that in trogons yellow carotenoids represent a derived condition relative to red.

The contour feathers of trogons typically are closely spaced and have well-developed (long and narrow) aftershafts, but trogon feathers are unusually easy to detach, perhaps because the skin of trogons is extremely thin. Nestling down is either absent or at most highly transient and confined to the tips of the expanding contour feathers; and adult down feathers are likewise either sparse or entirely absent. The oil gland (preen gland) of trogonids is bare of feathers and at least in the resplendent quetzal is so small as to be seemingly nonfunctional. In that species, as well as in at least some other trogonids, there is a narrow band of bare skin around the neck, just below the head (Wetmore 1968).

Juveniles and young immatures usually have female-like plumages but are variously spotted or edged with white or buffy markings on the wing-coverts and sometimes the remiges and are generally more dull colored than adults. At least some and perhaps most trogon species are believed to assume their initial adult (definitive) plumage before the end of the first year, when breeding probably becomes possible (Dickey and van Rossem 1938). However, some species, such as the black-throated trogon and perhaps all quetzals, probably require two years to attain the definitive breeding plumage and reproductive condition (Skutch 1983). In the resplendent quetzal, at least, the extremely elongated and ornamental tail-coverts may not reach their maximum length until the third year, but this is probably an exceptional case.

The plumage that follows the juvenal plumage is usually called the first nonbreeding (or first basic) plumage. By the following breeding season, the birds are usually in their first nuptial plumage (also termed the first alternate plumage, which they reportedly acquire by means of a partial prenuptial (or first pre-alternate) molt. They acquire their subsequent nonbreeding plumages (which in trogons closely resemble the breeding plumages) by means of a complete postnuptial (or prebasic) molts. (Molting of the primaries occurs in the usual outward sequence, but tail molt sequences seem to be rather irregular.) The usual pattern of adult molt seems to involve a complete molt of body, wing, and tail feathers shortly after breeding and, several months later, a partial molt of some of the anterior body and head feathers prior to the next breeding season (Dickey and van Rossem 1938; Oberholser 1974). However, these two molts do not produce obvious seasonal changes in overall appearance.

Bill colors of adult males vary considerably among species and may be red, yellow, bluish, or grayish black; females and immatures have variably duller bills. The eyes of trogons are relatively large, perhaps reflecting their usually rather dark environs. The iris is usually brown among both adults and young but is sometimes ivory-white. Colorful eyerings (yellow, red, blue, and so on) occur in adults of most species but not among those having white irises. Toe and tarsal colors are generally not pronounced. Sexual dichromatism among adults may be slight or moderate (but is entirely lacking in the Cuban trogon).

Sexual dimorphism in adult body mass is negligible, with females averaging very slightly (but prob-ably not significantly) larger in a few instances, although males of at least species such as the elegant trogon seem to have slightly wider gapes (Kunzmann, Hall, and Johnson 1998).

The syrinx of trogons is tracheo-bronchial in location, as it is in most coraciiform birds; its musculature has apparently not been described. During calling, the throat may expand considerably, presumably through air-sac inflation; and calling may expose bare and colorful throat skin, at least in the African species. Males of many species also raise the bill almost to the vertical when uttering loud advertisement calls, the neck stretching perhaps serving to modify tension on the syringeal membranes.

The tongue is typically short and triangular in shape, with backwards-pointing projections that probably aid in holding and swallowing prey. In at least the resplendent quetzal, and probably in all other species, a crop is lacking, but the esophagus is muscular and capable of regurgitating surprisingly large undigested seeds (Wheelwright 1983). These seeds may be nearly 20 mm (0.8 in) wide and 35 mm (1.4 in) long (Skutch 1983). The gizzard is large, and at least in the predominantly fruit-eating species intestinal caeca are well developed, suggesting that some digestion by fermentation may occur (see the next section).

GENERAL BEHAVIOR AND ECOLOGY

Trogons and quetzals are highly arboreal, often sitting motionless and silent on branches for prolonged periods. They forage mostly on the fruits of trees (quetzals) or on both fruits and insects (trogons), often plucking insects, fruits, or berries from vegetation while in flight or capturing insects from the air (hawking). Small vertebrates such as lizards or frogs are taken on occasion, especially when the birds are capturing food for well-grown chicks.

Caterpillars, especially smooth-skinned ones, seem to be a favorite food of the more insectivorous species and are often fed to young. Other commonly consumed prey include stick insects, locusts, cicadas, and beetles. Most of these prey species are fairly large and slow moving and are generally not proficient fliers. Some, such as the stick insects and various locusts, are well camouflaged, many of them leaf-green or twig-brown, attesting to the excellent color vision and visual acuity of trogons. Adult moths and butterflies as well as dragonflies seem to be eaten only rarely, and hairy or spiny caterpillars are also taken only infrequently. The bristle-like feathers that extend forward from around the base of the bill probably serve as a funnel for catching flying insects, and the gape of nestling chicks is remarkably wide, sometimes even wider than that of the adults in the case of well-grown young.

Remsen, Hyde, and Chapman (1993) examined 246 trogonid stomachs involving 17 species, and of these about 41 percent contained arthropods only, 24 percent contained mixed fruit and arthropods, 33 percent contained fruit only, and 2 percent contained unidentified materials. Drupes, berries, and arillate fruits are all eaten by quetzals and are certainly their primary foods. The four species of quetzals (19 specimens) examined by Remsen, Hyde, and Chapman had much higher percentages of fruit present than did the typical trogons, and in general the larger species of trogons showed a higher proportion of stomachs containing fruit than was true of smaller species. These investigators also found that quetzals have bills that are relatively flatter than those of the more insectivorous species, a trend that is also typical of other frugivorous groups of birds.

Flights by trogons tend to be rapid and undulating but are usually not lengthy. Short-range movements and limited seasonal migrations are typical of these generally tropical birds; some montane species, such as quetzals, exhibit rather substantial altitudinal migrations between breeding and nonbreeding seasons.

Trogons only rarely descend to the ground, and their small leg muscles seemingly make them poorly adapted to reaching for fruit or other food while perched. Preening occurs frequently, but mutual preening (allopreening) has not yet been reported for trogons. Bathing in water has been observed occasionally in young trogons and less frequently in adults. Drinking has been observed only rarely

(Kunzmann, Hall, and Johnson 1998), so a physiological dependence on surface water is doubtful.

Taylor (1994) reported several interesting cases of trogon memory in the elegant trogon. Not only did he observe repeated use of the same nest site (nest-site fidelity) in these migratory birds, but he also observed yearly birds returning to the nest where they were reared and trying to claim it for their own. The birds also gradually habituate to (learn to ignore) tape recordings of their species' calls when repeatedly exposed to them.

Vocalizations are evidently important in social interactions and tend to be loud, repetitive, and acoustically simple in both sexes (figure 9). Advertisement calling (or singing) by the male often occurs during morning and evening at territorial boundaries. It is questionable whether any trogon vocalizations qualify as songs in terms of their acoustic complexity alone, but in at least some of the other usual attributes of song (species- and sex-specificity, temporal association with pair formation or breeding, presumed hormonal controls) some of the utterances of male trogons seem to qualify as such. Male "songs" typically consist of multiple repetitions (sometimes as many as 100 or more) of the same note at fairly regular intervals; there is little change in pitch, but sometimes the series speeds up or becomes louder or softer toward the end. In larger species these vocalizations are generally lower in pitch, louder, more resonant, and often more wooden in quality, sometimes sounding rather cuckoo-like or rail-like. In smaller species these utterances become faster, higher, more rapidly repeated, and more musical.

Other trogon vocalizations, which seem to quality both acoustically and functionally as calls, are often seemingly associated with disturbances and typically are brief and rapid chattering or chirring sounds that sound rather woodpecker-like or kingfisher-like. Squealing notes of uncertain function occur in at least one species (the eared trogon). Mechanically produced sounds are seemingly lacking as social signals, but hissing and bill-snapping noises are produced by the nestlings of some species. Group calling by males has been reported in a few

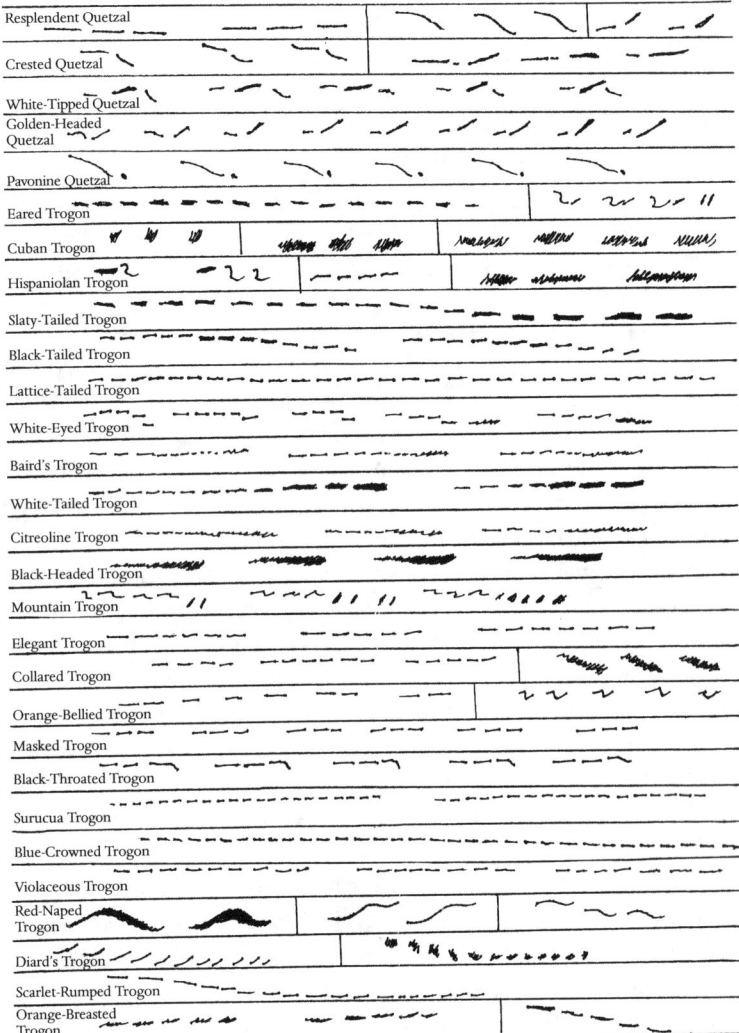

Figure 9. Sonosketches of vocalizations of trogons, based on subjective interpretations. Pitch (vertical axis) and duration (horizontal axis) are not drawn to scale. Vocalizations of Asian species after MacKinnon and Phillipps 1993.

species (see part 2); these calls apparently serve neither as territorial proclamations nor as specific courtship competitions. Exhibition of the tail by repeated spreading and closing, as well as tail raising and lowering, may also be a significant visual display during vocalizing in at least some species. Many if not most of the undertail patterns of adult male trogons are species specific (see figures 10 and 11). In this general way the birds resemble many other rather vocal and long-tailed species having distinctive voices and tail patterning, such as doves and cuckoos.

In many trogon species the undertail feather pattern (produced mainly by the three outer pairs of

rectrices) is also sex specific and probably supplements similarly specific male vocal signals. Compared with males, females are less vocal, especially outside the breeding season; but after pairing has occurred females may participate in advertising vocalizations, sometimes by duetting. Courtship chases are certainly a part of the prebreeding activity of many trogons, as is singing by the male while perched near a prospective nest site. Aerial display flights, accompanied by calling of the male as he flies above the forest canopy, have been described for the resplendent quetzal (LaBastille, Allen, and Durrell 1974; Skutch 1983); and male display flights have also been seen in

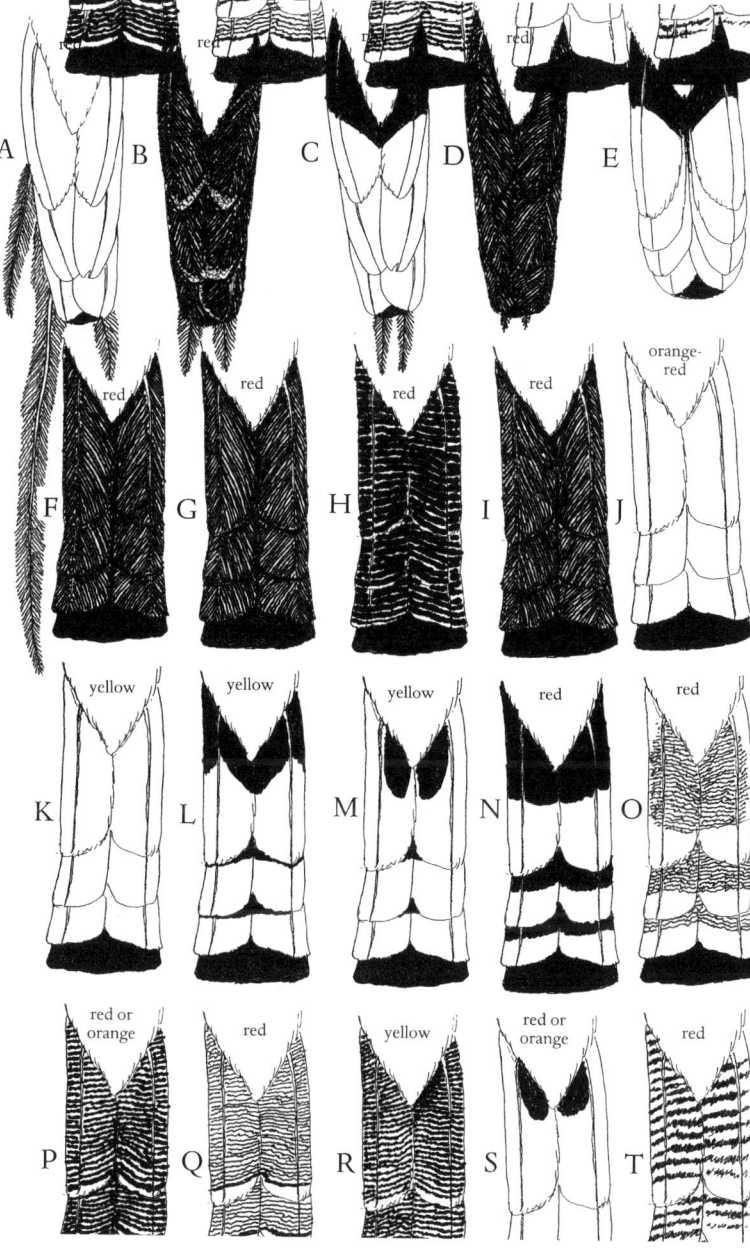

Figure 10. Undertails of some adult male New World trogonids, including *Pharomachrus mocinno* (left) and *P. antisianus* (right) (A), *P. auriceps* (B), *P. fulgidus* (C), *P. pavoninus* (D), *Euptilotis neoxenus* (E), *T. massena* (F), *T. melanurus* (G), *T. clathratus* (H), *T. comptus* (I), *T. bairdii* (J), *T. viridis* (K), *T. citreolus* (L), *T. melanocephalus* (M), *T. mexicanus* (N), *T. elegans* (O), *T. collaris* and *T. aurantiventris* (P), *T. personatus* (Q), *T. rufus* (left) and *T. violaceus* (right) (R), *T. surrucura* (S), and *T. curucui* (T). Not drawn to exact scale.

the golden-headed quetzal (Lee Schoen, pers. comm. with author). Similar display flights have not yet been described in other trogon species. However, details of both vocal and visual displays are woefully limited or lacking for nearly all trogons.

One of the most remarkable aspects of trogon plumages is the high level of adult plumage iridescence and, in most, the usually moderate degree of sexual dichromatism, which exists in spite of a seemingly obligatory monogamy (except, as we

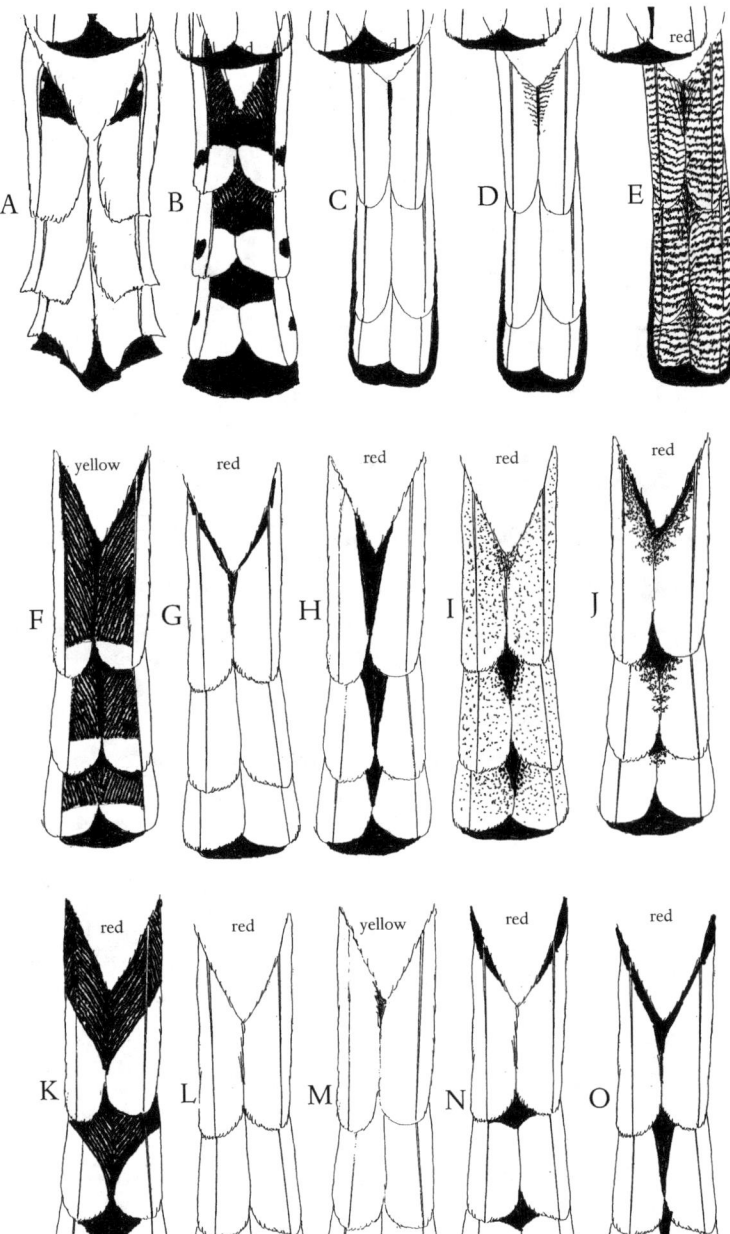

Figure 11. Undertails of some adult male West Indian and Old World trogonids, including *P. temnurus* (A), *P. roseigaster* (B), *A. narina* (C), *A. aequatoriale* (D), *A. vittatum* (E), *H. reinwardti* (F), *H. fasciatus* (G), *H. kasumba* (H), *H. diardii* (I), *H. ardens* (J), *H. whiteheadi* (K), *H. duvaucelii* (L), *H. orrhophaeus* and *H. oreskios* (M), *H. erythrocephalus* (N), and *H. wardi* (O). Not drawn to exact scale.

have seen, in the Cuban trogon). Many other groups with comparably bright male plumages (hummingbirds, pheasants, and so on) have nonmonogamous mating systems and correspondingly high sexual-selection pressures that help explain these features.

It is true that many species of bee-eaters, motmots, and kingfishers are also monogamous and quite brightly colored, but few if any exhibit the degree of sexual dichromatism that most quetzals and trogons do. Like the trogonids, however, the young of the former groups are raised to fledging in rather dark cavities, and they often acquire some iridescent feather colors with the acquisition of their juvenal plumages. Skutch (1987) suggested that the brilliant coloration of trogonids and other frugivorous and nectarivorous birds is related to their foraging needs, the birds initially becoming highly color sensitive as an adaptation for locating colorful fruits and flowers. Later, they employed such coloration as a social and sexual signal: more colorful birds may be preferred as mates (sexual selection), and bright colors may prevent mismatings with other species (species recognition mechanisms). Thus, in Skutch's view, monogamous but frugivorous or nectarivorous groups, such as tanagers and sunbirds, are more inclined to become brightly plumaged than, for example, are seed and grain eaters, such as finches.

Among the twenty-five New World species of trogonids analyzed ecologically by Stotz et al. (1996), thirteen utilize lowland tropical evergreen forests, nine use montane evergreen forests, at least eight use secondary forests, four use pine forests, three use tropical deciduous forests, three use gallery forests, and two use pine-oak forests. All species confined to a single forest type use either lowland tropical evergreen forests or montane evergreen forests. Eight or nine species occur in two forest types, three in three types, and three in four types. Most of the species were classified as having a medium sensitivity to disturbance, but three were judged to have a low level of sensitivity. All are mid-understory to canopy-level foragers.

BREEDING BIOLOGY AND POPULATIONS

Pairing among trogons and quetzals is monogamous, and presumably (but not certainly) pair bonds are held indefinitely, at least for species with prolonged breeding seasons. However, some species seem to be solitary and quiet outside the breeding season, with no clear indication of continued pair bonding. For many tropical species that occur primarily in wet habitats such as rain forests or cloud forests, breeding seems to be centered on the dry season,

which is when fruits ripen and other foods, such as insects, are also abundant. Spring or summer breeding occurs in species that range into latitudes that are distinctly more temperate. Wet-season breeding occurs in those forms that extend into rather arid habitats; this wetter period often occurs during spring or summer months. Of the eighteen species of New World trogonids breeding north of the equator, most (ten species) have breeding times centering on May, but the range is from March (one species) to August (one species). For two species breeding in temperate South America, the probable center of the breeding period is December.

Assemblages of singing males have been observed in several species, especially among two of the African trogons. Although females may be attracted to such groups, trogons do not appear to use the groups predominantly as pair-forming competitions or as precopulatory mechanisms, as classic lek-forming species do. Even already paired males have been observed to leave their territories to participate in these group activities, making their possible function even more uncertain (Brosset 1983). Pair-forming, pair-maintaining, and copulatory behaviors of trogons are all poorly understood and, for most species, remain completely undescribed.

In one of the important early phases of breeding behavior, the male locates a suitable nest site, begins to excavate it if needed, and starts advertising his and its location by singing persistently. Besides using self-excavated cavities in rotting wood, trogons may also excavate nests in arboreal termitaria or in vespiaries made by paper wasps. Sometimes old woodpecker holes may be claimed and used with little or no modification (Skutch 1976). The nest may even be enclosed among a mass of epiphytic roots or at times may consist of a well-enclosed chamber with a tunnel-like entrance. Skutch (1983) has suggested that there are two basic types of excavated wood cavities, classification being based on the relative depth of the cavity and the resultant degree to which the incubating or brooding bird is hidden from view from outside (see figure 12). Although earthen nest cavities comparable to those of many kingfishers have not been reliably reported, they

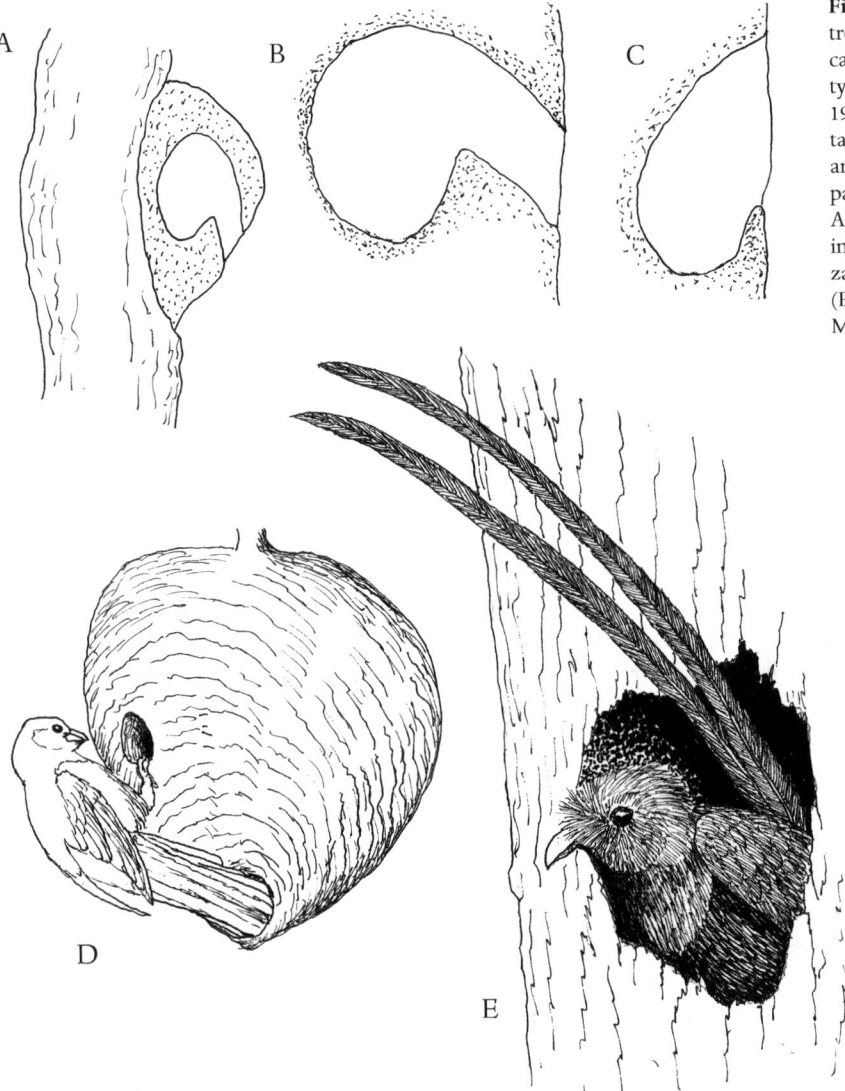

Figure 12. Nest types of trogons, including deep excavation type (A) and niche type (B), both after Skutch 1983; excavation in termitarium (C), after Sick 1993; and in vespiary (D), after a painting by D. Eckelberry. Also shown is an incubating male resplendent quetzal in a niche-type tree nest (E), after a photograph by Marco Saborio.

have been suggested for at least one Asian trogon species. Reuse of the same nest site in successive or in later but nonsuccessive years has been reported for some New World species, such as the eared trogon and the elegant trogon (Taylor 1994). However, in many tropical situations nests are built in trees that are so rotted that they are prone to collapsing between successive nesting seasons. In addition, the lack of nest sanitation often leaves the nest bottom a mass of feces, regurgitated seeds or undigested food remains, and maggots. Competition for suitable

sites with other hole-nesting species is sometimes severe; competitors include small hole-nesting owls, woodpeckers, other trogonids, and various cavity-nesting passerines (Taylor 1994).

Although territoriality is unstudied for most species, the defended area of the elegant trogon seems to include the entire home range of the pair. In Arizona the home ranges of males varied from 76–205 hectares (188–506 acres), averaging 123 hectares (304 acres), and were about a kilometer in length, although in some locations a pair may advertise over a distance of 1.6 km (1.0 miles; Taylor 1994). At least one radio-tagged male returned the following summer to its place of tagging, suggesting that a degree of territorial fidelity exists in this migratory population. During the breeding season male elegant trogons may spend 7–20 percent of their total activity time vocalizing, as compared to 7–11 percent for females (Kunzmann, Hall, and Johnson 1998).

Both sexes participate in excavation, either working in turns or simultaneously. The nest is left unlined during both egg laying and incubation, although debris typically accumulates at the base of the nest. The two to four unmarked, nearly globular eggs (23–39 mm long [0.9–1.6 in] and 20–30 mm in diameter [0.8–1.2 in] and weighing 7–20 g [0.2–0.7 oz]) are smooth and dull white or, in a few cases, tinted brownish, bluish, or greenish. It is unusual for hole-nesting birds to have pigmented eggs; perhaps this tinting is an indication of an earlier time when cavity nesting was not universal among trogons. Similarly, hole-nesting American bluebirds (*Sialia* spp.) continue to lay blue eggs, like those of their non-cavity-nesting relatives. Eggs are usually laid on a daily basis, but in at least some species a 2-day interval seems more typical. Incubation may begin with the first egg or perhaps somewhat later in species with larger clutches; in any case the hatching of the last-laid egg is often delayed by a day. Among eighteen species of New World trogons, two-egg clutches appear to be typical in twelve species, three-egg clutches are typical in five, and four-egg clutches may be most common in one. Such small clutch sizes are not rare in nidicolous Neotropical birds and are often accompanied by a high rate of nest failures (Skutch 1985).

The 16–21-day incubation period is shared by both sexes. In several if not most trogon species the male incubates starting in the morning for 6 to 9 hours, usually remaining until late afternoon, whereupon he is replaced by the female. However, in the resplendent quetzal the pair members typically each take two turns on the nest per 24-hour period. In either case, the eggs are almost constantly covered, and occasionally the sitting bird seemingly refuses to be relieved (Skutch 1976). A similar pattern seems to occur in the elegant trogon: the female incubates all night and is then replaced by the male early in the morning for a few hours. After the female returns for a few hours in midmorning, the male takes over for a second stint again in the afternoon (Taylor 1994). According to Taylor, responsibility for defense of the territory shifts from the male elegant trogon to the female after incubation begins, presumably because she is away from the nest for much of the daytime period. By late afternoon, a time of very hot temperatures in Arizona, the male may desert the nest, leaving it unattended until the female returns. Nevertheless, the nest is typically occupied 85–95 percent of the daytime period.

Unhatched clutches may be incubated by the parents well beyond the normal duration of the incubation period. There is even a case of a pair of Baird's trogons that incubated their unhatched eggs for 51 days, long beyond their normal 16–17-day incubation period (Skutch 1983).

The young are hatched naked and blind and have no immediate downy feathers (neossoptiles) preceding their juvenal feathers. However, species such as the Narina and eared trogons reportedly have brief coats of dark down just prior to the emergence of the contour feathers from their long sheaths, and the tips of many of the developing juvenal feathers of the resplendent quetzal are similarly downy. Presumably these down-like coats provide some thermal protection as the brooding tendency of adults begins to wane. Nestlings may gape, hiss, bill-snap, or move the tongue from side to side when threatened; and they also vocalize to some extent. Both adults feed the young bill-to-bill or regurgitate recently captured prey, often nonhairy caterpillars and other crawling, jumping, or slow-flying insects, which frequently are well camouflaged. Like kingfishers and other coraciiform groups, trogons do not perform nest sanitation (feces removal by adults). Accumulated wastes (including regurgitated seeds, chitinous insect remains, and other indigestible materials) may eventually reach a depth of nearly 10 cm (3.9 in) in nests of such large species as the resplendent quetzal (Skutch 1983).

Table 3. Relationship of Egg Volume to Incubation and Fledging Periods

Species	Egg Size (mm)	Egg Volume (ml)	Incubation Period (days)	Fledging Period (days)	Total (days)	Source
Bare-cheeked trogon	26.5 × 23.5	6.9	16	16	32	Fry et al. 1988
Black-throated trogon	28 × 22	6.9	18	14–15	32-33	Skutch 1976
Narina trogon	28 × 23	7.6	16–21	25–28	41–49	Fry et al. 1988
Black-headed trogon	30 × 23	8.1	19	16–17	35–36	Skutch 1983
Mountain trogon	29 × 24	8.5	19	15–16	34–35	Skutch 1976, 1983
Elegant trogon	29 × 23	8.6	17–19	15–23	32–42	See text
Baird's trogon	33 × 25	10.5	16	25	41	Skutch 1983
Golden-headed quetzal	37 × 29	16.4	17–20	24–25	41–45	Captives, see text
Resplendent quetzal	39 × 32	20.4	17–18	31	48–49	Skutch 1976
Blue-throated green motmot	29 × 23	7.9	21–22	29–31	50–53	Skutch 1976
Pied kingfisher	29 × 23.5	8.2	18	24–29	42–47	Fry et al. 1988
White-collared kingfisher	31 × 25	9.9	?	?	44	Fry et al. 1988
Lilac-breasted roller	32 × 26	11.1	17–18	35	52–53	Fry et al. 1988
Amazon kingfisher	32 × 27	11.9	22	29–30	51–52	Skutch 1976
European roller	36 × 28	14.5	18–19	26–30	44–49	Fry et al. 1988

Note: Arranged in order of increasing estimated egg volume. Trogon egg measurements are usually those of Skutch.

Like other hole nesters such as kingfishers, jacamars, and woodpeckers, trogonid chicks have papilla-like calluses on the undersides of their tarsi, which seem to function as nonslip cushions for resting on the hard nest substrate and perhaps serve to raise the birds slightly above it. Fledging requires 14–30 days, the fledging interval often being about the same length as the incubation period, but sometimes longer. The combined incubation and fledging periods tend to be shorter than those of kingfishers, motmots, and bee-eaters, species that lay eggs of about the same size; but these three species also have somewhat larger average clutch sizes (Skutch 1976; see table 3).

In the first few days following fledging young trogons fly quite poorly and attempt few flights. The rectrices are usually only partially grown, and as a result, the young sometimes spend much of their time on low branches or even the ground, where they are fed by their parents and thus are highly vulnerable to both ground and aerial predators. When there are only two fledglings, each may be attended by a single adult. The first foods taken by young elegant trogons are often berries, although their parents may supplement these foods with insect prey (Taylor 1994). The parents attempt to defend their young against various predators such as snakes, squirrels, and hawks, including even such serious threats as Cooper's hawks (*Accipiter cooperi*).

Dependence on the adults for food probably continues for a considerable period following fledging, given the rather specialized food-catching behavior of trogons. This dependence may continue for at least two months, and the young are likely to follow their parents for some time thereafter. Fogden (1972) observed adult scarlet-rumped trogons still feeding their young 17 weeks after the chicks had fledged. However, second broods within a single breeding season do often occur in some areas, such as with the resplendent quetzal and elegant trogon in Costa Rica (Stiles and Skutch 1989); although in southern Mexico one brood per year seems typical. Skutch (1983) noted that two of three cases of second nestings by the resplendent quetzal involved pairs that were successful in fledging their first brood. Second efforts, however, may most frequently involve situations in which the first attempt was unsuccessful in fledging any chicks. The relatively few observations so far available seem to indicated that nesting success is generally very low in most of the tropical species of trogons; Skutch (1983) noted that only three of twelve black-throated trogon nesting attempts he studied were successful, and similarly low success rates of 50 percent or less have been reported for the resplendent quetzal and the Narina trogon. For 42 species of birds in Costa Rica, Skutch (1985) reported an average nesting success of 39.7 percent for 237 nests. However, average fledging-success rates for the elegant trogon in Arizona have been substantially better, as high as 78 percent during the best years (Kunzmann, Hall, and Johnson 1998).

In Arizona, breeding territories begin to dissolve after the young have fledged, and the adults begin to range widely in search of food. At this time both adults and young feed voraciously on fruits and berries, and by the time of fall migration the young birds are as adept as their parents in plucking fruits and berries from plants while in full flight (Taylor 1994). Almost nothing is known of the duration and destinations of these postbreeding migrations in the elegant trogon, but in cloud forests the birds typically move to lower elevations after they complete their breeding activities.

PART TWO

Species Accounts

AFRICAN TROGONS
(Subfamily Apaloderminae)

Bare-Cheeked Trogons (Tribe Apalodermini)

The subfamily Apaloderminae is the first of two recognized trogon subfamilies and may be characterized as having the two anterior toes unfused basally and the marginal teeth on the cutting edges of the maxilla weak to vestigial. The bill is yellow in adults of both sexes but more grayish in immatures. Adult birds have one or two bare and colorful (often yellowish to greenish-blue) areas of skin between the base of the eye and the posterior edge of the maxilla and also have colorful (yellow to bluish) bare eye-rings. A small area of bare greenish or bluish skin is sometimes present directly above the eye-ring of males, and in at least one species (*Apaloderma narina*) the gular skin is also blue and is exposed during calling by males. The commisural junctions of the bill are also tinted with yellow to greenish or bluish. The tarsi are partly feathered, and the oval nostrils are well concealed by antrose, bristle-like feathers. Adults of all three species in the family are strongly iridescent green dorsally and brilliant carotenoid red ventrally; there is moderate sexual dichromatism and almost no sexual dimorphism. The upper wing-coverts are finely barred or vermiculated with black-and-white patterning, more noticeably so in females and especially in immatures; and the tertials of immatures are also conspicuously spotted and edged with white. The central pair of rectrices in adults of both sexes is iridescent greenish, without the terminal black band that occurs in the New World trogons. The outer three pairs of rectrices are either white (two species) or narrowly and uniformly barred with black and white (one species). A single genus (*Apaloderma*) is now generally accepted, but the two subgenera recognized in this work have often been considered as representing separate genera. The three included species are all of small to moderate size and mass (total length 25–32 cm [9.9–12.6 in], adult mass 50–100 g [1.8–3.5 oz]) and seem to be entirely insectivorous.

Fruits and berries are reportedly never eaten, which makes the source of the carotenoid feather pigments unknown (presumably they come secondhand, from consumption of plant-eating insects).

The African trogons are rather widely distributed through the forested portions of Africa, including both lowland forests and montane forests and thus are sensitive to destruction of these habitats (figures 13 and 14). In the 1960s, tropical rain forests in the Congo Basin comprised nearly 50 percent of the land areas of the regions then known as the Léopoldville and Brazzaville divisions of the Congo (Heske 1973). (Later known as Zaire, the region, as of 1997, is now the Democratic Republic of the Congo.) These rain forests have since been subjected to extensive cutting, burning, and disturbances associated with tribal warfare, making the conservation status of forest-dependent species such as trogons rather uncertain. According to 1998 World Resources Institute data, deforestation rates of tropical forests in

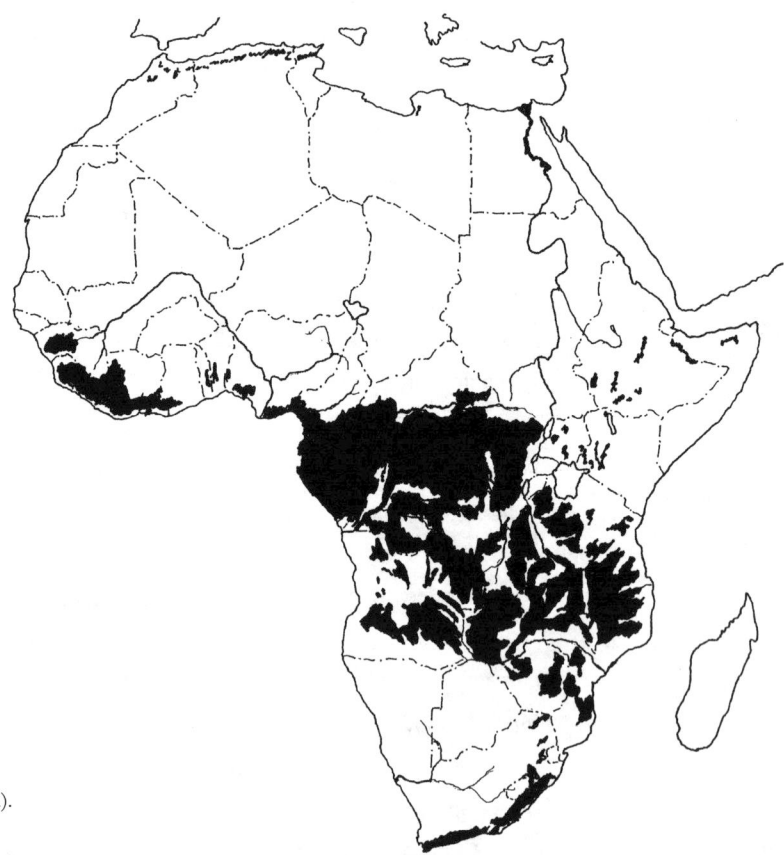

Figure 13. Recently (ca. 1960) forested regions of Africa (inked). After various sources, including Heske 1973.

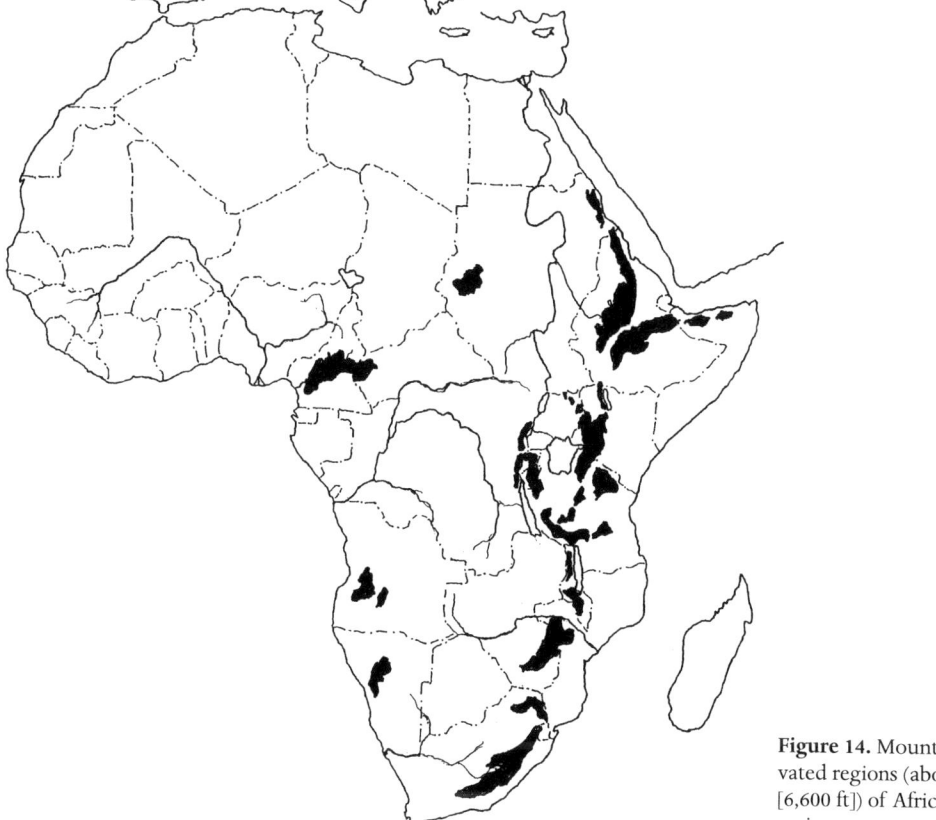

Figure 14. Mountainous and elevated regions (above 2,000 m [6,600 ft]) of Africa (inked). After various sources.

Africa reportedly averaged about 5.5 percent in the decade 1961–1970, versus about 8 percent during the period 1981–1990. Of Africa's 527 million hectares (1.30 billion acres) of tropical forests, nearly half (207 million hectares, 511 million acres) is located in central Africa, where, on average, 1.1 million hectares (2.7 million acres) were lost to logging annually during the 1980s. A total of 4.7 million hectares (12 million acres) of tropical rain forest (out of an estimated total 91 million hectares [220 million acres]) was believed lost to logging in Africa during this period, with the highest deforestation rates occurring in moist deciduous forests (FAO 1993). However, quantitative data on deforestation rates in Africa are rather sparse, and their reliability is suspect.

Of 26 endangered and vulnerable bird species in Zaire, 17 (65 percent) are specifically forest adapted, and 3 more include forests within their habitat range (Collar, Crosby, and Stattersfield 1994). In Zaire approximately 7 percent of the land area was classified as protected, but this designation currently means little or nothing in terms of actual present-day conservation measures, considering the chaotic state of this war-ravaged region. Perhaps as many as 600 million hectares (1.48 billion acres) of lowland forest may still exist in this vast and politically unsettled region, making it ripe for massive logging operations in the future (Wheatley 1996b).

BARE-CHEEKED TROGONS (GENUS *APALODERMA* SWAINSON 1832-1833 [1833])

The genus *Apaloderma* has the same distinguishing characteristics as the subfamily Apaloderminae. Espinosa de los Monteros (1998) regarded this genus as representing the basal lineage of all the extant Trogonidae.

Subgenus *Apaloderma* Swainson 1832-1833 (1833)

The subgenus *Apaloderma* differs from the other subgenus (*Heterotrogon*) of this taxon in that the outer rectrices lack black barring and the lower edge of the maxilla is somewhat serrated.

NARINA TROGON

Apaloderma narina (Stephens) 1815

OTHER VERNACULAR NAMES

Gold Coast Narina trogon (*constantia*), West African trogon; couroucou narina (French).

RANGE

Widespread in sub-Saharan Africa: from Sierra Leone to Ghana and from Nigeria east to southeastern Sudan, Ethiopia, and north-coastal and southern Somalia, and south through the Congo Basin and Rift Valley to Angola, Zimbabwe, and the coastal forests of southeastern South Africa. See map 1.

SUBSPECIES

Apaloderma narina constantia Sharpe and Ussher 1872: Sierra Leone and Liberia to Ghana.

Apaloderma narina brachyurum Chapin 1923: southern Cameroon and southern Nigeria to Uganda and Zaire.

Apaloderma narina narina (Stephens) 1815: Sudan and Ethiopia south to Angola and the former Cape Province, South Africa.

Apaloderma narina littoralis van Someren 1931: southern Somalia to the former Natal Province, South Africa, including coastal woods of eastern Kenya and eastern Tanzania, inland to the lower Tana River; also Zanzibar Island.

MORPHOMETRICS

Measurements Wing, males (of *narina*) 122–140 mm (4.8–5.5 in; average of 17, 132 mm [5.2 in]); females 124–139 mm (4.9–5.4 in; average of 12, 132 mm [5.2 in]). Tail, males (of *narina*) 145–170 mm (5.7–6.7 in; average of 17, 158 mm [6.2 in]); females 155–170 mm (6.1–6.7 in; average of 12, 161 mm [6.3 in]). Wing, five birds of both sexes of *littoralis* average 124.5 mm (4.9 in); seven of *constantia* average 125 mm (4.9 in); thirty-four of *brachyurum* average 129 mm (5.1 in). Tail (same sample sizes) of these forms average 148 mm, 156 mm, and 148 mm (5.8, 6.1, and 5.8 in), respectively (Fry, Keith, and Urban 1988). Egg of *narina* 28.2 × 22.5 mm (1.1 × 0.9 in; Schönwetter 1966).

Weights Eight males of *narina* 55–72 g (1.9–2.5 oz; average 61.6 g [2.2 oz]), seven females 51–71 g (1.8–2.5 oz; average 60.4 g [2.1 oz]). Fourteen males of *brachyurum* 60–77 g (2.1–2.7 oz; average 69.9 g [2.4 oz]), ten females 60–95 g (2.1–3.3 oz; average 68.8 g [2.4 oz]). Five males of *constantia* 63.3–71.4 g (2.2–2.5 oz; average 68.8 g [2.4 oz]; Fry, Keith, and Urban 1988). One male of *narina* 65 g (2.3 oz; specimen, FMNH). Estimated egg weight 7.5 g (0.3 oz; Schönwetter 1966).

Narina trogon, adult male. Photograph by author.

Map 1. Distribution of the Narina trogon. Two peripheral locality records are indicated by dots.

DESCRIPTION (of *narina*, mainly after Fry, Keith, and Urban 1988; see plate 2)

Adult Male Crown, nape, neck, face, chin, and throat iridescent bronzy green; scapulars and back to upper tail-coverts bronzy green (the latter with slight blue iridescence); middle of tail dark blue above, middle rectrices with narrow green edges, undersides dull blue-black; outer three rectrices graduated; outermost rectrices half the length of middle ones, each white with a gray base; resulting undertail pattern appears predominantly white, with narrow black tip (the black representing the exposed ends of the longer central rectrices), grayish toward bases of outer rectrices; breast bronzy green; abdomen, flanks, and under tail-coverts crimson; primaries dull blackish with narrow white outer vanes; secondaries pale gray with finely vermiculated black-and-white outer vanes; greater and median wing-coverts gray, finely vermiculated, with

narrow green margins; bases of remiges white; underside of wing mostly blackish, except for white at bases of primaries and secondaries, this forming a broad white stripe on underside of extended wing; bare areas on lores and middle cheeks (two discrete areas, below rear edge of eye and behind base of bill, separated by a narrow line of green feathers) yellow to greenish yellow; gular skin bright blue; maxilla pale bluish toward tip, becoming yellow at base and toward gape; mandible yellow, becoming bluish toward tip; bare supraorbital area blue, a narrow bluish eye-ring (more greenish in *littoralis*) encompassing remainder of eye; iris chestnut-brown; feet and toes flesh-pink to dusky purple. (The head details in the Gould color plate are incorrect and fail to show the areas of bare facial skin. See figure 1 for location of bare areas.)

Adult Female Crown to scapulars, back, rump, upper tail-coverts, and tail as in male; forehead and ear-coverts cinnamon-brown; chin, throat, and breast light cinnamon-brown, foreneck sometimes tinged with green; abdomen grayish pink, vent area and under tail-coverts crimson; wings as in male but secondaries green on outer vanes, over vermiculated patterning; bare lores blue or blue-green; other soft-parts similar to adult male; similar bare facial areas.

Immature   Head and scapulars to upper tail-coverts dull bronzy green; tail dull blue-black; three outer rectrices white; primaries dark gray with narrow white outer vanes; secondaries (especially tertials) and their coverts greenish, with large terminal or subterminal buffy white spots, these forming rows of contrasting spots on otherwise blackish upper wing; throat to abdomen gray with narrow light-brown barring. Immature females less pink on underparts than adults. Older immatures much like adults but with faint dark gray-pink horizontal bars still visible on abdomen and flanks; bill dull yellow to grayish yellow, gape area at base of bill orange; bare eye-ring whitish; iris dark brown.

Juvenile   Upperparts dull bronzy green; underparts gray with fine buff barring, throat to breast mottled brown; wing-coverts and innermost secondaries white-tipped; under tail-coverts white with brown and blackish markings; eye-ring white; bill dull yellow, becoming orange at gape; iris dark brown.

IDENTIFICATION

In the Hand   Like the bare-cheeked trogon, the Narina trogon has small areas of bare skin above and below the eyes, but in the latter the subdivided bare-cheek area below the eye is greenish yellow to greenish blue (in the former, the area is instead yellow and not obviously subdivided by feathers). The small patch of skin above the eye-ring is bright yellow to bluish green. The eye-ring may be blue (in *narina*), green (in *constantia*), greenish yellow (in *littoralis*), or even whitish (in younger birds); and a small, usually hidden area of bare skin on the upper throat is blue. There are some racial variations in these skin colors, as well as some variations by sex or age or both.

In the Field   The usual (territorial advertisement) vocalization of the Narina trogon is a sequence of double-noted *hoo-hoo', hoo-hoo'* . . . utterances. Sequences of four to fourteen such dove-like vocalizations, lasting 4–7 seconds, are typical. The sequence progressively rises in pitch and volume and is repeated at irregular intervals. An area of blue gular skin is expanded and exposed with each utterance. Duetting, or synchronized calling, among nearby birds may occur. Sympatry with the very similar bare-cheeked trogon in some areas may make field identification difficult, but the vocalizations of the two species are distinctive. Whereas the Narina trogon has a soft double-noted cooing call, with the second note louder and higher pitched, the bare-cheeked trogon utters a descending series of six to eight *chuuu* notes that have a melancholic timbre The male of the less similar bar-tailed trogon (which usually occurs at higher elevations and has distinctive undertail barring) has a higher pitched, more yelping and repeated *yow* note that also gradually increases in volume. See also the vocalizations section.

GEOGRAPHIC VARIATION

Geographic variation in the Narina trogon mostly involves minor plumage-color variations, especially of the iridescent upperparts, tail, and breast; but some differences in the intensity of scarlet under-part coloration also occur. The races *constantia* and *littoralis* average slightly smaller in wing measurements than the other two races. The nominate race has a fairly long tail, but *brachyurum* is distinguished by its relatively short tail. The bare areas of skin near the eye are green or greenish blue in males of *littoralis* and *brachyurum* but are yellow to greenish yellow in *narina* and *constantia*. Some sex, age, and individual variations in these colors also apparently exist.

ECOLOGY

Habitats   The Narina trogon is the most widespread and ecologically tolerant of the African trogons. It occurs in rain forests and gallery forests and occasionally penetrates into savanna woodlands. It also sometimes occurs in well-treed gardens. It is found mainly in low-altitude forests, below 2,500 m (8,200 ft); but it reaches 3,500 m (11,500 ft) in Ethiopia (Fry, Keith, and Urban 1988) and extends into highland forests in northern and western Kenya

(Zimmerman, Turner, and Pearson 1996). In northern Somalia the nominate race occurs in wooded hills, especially in dense junipers, whereas the *littoralis* race occurs in lowland forests of southern Somalia (Ash and Miskell 1998).

Foods and Foraging Ecology   In contrast to the New World trogons, the African species, including the Narina trogon, apparently never consume fruit or berries, even when these are abundant. Their diet consists mostly of spiders and insects, especially nonhairy caterpillars (of both butterflies and moths, including sphinx moths), orthopterans (grasshoppers, crickets, mantids), cicadas, beetles, winged termites, and sometimes flying moths. Small arboreal lizards are also eaten at times (Fry, Keith, and Urban 1988). Chapin (1939) examined the contents of 20 stomachs and found caterpillars most frequently (present in 16 stomachs), followed by orthopterans in 10 (mostly large, green jumping species, plus two mantids). Like other trogons, the African species capture food during short, often swooping flights and return to the perch to subdue and eat their prey.

BEHAVIOR

General, Social, and Sexual Behavior   The Narina trogon is said to be territorial throughout the entire year, with strong monogamous pair bonds. The territory may cover 1–10 hectares (3–25 acres), but at times up to seven males will congregate during the rainy (breeding) season and form a slowly mobile group, continuously singing and displaying. These singing assemblages are male-only groups, but females are evidently attracted to them. At times both the Narina trogon and the bare-cheeked trogon have been seen displaying together in such groups. The groups resemble leks, but because these trogons are both monogamous and territorial, it is doubtful that the groups serve as pair-forming devices. Even paired birds will participate, the males temporarily leaving their mates and territories and crossing the territories of others to participate (Brosset 1983; Cunningham-van Someren 1973).

Vocalizations   Like cuckoos, Narina trogons often call before or during rain. Calling is done from open branches 7–20 m (23–66 ft) above ground but not from the upper canopy. The usual male territorial

vocalization is a soft and dove-like *hoo-hoo'*, *hoo-hoo'* or *oh-coo'* that is repeated many times, the series gradually becoming louder. Bare blue gular skin is exposed as the throat is expanded with each call and the head is jerked upward. The sequence usually consists of four to fourteen doublet calls and lasts about 7 seconds, with irregular intervals between calls. Single low *kuk* notes may be uttered two to twelve times during courtship, and the same note is used by the male when relieving its mate at the nest. There are also aggressive *kwaurr* or *koorr* notes uttered in the presence of other males. Females are evidently fairly quiet compared with males and are silent outside the breeding season (Fry, Keith, and Urban 1988). Nestlings may hiss when alarmed, simultaneously exposing their pink gapes and making lateral tongue movements. The sound and visual effect thus produced are distinctly snake-like (Harcus 1976).

BREEDING BIOLOGY

Chronology of Breeding   Breeding occurs during a broad time period over the Narina trogon's range. In the Congo Basin breeding probably occurs throughout the entire year, or at least from January through September (Chapin 1939). In Kenya the peak of laying is from April to June, during the wet season; but laying has been recorded in all months except January, July, and September. In Ethiopia breeding occurs in spring, from March to June; and south of the equator in Malawi, Zambia and Zimbabwe, breeding occurs during September. Likewise in South Africa breeding extends from October to January, in the austral spring and early summer (Fry, Keith, and Urban 1988).

Nest Sites   Narina trogon nests are in natural cavities, such as rotted holes in dead limbs or stumps, from 1–12 m (3–40 ft) above ground. The entrance is typically only large enough to let the bird pass through, but the nest cavity may be fairly large. Like other trogons, Narina trogons add no nest lining. Nests may be used from year to year; in one case a site supported four clutches in 7 years (Brown 1975).

Eggs and Incubation   Although from one to four eggs have been found in Narina trogon nests, the usual number is two or three (the average in South

Africa is 2.6 eggs). The eggs are laid at daily inter-vals, and incubation begins with the first egg. Like most New World trogons, the male Narina incu-bates during daytime hours for about 9 hours, and the female at night. Incubation periods have been estimated variously at 16–17, 18, and 21 days (Fry, Keith, and Urban 1988).

Brood Rearing   Newly hatched Narina trogon chicks are reportedly downy (Fry, Keith, and Urban 1988), although this seems unlikely on the basis of what is known of other trogons; Zimmerman, Turner, and Pearson (1996) state that the young are initially naked. By 10 days the juvenal plumage breaks free of feather sheaths. Both sexes feed their young, the male more actively than the female. Caterpillars are among the items fed to chicks. There are two estimates of the fledging period, 25 and 28 days, but the flight feathers may not be fully grown until about 60 days. The young may remain with their parents and receive food even after this stage. Nesting success is evidently rather low; in one case eight nests produced four fledged young (Fry, Keith, and Urban 1988).

CONSERVATION AND EVOLUTIONARY RELATIONSHIPS

The Narina trogon is easily the most common and most widespread trogon in Africa, with population densities of five pairs in 8 hectares (1.25 individuals per hectare) reported in Kenya.

Although the Narina is widely sympatric with the bare-cheeked trogon, the two differ most markedly in vocalizations, as well as in their bare facial colors; they also differ slightly in size. The bare-cheeked trogon is, however, restricted to primary evergreen forests, whereas the Narina trogon has a much broader habitat spectrum.

Relationships between *Apaloderma* and other trogon genera remain speculative; one might imag-ine that an ancestral African trogon gave rise, at different times, to both the Asian and the New World groups. On the basis of DNA-sequence data, Espinosa de los Monteros (1997) recognized the African trogons as the most basal trogon clade, with the New World trogons being a sister taxon to the Asian group.

BARE-CHEEKED TROGON

Apaloderma aequatoriale Sharpe 1901

OTHER VERNACULAR NAMES

Bare yellow-cheeked trogon, yellow-cheeked trogon; couroucou à joues jaunes (French).

RANGE

Lowland rain forests, with two seemingly geographically isolated populations, one in Cameroon and Gabon and another in northeastern Zaire; some specimens help to connect these areas (Louette 1987). See map 2.

SUBSPECIES

None recognized.

MORPHOMETRICS

Measurements Wing, males 115–125 mm (4.5–4.9 in; average of 33, 120.5 mm [4.7 in]); females 110–125 mm (4.3–4.9 in; average of 9, 120.5 mm [4.7 in]). Tail, males 124–161 mm (4.8–6.3 in; average of 33, 142.5 mm [5.6 in]); females 135–156 mm (5.3–6.1 in; average of 9, 145.5 mm [5.7 in]; Fry, Keith, and Urban 1988). Wing, twenty-eight males 115–126 mm (4.6–5.0 in); seven females 110–125 mm (4.3–4.9 in). Tail, twenty-eight males 136–161 mm (5.3–6.3 in); seven females 140–156 mm (5.5–6.1 in; Chapin 1939). Egg 26.7 × 23.1 mm (1 × 0.9 in; Schönwetter 1966).

Weights No information on body mass available, but linear measurements suggest a very slightly smaller body mass than in *narina*. Estimated egg weight 7.7 g (0.3 oz, Schönwetter 1966).

DESCRIPTION (mainly after Fry, Keith, and Urban 1988; see plate 3)

Adult Male Entire upperparts (from head to upper tail-coverts), chin, throat, and upper breast iridescent green with golden and blue gloss; tail long; middle rectrices violet-blue, inner pair with narrow green margins, three outer pairs graduated in length, white at tips, brown at bases, glossed purplish and green; outermost rectrices almost wholly white; underside of outer rectrices largely white, except for grayish or brownish freckling or vermiculation visible near base; lower breast to under tail-coverts pinkish red; primaries mostly black, but white basally and narrowly white on outer vanes; secondaries vermiculated on outer vanes and black on inner vanes; innermost secondaries vermiculated on both vanes; upper wing-coverts vermiculated black and white or black with green margins; anterior cheek area mostly bare yellowish skin, extending from below eye forward through lores to base of maxilla, a second area of similar yellowish skin extending below and behind eye, continuous with a narrow eye-ring of bare skin; gular skin light blue; maxilla dull grayish to greenish gray, with cadmium yellow base; mandible yellow, brighter toward gape; iris dark reddish brown; feet and toes brownish flesh to light red.

Adult Female Crown and upperparts as in adult male; forehead, lores, chin, and throat rusty brown becoming dull brown on upper breast and gradually merging on lower breast to gray with rosy wash and into blood-red on abdomen and under tail-coverts; vermiculated patterning of secondaries and upper wing-coverts less distinct than in male; maxilla with deep yellow base, tip dull pale green, shading to dusky on culmen; mandible yellowish; bare skin areas near eye variable, usually lemon-yellow; iris brownish red; feet and toes pale red to flesh colored.

Immature Female like adult, but duller. Male like adult above but upper breast fulvous; lower breast narrowly barred with green; inner secondaries and larger wing-coverts with large white or buff subterminal spots. Maxilla of both sexes dusky, with greenish-yellow base; mandible light yellowish green with light gray tip; iris brown; feet and toes pinkish.

Juvenile Similar to immature, but forehead and underparts pale tawny, mottled with black.

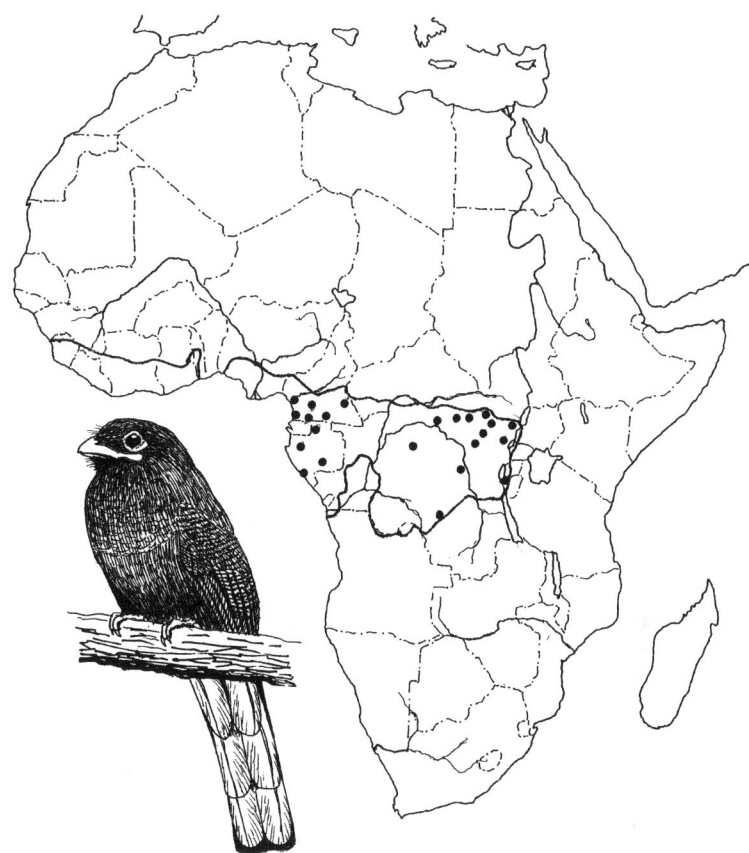

Map 2. Potential maximum distribution of the bare-cheeked trogon (lines show the limits of the equatorial forest). Specific locality records (after Louette 1987) are indicated by dots.

IDENTIFICATION

In the Hand   The bare-cheeked trogon is locally sympatric with both of the other African trogons, but its bare facial skin is more extensive than that of the others, is bright lemon-yellow to cadmium yellow, and is undivided. In contrast to the undertail pattern of the bar-tailed but in common with that of the Narina, the visible undertail pattern of the three outer pairs of rectrices is unbarred white with some inconspicuous brown or grayish and white freckling basally.

In the Field   In addition to the color differences of the head and undertail just mentioned, the bare-cheeked trogon has a distinctive quavering vocalization, repeated many times until it fades away. It consists of six to eight notes, *chuu-chuu-chuu* . . . , melancholy in tone, uttered as if in pain, and repeated at intervals of 15–20 seconds. Chapin (1939)

has interpreted the same call as a repeated *owng* or *oodle*, with all but the first note being nearly disyllabic. With each note the throat is expanded, exposing blue gular skin, and the tail is dropped down to the vertical. Calling is performed by both sexes, sometimes in antiphonally, the female having a lower-pitched call. See also the vocalizations section.

GEOGRAPHIC VARIATION

None described.

ECOLOGY

Habitats   Chapin (1939) noted that although the Narina trogon sometimes was seen in areas of tall secondary growth, the bare-cheeked was never observed outside primary forest. They were usually found on lianas or bare branches about 5 m (16 ft) above ground.

Foods and Foraging Ecology   Of twelve bare-cheeked trogon stomachs examined by Chapin (1939), ten contained caterpillar remains (in one case hairy caterpillars). Orthopterans were present in seven, the latter primarily consisting of large, green jumping species such as tettigonids. Minor contents included a snail and a chrysalis. No evidence of fruit eating was found.

BEHAVIOR

General, Social, and Sexual Behavior   Calling assemblages of male bare-cheeked trogons have been reported, as in the Narina trogon. In one case six birds were joined in a singing assembly by a male Narina trogon. As with that species, singing is more or less continuous, with no overt aggression apparent among participants. However, some supplanting behavior involving perch sites does occur. The members of the group may sing simultaneously or sometimes alternately for an hour or two each morning, and then the males return to their own territories and mates (Brosset 1983).

Vocalizations   Little about bare-cheeked trogon vocalizations is known beyond what is mentioned in the identification section. Besides the melancholic, repeated *chuu-chuu-chuu* series of notes, the birds also utter raucous whining notes similar to those of the Narina trogon, as well as an aggressive "fighting song." Calling is done by both sexes (Fry, Keith, and Urban 1988).

BREEDING BIOLOGY

Chronology of Breeding   Bare-cheeked trogons evidently breed during the rainy season; in the Congo Basin breeding-condition birds have been taken in April and August, and nestlings or immatures in August and October. Most singing occurs between July and September. Nonbreeders have been taken between December and July (Chapin 1939). In Cameroon breeding has been reported in January (Bates 1927), and in Gabon breeding has been reported for January and February (Fry, Keith, and Urban 1988).

Nest Sites   Bare-cheeked trogon nest sites are reportedly chosen by the female and consist of natural cavities 4–8 m (13–26 ft) high in rotting trees. The cavity may be up to 30 cm (11.8 in) deep (Fry, Keith, and Urban 1988).

Eggs and Incubation   The bare-cheeked trogon clutch consists of two pinkish-buff and glossy eggs. The incubation period has been estimated at about 16 days (Fry, Keith, and Urban 1988). Although only the female has been seen incubating, with stints of 2.5 hours of sitting alternating with 30-minute periods away from the nest, this apparently single-sex incubation behavior would be an unusual exception to the trogonid pattern.

Brood Rearing   No information available.

CONSERVATION AND EVOLUTIONARY RELATIONSHIPS

As noted above, the bare-cheeked and Narina trogons clearly are closely related, and at times it has been suggested that their differences do not warrant species separation. However, whereas the Narina trogon is easily the most highly adaptable of all three African trogon species, the bare-cheeked trogon seems to be highly restricted to undisturbed tropical forests. It has the most restricted range of all the African trogons and is thus the most vulnerable species.

Subgenus *Heterotrogon*
Richmond 1895

The single species of the subgenus *Heterotrogon* differs from the typical *Apaloderma* in having barred rather than white outer rectrices, a bill that is smaller and more depressed, and smaller bare facial areas; in addition the cutting edge of the maxilla is sometimes slightly serrated but often smooth.

BAR-TAILED TROGON

Apaloderma vittatum (Shelley) 1882

OTHER VERNACULAR NAMES

Cameroon bar-tailed trogon (*camerunensis*); couroucou à queue barrée (French).

RANGE

Local in montane forests, with separate populations in Nigeria and Cameroon; Angola; eastern Zaire; the highlands of Kenya and Tanzania; Malawi; and Mozambique. See map 3.

SUBSPECIES

Apaloderma vittatum camerunensis Reichenow 1902: Nigeria and Cameroon, Angola, and Zaire and probably western Uganda. Also on Bioko (formerly Fernando Póo) Island, Gulf of Guinea. This race was not recognized by Fry, Keith, and Urban (1988).

Apaloderma vittatum vittatum (Shelley) 1882: Kenya, Tanzania, Malawi, and Mozambique.

MORPHOMETRICS

Measurements   Wing, males (races combined) 116–123 mm (4.6–4.8 in; average of 5, 119 mm [4.7 in]); females 118–126 mm (4.6–5.0 in; average of 2, 122 mm [4.8 in]). Tail, males 154–165 mm (6.1–6.5 in; average of 4, 157 mm [6.2 in]); females 150–170 mm (5.9–6.7 in; average of 5, 158 mm [6.2 in]; Fry, Keith, and Urban 1988). Wing, eleven males (of *camerunensis*) 111–120 mm (4.4–4.7 in); five females 115–119 mm (4.5–4.7 in; Bannerman 1951). Wing, both sexes of *vittatum* 123–131 mm (4.8–5.2 in), of *camerunensis* 115–125 mm (4.5–4.9 in; Chapin 1939). Average of four eggs 25.9 × 21.8 mm (1.0 × 0.9 in; Fry, Keith, and Urban 1988).

Weights   Five birds of both sexes 50–57 g (1.8–2.0 oz; average 55.1 g [1.9 oz]; Dowsett 1970). Eight males from Uganda 49–58 g (1.7–2.0 oz; average 53.8 g [1.9 oz]), three females 54–59 g (1.9–2.1 oz; average 56.8 g [2.0 oz]; Fry, Keith, and Urban 1988). One male 56.5 g (2.0 oz), one female 56.4 g (2.0 oz; Sclater and Moreau 1932). Estimated egg weight 6.3 g (0.2 oz).

DESCRIPTION (of *vittatum*, mainly after Fry, Keith, and Urban 1988; see plate 4)

Adult Male   Lores black; crown, ear-coverts, and nape iridescent blackish; hindneck, scapulars to upper tail-coverts iridescent dark green, sometimes with blue gloss; chin and throat black with blue-green gloss; middle rectrices purple-blue, narrowly edged with green; outer rectrices graduated, barred white and black, basally black above, with blue gloss; outermost rectrices half length of middle ones; undersurface of three outer pairs of rectrices almost entirely regularly barred with gray and white, these seemingly narrowly tipped with black when closed tail is seen from below (these "tips" being the exposed underside ends of blackish central rectrices); upper breast dark bronzy green with narrow band of iridescent purple below, becoming bright blue, then dark blue-green; lower breast, flanks, abdomen, and under tail-coverts deep crimson; primaries black with very narrow white edging on outer vanes; inner secondaries, tertials, and greater and median wing-coverts with overall green tinge, vermiculated very narrowly with black and white; lesser wing-coverts black, each feather edged green; underside of wing entirely blackish except for white bases of primaries and secondaries, these forming a broad white stripe on extended wing; two areas of bare yellow or orange facial skin, these located below eye and immediately behind bill and separated narrowly by a line of blackish feathers; also a small area of yellow or gray skin above eye; bill entirely yellow or greenish yellow, brighter toward gape; iris reddish brown to dull orange; feet and toes pale brown to dusky orange.

Adult Female   Head to nape chocolate-brown with variable green sheen; ear-coverts darker; neck to upper tail-coverts iridescent dark green with dark blue reflections, particularly on upper tail-coverts; chin, throat, and upper breast cinnamon-brown; lower breast rich orange-cinnamon; sides of breast somewhat glossed with green; flanks and abdomen to under tail-coverts bright pinkish crimson; wings and tail as in male; two bare areas of skin below eye yel-

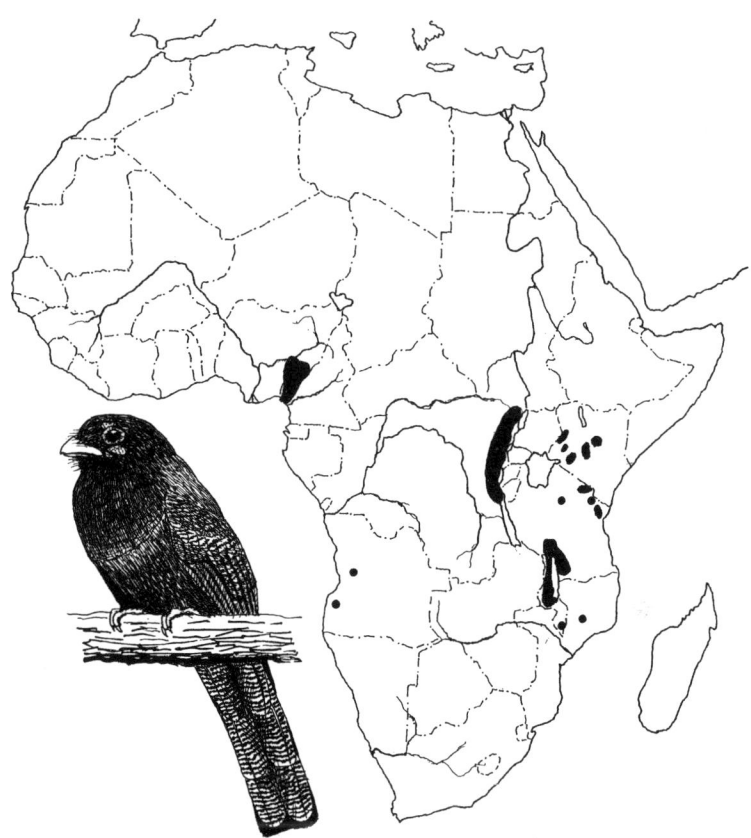

Map 3. Distribution of the bar-tailed trogon. Six peripheral locality records are indicated by dots.

low or orange; maxilla dark horn-brown with yellow cutting edges; mandible greenish yellow to orange; iris brown; feet and toes brown.

Immature   Like adult female but inner secondaries, tertials, and greater and median wing-coverts with white or buff subterminal spots; abdomen whitish. Immature males gradually acquire green gloss to blackish feathers on head and breast.

Juvenile   Head, wings, and remiges dark brown; scapulars, back, and rump brown glossed with green; buffy spots on wing-coverts and some secondaries; middle rectrices glossy blue, with rounded tips; outer three pairs of rectrices white, with a few brown bars; underparts mostly a mixture of dark brown and cinnamon, barred with bluish gray; bases of breast feathers yellow, of abdominal feathers gray. Yellow-orange breast and beige-colored abdominal feathers of nestlings similarly barred with grayish blue.

IDENTIFICATION

In the Hand   The bar-tailed trogon differs from the other two African trogons in that its outer three rectrices are barred conspicuously with black.

In the Field   The barred undertail pattern of the bar-tailed trogon is usually visible in perched birds when facing the observer. When the bird is calling, the tail is lowered to the vertical with each note. The male's territorial vocalization is a distinctive series of sharp, high-pitched *wup-wup-wup-wup* notes, repeated seven to fourteen times, the series gradually increasing in volume and lasting 5–10 seconds. See also the vocalizations section.

GEOGRAPHIC VARIATION

Bar-tailed trogons from western Africa (*camerunensis*) average slightly smaller than those of the nominate race.

ECOLOGY

Habitats   In the eastern edge of the Congo Basin, the bar-tailed trogon is limited to montane forests above 1,500 m (4,900 ft) and is thus altitudinally separated from the more lowland-adapted Narina trogon. It may reach altitudes of more than 2,500 m (8,200 ft) in eastern Africa, in moist montane forest. Color plate 3 illustrates a moss- and lichen-draped forest on Mount Elgon in western Kenya, at about 3,000 m (9,800 ft).

Foods and Foraging Ecology   Chapin (1939) found only smooth-skinned caterpillars in the stomachs of the three bar-tailed trogons that he examined. Insects caught in flight seem to be commonly taken. There is no evidence that fruits or berries are eaten.

BEHAVIOR

General, Social, and Sexual Behavior   Although reportedly territorial, with rather large territories of 1–7.5 hectares (2–19 acres) being estimated, the bar-tailed trogon may make occasional excursions outside the territory to forage. In one area a density of 12 pairs in 25 hectares (62 acres) was estimated (Dowsett-Lemaire 1983). There have been no definite observations of courtship, but in one case an individual was seen repeatedly jumping over another, apparently engaging in premating behavior. However, no copulation occurred, and the birds instead then began what seemed to be a mating song: a sequence of high-pitched notes similar to the yelping of a small puppy or the squeaking of a rubber toy. Each song sequence consisted of about eight notes, repeated at half-second intervals, with one bird starting its sequence when the other was about halfway through its own (Moreau and Moreau 1939).

Vocalizations   Little about bar-tailed trogon vocalizations is known beyond what is mentioned in the identification section. No singing assemblies of males have been reported. In the presence of a female, the male produces an "excitement," or "courtship," call; and the female utters an extended, descending *chee-uu* note. The male's call has also been variously written as *chuiwi*, as *chur-chur*, or even as a low *meow*

note (Fry, Keith, and Urban 1988). Both sexes utter mournful *wheeee-oh* notes as well.

BREEDING BIOLOGY

Chronology of Breeding   Bar-tailed trogons in breeding condition have been taken in Cameroon from October to December, on Bioko Island in November, and in the eastern Congo during February and March (Mackworth-Praed and Grant 1970–1973). Breeding in Zaire occurs from February or March to May, and breeding-condition birds have been taken in Tanzania from November to February (Fry, Keith, and Urban 1988). Breeding in southeastern Africa probably occurs between September and November, during the austral spring (Mackworth-Praed and Grant 1957).

Nest Sites   Bar-tailed trogon nests are located in natural tree cavities, from 1.5–5 m (5–16 ft) above ground (Fry, Keith, and Urban 1988). One measured cavity was about 20 cm (7.9 in) deep and 40 cm (15.7 in) in diameter (Prigogine 1953).

Eggs and Incubation   Two or three eggs constitute the bar-tailed trogon clutch. Nothing is known of the incubation or prefledging phases (Fry, Keith, and Urban 1988). The nestling shown in color plate 3 is based on a specimen in the Field Museum of Natural History; nestlings of this species have not been illustrated previously.

Brood Rearing   No information available.

CONSERVATION AND EVOLUTIONARY RELATIONSHIPS

Clearly a close relative of the other two African trogons, the bar-tailed trogon's barred outer tail feathers are perhaps a species-level characteristic, and the species' less serrated maxilla may have some limited ecological significance. Its montane distribution places it out of contact with most trogon populations, but it probably has some limited contact with the montane (nominate) populations of the Narina trogon. Being dependent on generally isolated montane forests makes it potentially subject to the loss of local populations through deforestation.

NON-AFRICAN TROGONS
(Subfamily Trogoninae)

New World Trogons and Quetzals
(Tribe Trogonini)

The species of the subfamily Trogoninae are all of subtropical to tropical distributions either in the Americas or in Southeast Asia and the Greater Sunda Islands. Their two anterior toes (the third and fourth) are fused for as much as half their length, and there are no isolated patches of colorful skin on the lores and the sides of the head, although variably colorful bare eye-rings are present in adults of nearly all species.

The tribe Trogonini, which consists entirely of New World forms, may be additionally characterized as having upper mandible edges that are either serrated or notched only near the tip, and upperparts that are predominately iridescent green to blue. Colorful and bare eye-rings are usually present, and the head, throat, and upper breast regions are often uniformly iridescent green, bluish, or even glossy blackish, whereas the underparts are usually a strongly contrasting vivid red to yellow. In many species (nearly all of those in the genus *Trogon*) the upper wing-coverts are finely vermiculated or more coarsely barred with wavy markings in both sexes, but in some genera (*Priotelus*, *Pharomachrus*, and *Euptilotis*) they are unpatterned and iridescent green in one or both sexes. It is believed that sexual maturity typically occurs by the end of the first year, although several years may be required for maximum plumage development among male quetzals. The three outer pairs of rectrices are often extensively patterned with black and white, the females and immatures usually having more such patterning than the adult males. The 25 included species are insectivorous (spiders and sometimes even small vertebrates may also be consumed) or largely frugivorous (berries and relatively large arboreal fruits are eaten) or both. The quetzals (*Pharomachrus*) are the largest and most frugivorous, whereas the group of typical trogons (*Trogon*) includes the smallest and most insectivorous species. The aberrant eared trogon (*Euptilotis*) is somewhat

transitional between these two morphological and ecological types and has at times been taxonomically associated either with the quetzals or with the typical trogons. The two West Indian trogons (*Priotelus*) differ only in minor morphological respects from the typical New World trogons and seem likely to have diverged from the latter more recently than did the quetzals. Only four genera are accepted here, but several additional subgenera (previously often considered distinct genera, as by Ridgway [1911]) are also recognized.

On the basis of gene-sequencing studies, Espinosa de los Monteros (1998) considered the New World trogons to be monophyletic, with the Asian trogons being their sister group. Within the New World assemblage, *Priotelus (temnurus)* was regarded as the sister taxon to the other New World species.

The New World trogonids are widely distributed in Central and South America, inhabiting both lowland and montane forests (see figures 15 and 16). They also occur in two of the larger West Indian islands. Only one American species of trogonid (the resplendent quetzal) is currently considered endangered, but three others (Baird's, white-eyed, and orange-bellied) have very restricted ranges. All of these are close relatives of more widely ranging species, and the validity of at least one (the orange-bellied) is distinctly dubious. The greatest species diversity of trogonids occurs in Colombia and Ecuador (see table 1), and there nearly all of the

species are primarily associated with fairly undisturbed forests.

Estimates of current forest-destruction rates are unavailable for these countries, but of 62 critical, endangered, or vulnerable bird species in Colombia, 48 (77 percent) are forest dependent, and 8 more include forests in their habitat range. Of 50 such threatened species in Ecuador, 34 (68 percent) are forest dependent, and 6 more include forests within their habitat range (Collar, Crosby, and Stattersfield 1994). In Colombia less than 5 percent of the land area has been designated as protected, and in Ecuador the proportion of similarly designated land is about 10 percent, although as in many other Latin countries such designations often mean very little.

As of the 1960s, about 57 percent of Colombia's land area and about 74 percent of Ecuador's land area was forested (Heske 1973). According to 1998 World Resources Institute data, deforestation rates of tropical forests in South America averaged about 5.5 percent during the period 1960–1970, versus about 7.5 percent during the period 1981–1990; and logging rates during the 1980s were highest in Brazil (1.9 million hectares [4.7 million acres]), Ecuador (152,000 hectares [376,000 acres]), and Colombia (108,000 hectares [267,000 acres]). In 1990 the total tropical forest cover in Latin America consisted of about 918 million hectares [2.27 billion acres]; and of this area 87 percent was in South America and 474 million hectares (1.17 billion acres) consisted of tropical rain forest. Annual Latin American losses through deforestation during the 1980s averaged 7.4 million hectares (18 million acres), with the highest rates occurring in moist deciduous forests (FAO 1993).

Brazil's vast Amazon Basin once contained the largest unbroken tropical forest in the world (covering about 500 million hectares [1.24 billion acres]), with about a quarter of this area (130 million hectares [320 million acres]) represented by the moist Atlantic forests of eastern Brazil. Nearly 90 percent of these moist Atlantic forests have already been destroyed (Viana, Tabanez, and Batista 1997), and the interior lowland forests are also under concerted attack. As much as 5 million hectares (12 million acres) of Brazilian forest have been lost in a single year (Wheatley 1995). Brazil has 102 bird species listed as critical, endangered, or vulnerable, of which 69 (68 percent) are forest dependent and another

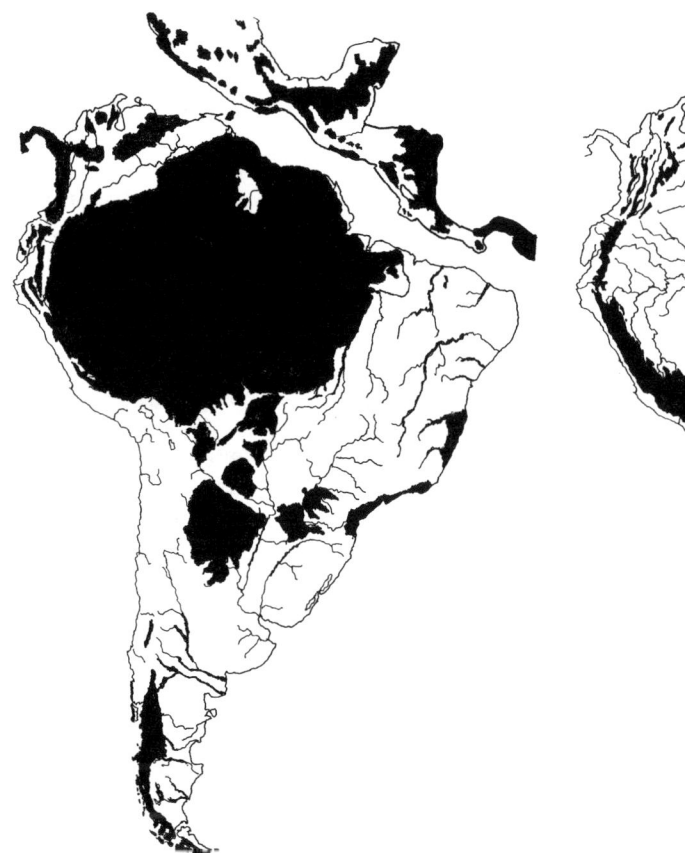

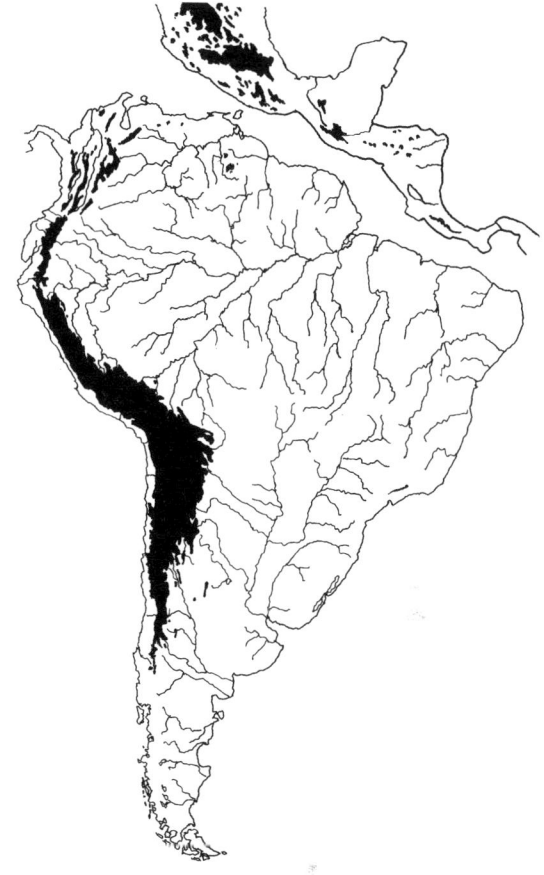

Figure 15. Recently (ca. 1960) forested regions of Central and South America (inked). After various sources, including Heske 1973.

Figure 16. Mountainous and elevated regions (above 2,000 m [6,600 ft]) of Central and South America (inked), after Fjeldså and Krabbe 1990.

10 include forests in their range of habitats (Collar, Crosby, and Stattersfield 1994). Among 290 threatened bird species of the non-Caribbean Neotropics, nearly 55 percent are associated with wet forests, and 97 are Brazilian species, with the Atlantic coastal forests having the highest number (about 30 percent) of the threatened forms (Wege and Long 1995).

The Central American countries support a substantial proportion of the New World trogonids. In 1990 the percentages of "natural" (including disturbed and secondary forests) forest cover relative to original forest totals were estimated as follows: Panama 41 percent, Costa Rica 28 percent, Nicaragua 50.6 percent, Honduras 41 percent, El Salvador 5.9 percent, Guatemala 39 percent, Belize 87 percent, and Mexico 25.5 percent. In the area comprising

Mexico and the Central American countries, Mexico has the largest number of threatened bird species (29), and 86 percent of these are ecologically associated with wet or dry forests. Mexico supports a total of 58 identified "key areas" that are known to support threatened species of birds (the greatest concentration of these occurs in Oaxaca and Chiapas, which contain 14 such areas). Of the 58 total, only 19 areas provide some degree of protection for these threatened birds (Wege and Long 1995).

Deforestation has been rampant in Central America; by 1980 about 70 percent of the region's primary forests had been removed, with most of that percentage being lowland forests (FAO 1984). Logging rates have been especially high in Nicaragua and Costa Rica (FAO 1993), and El Salvador has lost about 95 percent of its few remaining forests since the 1980s. Forest depletion in southern Mexico (mainly Chiapas) and Central America has also been especially severe recently, as a result of forest fires associated with the 1997–1998 El Niño phenomenon. Chiapas had been one of the most heavily forested states in Mexico (with forests covering about 45 percent of its land area as of the 1960s) and is the state with the largest amount of tropical evergreen and montane forests as well as the greatest botanical and mammalian diversity. It also has extremely high avian diversity; over 640 bird species have been recorded from Chiapas (Hernandez 1998), which represents about 60 percent of Mexico's total species. It also has the largest number of trogonid species of any Mexican state (see table 1).

The five Central American countries of Belize, El Salvador, Guatemala, Honduras, and Nicaragua support fourteen locations that have been identified as key areas, six of which enjoy some level of protection. Of the total fourteen sites, six are located in Guatemala, five are in Honduras, two are in Belize, and one is in El Salvador; Nicaragua has none. Costa Rica has fourteen key areas, and eight of these have some level of protection. Panama has eleven such key areas, of which eight enjoy some protection status (Wege and Long 1995).

Of all the Central American countries, Costa Rica has by far the highest percentage of effectively pro-tected lands (about 12 percent of the country represented by national parks alone), and it also has the greatest diversity of trogonids in Central America (see table 1). By the 1990s more than 20 percent of Costa Rica's land area had been set aside as national parks, biological reserves, protected zones, forest reserves, and other sanctuaries. The economic payoff, in terms of Costa Rica's booming ecotourism industry, has been enormous. By the same period Mexico had established 64 federally protected areas (including 44 national parks and 20 biosphere reserves). Similarly, Honduras had 54 areas (including 13 national parks and 37 areas of cloud forest). Guatemala had 44 actual or proposed protected areas (potentially representing 15 percent of its land area). The recently established Maya Biosphere Reserve in northern Guatemala is the largest Central American reserve and includes Tikal National Park. Nicaragua had 19 already protected areas plus 18 more potentially protected ones (collectively representing 3 percent of its area; Howell and Webb 1995). There are few well-protected areas in Belize (two national parks), El Salvador (two national parks, one in cloud forest), and Nicaragua (two biological reserves). In Belize a large percentage of the country's forest (comprising about half of its total land area as of the 1990s) remains essentially intact, and its eco-tourism business is now probably second only to Costa Rica's.

QUETZALS (GENUS *PHAROMACHRUS* DE LA LAVE 1832)

The genus *Pharomachrus* differs from the others of its tribe in that the maxilla lacks serrations on its cutting edges (it does, however, have a subterminal notch, and the mandible has undulate edges) and in that the nostrils are narrow and mostly hidden by an overhanging operculum. Bristly antrose feathers on the forehead, lores, and chin hide the base of the bill. The highly iridescent median wing-coverts are elongated, decurved, and pointed; and in adult males the four similarly iridescent middle upper tail-coverts are also greatly elongated and pointed, forming an ornamental train directly above and effectively obscuring or even extending well beyond the tail itself (thus the vernacular name train-bearers). These ornamental feathers are much less developed in females and immature males; up to three years may be needed for the coverts to reach maximum length among males. The three outer pairs of rectrices are graduated in length, tapered, and somewhat curved in shape; and they are rounded at their tips, unlike the six interior rectrices of adults, which have more-truncated tips. Nearly the entire upperpart plumage is iridescent green to golden green, and the underparts are at least partly brilliant red, the color produced by the carotenoid feather pigment zooerythrine, one of the cantaxanthins. The five species of the genus are strongly frugivorous and are relatively large and heavy (overall bill-to-tail length 28–37 cm [11.0–14.5 in] exclusive of longest tail-coverts; adult mass 150–250 g [5.3–8.9 oz]).

RESPLENDENT QUETZAL

Pharomachrus mocinno de la Lave 1832

See Eisenmann (1959) for a discussion of the proper spelling of *mocinno*.

OTHER VERNACULAR NAMES

Costa Rican quetzal (*costaricensis*), northern quetzal (*mocinno*), quesel; quetzaltototl (Nahautl); quetzal centroamericano (Spanish).

RANGE

Cloud forests of Central America: from eastern Oaxaca and Chiapas to western Honduras, eastern El Salvador (Honduran border region), and north-central Nicaragua; local in the mountains of Costa Rica (Tilarán and Talamanca Mountains and the central cordilleras) and western Panama (east, at least formerly, to Veraguas; now possibly limited to western Chiriquí). See map 4.

SUBSPECIES

Pharomachrus mocinno mocinno de la Lave 1832: Mexico, Guatemala, Honduras, and northern Nicaragua; also very local in El Salvador, where limited to the Honduras–El Salvador border area (Los Esesmìles, Montecristo).

Pharomachrus mocinno costaricensis Cabanis 1869: Costa Rica and western Panama.

MORPHOMETRICS

Measurements Wing, males (of *costaricensis*) 189–206 mm (7.4–8.0 in; average of 17, 199 mm [7.8 in]);

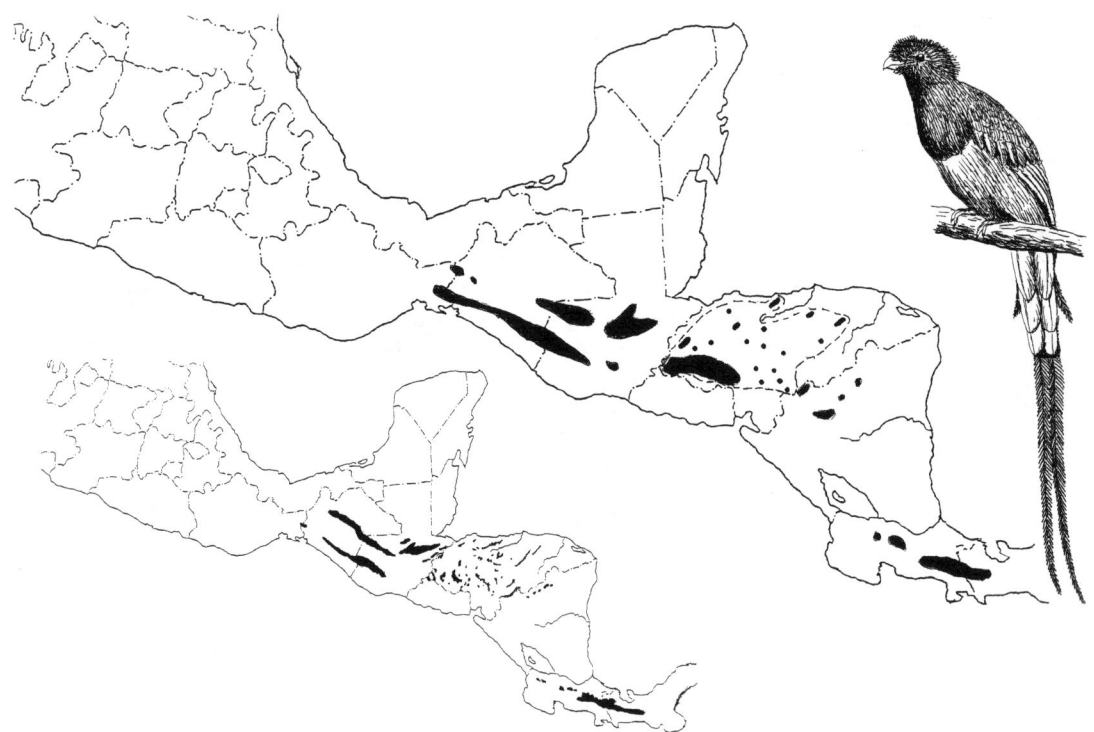

Map 4. Distribution of the resplendent quetzal (inked, but with Honduran range outlined and some locality records individually indicated; after Hanson 1982). Inset map shows distribution of wet montane forests (including cloud forests) within the range of the species.

Resplendent quetzal, adult male. Photograph by author.

195.5 mm (7.0–7.6 in; average of 17, 187.4 mm [7.3 in]); females 184–216 mm (7.2–8.4 in; average of 13, 196.8 mm [7.7 in]). Longest upper tail-coverts, males of seventeen *costaricensis* 480–825 mm (18.7–32.2 in), of fifteen *mocinno* 650–957 mm (25.4–37.3 in; Ridgway 1911). Egg of *mocinno* 34.9 × 29.1 mm (1.4 × 1.1 in), of *costaricensis* 38.9 × 30.2 mm (1.5 × 1.2 in; Schönwetter 1966).

Weights Seventeen birds of both sexes of *costaricensis* from Panama average 206.0 g (7.2 oz; Hartman 1961). Unspecified Costa Rican bird 210 g (7.4 oz; Stiles and Skutch 1989). One male of unknown origin 222 g (7.8 oz), one female 236 g (8.3 oz; specimens, MVZ). Two females of *mocinno* 136.8 and 225.5 g (4.8 and 7.9 oz; specimens, FMNH). One male from Honduras 233.9 g (8.2 oz); one male from Costa Rica 213 g (7.5 oz), one female 236.5 g (8.3 oz; specimens, LSU). Estimated egg weight of *mocinno* 16.2 g (0.6 oz; Schönwetter 1966), of *costaricensis* 19.6 g (0.7 oz). First clutch of two eggs of Costa Rican race average 18.85 g (0.7 oz), second clutch average 15.25 g (0.5 oz; Wheelwright 1983).

DESCRIPTION (of *mocinno*, adapted from Ridgway 1911; see plates 1 and 5)

Adult Male Upperparts, head, neck, and chest brilliant iridescent green or golden green, changing to bluish green or even greenish blue in certain lights, elongated greater wing-coverts with basal portion abruptly black; greater secondary and primary wing-coverts, alular feathers, and six middle rectrices uniform black; three outer rectrices white with black shaft, basal portion grayish black or slate; underparts posterior to chest intense carotenoid red, darkening to crimson or burnt carmine on upper breast; thigh feathers black, lower ones glossed with iridescent green; bill yellow to orange-yellow; iris dark brown; feet and toes olive to dull orange-brown.

Adult Female Crown and sides of head iridescent bronze-green, the cheek and laterofrontal feathers much less developed than those in male; back, scapulars, wing-coverts, rump, and upper tail-coverts bright, iridescent golden green, as in adult male; longer upper tail-coverts not reaching far, if at all, beyond tip of tail; remiges black, primaries broadly

females 193–208 mm (7.5–8.1in; average of 13, 198.4 mm [7.7 in]). The wings of *mocinno* males average 7 mm (0.3 in) longer than those of *costaricensis*, and the rectrices of the former average nearly 20 mm (0.8 in) longer. Tail, males (of *costaricensis*) 179.5–

edged with buff; tail black, three outer rectrices on each side broadly white distally and on outer vanes, white portion with dusky barring; chin and throat grayish brown; extreme lower abdomen, rear flanks, vent region, and under tail-coverts pure carotenoid red; thighs sooty blackish, the lower feathers glossed with iridescent green; maxilla blackish, streaked with yellow, sometimes entirely yellow; mandible dull yellow, tinged with green toward base; iris dark brown; feet and toes dull green to olive-brown.

Immature Male   Similar to adult female, but iridescent green of head and upperparts brighter; breast, abdomen, and sides gray instead of grayish brown; distal portion of rectrices broadly barred with white, outer rectrices barred only near base and more pointed than in adult; buff edges of outer vanes of primaries toothed with black; outer vanes of secondaries intermixed with buff; bill yellowish; other soft-part colors as in adult.

Immature Female   Similar to adult female but more bronzy to dull green above, and tail more coarsely patterned with black barring basally; tail-coverts (both upper and under) and outer three pairs of rectrices as in adult female, but feathers more pointed; bill yellowish, maxilla with brownish culmen.

Juvenile   Sooty brown above, scapulars and wing-coverts heavily mottled, wing feathers with buffy or tawny edges; breast dull cinnamon, edged with grayish brown; lower breast and abdomen white, barred with grayish brown on the sides; middle rectrices brownish black, outer ones white (Wetmore 1968). Skutch (1983) noted that at 23 days the head was mostly brown, with dull green feathers around the eyes. The sides of the neck and upper back were golden green, as were the central tail-coverts, these latter having black tips and brown subterminal spots. A bird near fledging had a mostly black upper head with a cinnamon-brown eyebrow stripe, pale cinnamon spots on the wing-coverts, grayish underparts, and some green iridescence at the base of the neck. Wheelwright (1983) noted that some green iridescence may appear on the contour feathers and wing-coverts as early as about 15 days after hatching; Skutch first observed green-tipped feathers on the neck and scapulars at 16–18 days and commented that these feathers were more cohesive than the

Resplendent quetzal, adult female. Photograph by author.

blackish, down-like feathers that preceded them. The bill was black and the iris brown, and the feet were lead colored. Somewhat older birds of undetermined sex resembled the female; but less green was visible on their plumage, and they had lighter breasts and abdomens and no red on the underparts. Younger nestlings with eyes still closed and early down-like feathers emerging have bills that are blackish only near the middle, with yellowish bases and tips (Bowes 1969).

IDENTIFICATION

In the Hand   Adult male resplendent quetzals are unmistakable, with their bushy crests and tail-coverts reaching well beyond their rectrices. Females' tail-coverts are not as long as those of the male, but like the male females have the feathers of both their crown and nape elongated into a bushy crest.

Resplendent quetzal, 21 days old. Photograph by
A. F. Skutch.

In the Field   No other quetzals occur in the range
of the resplendent quetzal (southern Mexico to
western Panama), so field identification is simplified.
The resplendent occurs in cloud forests, forest edges,
and areas of scattered trees near forests. In southeastern Panama (Darién) it is replaced by the golden-
headed quetzal, which somewhat resembles the
adult female resplendent quetzal but has a red lower
breast and a wholly dark undertail. The folded tail
of male resplendent quetzals has an exposed white
underside (comprising the three outermost pairs of
rectrices), and that of females is barred with black
and white. One typical vocalization is a loud and
repeated *hwaao* or *hwaco* (reputedly the onomato-
poetic basis for a native name for the quetzal in
Panama). In addition to the many variations of this
vocalization, there is also a throaty, repeated whistle,
keeeoooo, keeeooo. See also the vocalizations section.

GEOGRAPHIC VARIATION

The southern race (*costaricensis*) of the resplendent
quetzal is somewhat smaller in wing and tail meas-

urements, and the ornate upper tail-coverts of the
male are substantially shorter and narrower than
those of the nominate race. The southern birds are
also less golden toned.

ECOLOGY

Habitats   Collectively the resplendent quetzal
ranges from 1,000–3,300 m (3,300–10,800 ft) and is
essentially limited to montane evergreen forest (e.g.,
cloud forest), with nonbreeding habitats both lower
and more diverse. In Mexico this species is confined
to cloud forests between 1,400 and 3,000 m elevation
and is found mainly in Chiapas (Howell and Webb
1995). In Guatemala it occurs from 1,000 m to
2,300 m (3,300 to 7,500 ft), also in cloud forest (Land
1970); but LaBastille, Allen, and Durrell (1972) found
it at two cloud forest sites, one at 1,500–1,900 m
(4,900–6,200 ft) and another at 2,800–3,300 m (9,200–
10,800 ft). In Honduras it occurs in cloud forests
above 1,300 m (4,300 ft; Monroe 1968), and in Costa
Rica it extends from 1,200 m to 3,000 m (3,900 to
9,800 ft; Skutch 1983; Stiles and Skutch 1989). In
Panama it mostly occurs above 1,500 m (4,900 ft) in
humid montane forest but also uses forest edges and
park-like clearings (Ridgely and Gwynne 1989).
Skutch (1944, 1983) stated that preferred habitats
are cloud forest in which the trees are 30–45 m
(98–148 ft) tall, these cloud forests often being rich
in oak, laurel, mulberry, wild fig, and liquidamber
trees and having abundant epiphytic vegetation,
such as orchids and bromeliads. Also important are
fruit-bearing trees of the laurel family (Lauraceae),
such as wild relatives of the avocado (*Persea*), and
various other fruit-bearing trees such as *Nectandra*
and *Ocotea*. Outside the breeding season the birds
migrate to lower altitudes and may be found in park-
like pastures, on precipitous slopes, and at canyon
edges (Slud 1964). In the Monteverde area of Costa
Rica, which lies on the Continental Divide, the birds
may move down on either the Atlantic or Pacific
slope of the divide (Wille 1993). Powell and Bork
(1995) have commented on the fact that the establishment of nature preserves for species such as this
must take seasonal migrations into account if these
preserves are to be effective.

Foods and Foraging Ecology   Wheelwright (1983) and
Wheelwright et al. (1984) reported that resplendent

quetzals in the Monteverde area of Costa Rica consume fruits of 41 plant species (17 plant families), with nearly half (17–19 species) being drupes of the Lauraceae family. Other plant families for which more than one species was recorded included the Rubiaceae (four species) and the Moraceae, Symplocaceae, Rutaceae, and Verbenaceae (two species each). In Chiapas, Mexico, the birds were found to forage on 15 plant species, of which 6 were members of the Lauraceae family, with additional plants from the families Theaceae, Araliaceae, Verbenaceae, Solanaceae, Myrtaceae, Melastomataceae, Moraceae, and Clusiaceae (Avila, Hernandez, and Velarde 1996). Using evidence from regurgitated seeds, Wheelwright (1983) concluded that about half the total fruit intake at Monteverde was derived from lauraceous plants. At least at Monteverde, the quetzals apparently move according to the fruiting season of lauraceous plants, evidently "tracking" their favorite food resources. Quetzals typically pluck fruits in flight, but the birds sometimes land when pursuing insects or lizards. The birds can readily swallow fruits larger than one would predict from gape measurements (2.1 cm, 0.8 in), and their muscular esophagus is probably used in the regurgitation of seeds. There is no crop, but the gizzard has a diameter of 2.5 cm (1.0 in), which is somewhat larger than the mean diameter of lauraceous fruits (1.8 cm, 0.7 in) in the Monteverde area. Foods fed to the chicks during the first 10 days after hatching included insects (mostly beetles, lepidopteran larvae, and orthopterans, plus a few cicadas and a single example of Odonata), nine small lizards, and a few snails. In Chiapas 76 percent of the young were fed on insects (mainly orthopterans, coleopterans, lepidopterans, and various larvae) and small vertebrates during their first 10 days after hatching but received only 28 percent animal materials from then until fledging (Avila, Hernandez, and Velarde 1996). At Monteverde the adult birds remain at their food plants for a median duration of 3 minutes and have a processing time (ingestion-to-regurgitation period) that ranges from 19 to 67 minutes for nine species of lauraceous plants (Wheelwright 1991). Quetzals tend to gather their fruit by short sallies rather than by perching and hopping and usually forage in the lower and outer portions of trees (Santana and Milligan 1984).

BEHAVIOR

General, Social, and Sexual Behavior   During spring the male resplendent quetzal engages in occasional flight displays, circling above the forest canopy. While ascending he utters a *wac-wac* phrase similar to the possible alarm call, but in Skutch's view the call is simply a reflection of exuberance. Another apparent flight display, performed in a clearing with the long tail coverts lashing about, is accompanied by a similar *wac-wac-wac-wac-way-ho-way-ho*. The male's breeding territory is roughly spherical and averages 300–350 m (1,000–1,100 ft) in radius, extending up to about 4 m (13 ft) above the canopy. The overall home range during the breeding is larger and occupies 6–10 hectares (15–25 acres) in the upper third of the forest canopy. There are usually one or two giant fig (*Ficus*) trees present within the home range of each pair (Bowes and Allen 1969; LaBastille 1983).

Vocalizations   Skutch (1983) described five resplendent quetzal vocalizations, of which the most commonly heard is the *wac-wac, wac-wac* call, often uttered in flight, perhaps in alarm. During spring the birds produce a series of slurred, fused, and mellow notes that are deeper, fuller, and more powerful than those of smaller trogons; and in May a high soprano-like *whooo* note is uttered. LaBastille, Allen, and Durrell (1972) identified the five vocalizations described by Skutch, plus three additional ones. They believed that the two-note whistle of the male serves as a territorial advertisement call, which is uttered every 8–10 minutes with crest raised, cheeks and crest compressed, and breast feathers expanded. They considered Skutch's *wac-wac* flight call to be a pair-recognition call and described it as *wahc-ah-wahc*. They described the soprano-like note as a *coouee* whistle, either rising in scale or rising and falling like a wolf whistle. Its function was thought to be in mating, nest exchange, or nestling care. Another call, the *uwac*, that also rises in scale was heard during courtship chases.

Vocalizations of this species have been recorded (Hardy, Reynard, and Coffey 1995). The recording included songs and disturbance calls near the nest (recorded in Costa Rica by Gary Stiles) and calls of a captive female and male from Costa Rica (recorded by G. B. Reynard). The songs (examples one and five on the recording) consisted of rather loud *keeeeooo* notes of uniform pitch and volume uttered at about 1-second intervals, either uniformly spaced or (example one) in groups of three followed by brief pauses. As many as 12 such notes were uttered in a continuous series. A similar example (three) consisted of several down-slurred *peeeooo* notes similar in tonal quality to the extended song type but dropping in pitch. Another example (two) included chattery notes and a sharp *go-back* call (probably the *hwaco* call noted earlier) similar to a willow ptarmigan's (*Lagopus lagopus*) alarm or disturbance call. See also the identification section above.

BREEDING BIOLOGY

Chronology of Breeding   Three resplendent quetzal egg sets from Oaxaca in the collection of the Western Foundation of Vertebrate Zoology date from March 29 to April 29. In Guatemala laying occurs from March to May, the extreme dates for four nests being March 26 and May 21 (LaBastille, Allen, and Durrell 1972). In El Salvador there are several breeding records for May and June (Thurber et al. 1987). In Costa Rica, Skutch (1983) found eggs or incubating birds during April, May, and June. Wheelwright (1983) found that the peak of the breeding season at Monteverde (April and May) corresponds to the peak fruiting period for 12–15 species of Lauraceae. Breeding in Panama occurs from February to May (Ridgely and Gwynne 1989).

Nest Sites   Three resplendent quetzal nests found in Oaxaca were all in dead trees or stumps at heights of 2.5 to 4.7 m (8 to 15 ft) above ground. The nest cavity of one site was about 35 cm (13.8 in) in diameter but was very shallow, the bottom of the cavity only 5 cm (2.0 in) below the entrance. In another the bottom was about 30 cm (11.8 in) below the entrance (Rowley 1984). Thurber et al. (1987) saw a nest in a dead tree about 15 m (49 ft) tall, with the opening about 2 m (7 ft) below the top. Skutch (1983) described six nests, which ranged from 4.3–27 m (14–89 ft) above ground, with entrances 10–11.4 cm (10.4–4.5 in) in diameter. One nest had a cavity 28 cm deep by 15 cm wide (11.0 × 5.9 in), and in another the bottom of the cavity extended only 11.4 cm (4.5 in) below the entrance. It is possible that at

times the birds take over old woodpecker holes, enlarging them when necessary. Many of the trees in which the birds nested were in the last stages of decay, and two collapsed during inspection. As with other trogons, both sexes of the resplendent quetzal apparently help excavate. In Guatemala ten nest sites averaged nearly 10 m (33 ft) high and had been excavated in rotting stubs about 12.5 m (41 ft) in height. The nest entrances averaged 10 cm (3.9 in) in diameter, with interior cavities averaging 20 cm (7.9 in) wide and 30 cm (11.8 in) deep. The birds sometimes moved their sites downward as the snag rotted from above (Bowes and Allen 1969; LaBastille, Allen, and Durrell 1972).

Wheelwright (1983) noted that one pair began excavating five sites within a month before finally selecting one that was within 100 m (300 ft) of the earlier sites. These were mostly excavated in decaying *Ocotea* or other lauraceous trees. Furthermore, 74 percent of 43 nest sites were in forest, 12 percent were at forest or pasture edges, and 14 percent were in the open. The mean nest height was 8.8 m (29 ft), and the nests were in limbs or trunks of trees having a mean diameter of 0.63 m (2 ft) breast height.

Eggs and Incubation   The normal clutch size for the resplendent quetzal appears to be two eggs (LaBastille, Allen, and Durrell 1972; Rowley 1984; Skutch 1983). A few single-egg nest records probably represent incomplete clutches. Skutch (1983) found that in this species the male incubated in the morning for 2 to 4 hours and was then replaced by the female, who incubated for a few hours. The male then returned to the nest for an incubation stint in the afternoon. The female returned at about sundown to take over nighttime incubation. Sometimes the male performed a flight display upon being relieved. At one nest an incubation period of 17–18 days was established by Skutch, and at two nests observed by Wheelwright (1983) the incubation period was 18–19 days. Wheelwright did not observe the double male incubation shift during daylight hours described by Skutch, but LaBastille, Allen, and Durrell (1972) said that their observations of incubation rhythms roughly corresponded with Skutch's.

Brood Rearing   Skutch (1983) reported that nestling resplendent quetzals, like all other trogons studied so far, are hatched with tightly closed eyes. Again like other trogonids, they have small protuberances on their heels, which probably aid in gripping the substrate and perhaps also in rearing to obtain food. Pinfeathers began to appear at 2 days and began to break open to expose contour feathers on the 7th day, although the rectrices and remiges did not expand until the 10th day. Until their 10th day the chicks were fed only animal foods, but from 11 days on the adults began to feed them fruits, especially fruits of laurels. Toward the end of the brooding period only the male fed the chicks, which fledged at 23 days. At two other nests the chicks fledged at 29 days and 31 days. Two chicks 10–11 and 13–14 days old were described by LaBastille, Allen, and Durrell (1972) as being only partly feathered out and having their eyes still closed. Most of the developing feathers were blackish, and some were distinctly "fluffy," temporarily producing a rather down-like appearance. One of these chicks was illustrated photographically (see figure 17, bottom, for a drawing based on the photo), and both chicks were also illustrated in a color plate (Bowes 1969). Older chicks were illustrated by Wille (1993) and Winter (1998).

In Wheelwright's (1983) observations, the male parent brought more food items to the nest than did the female, and these items included more insects; whereas the female delivered relatively more fruits. The nestlings took rather large drupes when still very young; at 10 days one nestling's gape was even wider than the adults'. Parents bringing fruit returned at shorter intervals than did parents bringing insects (averages 15 minutes versus 24 minutes). LaBastille, Allen, and Durrell (1972) observed the adults bringing fruits, termites, spiders, a moth, and a beetle to the nest.

Each of the three pairs studied by Skutch attempted to produce a second brood, apparently within a month after the first brood fledged. Two of the pairs laid in the same nest hole as previously, whereas the third, whose nest tree had fallen, began in a new site. Wheelwright (1983) noted that one pair laid its first egg in its original nest only 14 days after loosing the chicks of its first effort, and a second egg was laid the following day. In two other cases the pairs chose new sites after losing their eggs to predators. There was a high rate of nest failure, estimated at 67–78 percent of nine nesting efforts. Weasels (*Mustela frenata*) were believed to be a ma-

jor predator of eggs and nestlings, and an unidentified hawk and a margay (*Felis weidii*) each took an adult. All three nests studied by LaBastille, Allen, and Durrell failed: two were abandoned, and one was destroyed when the stub supporting the nest collapsed. Both of the failed pairs studied by Bowes and Allen (1969) renested. Winter (1998) reported that 80 percent of young quetzals die before fledging and that less than 20 percent of the remainder survive to adulthood.

CONSERVATION AND EVOLUTIONARY RELATIONSHIPS

The resplendent quetzal is classified as a near-threatened species of trogonid by the World Conservation Union and is listed (appendix II) by CITES. Although officially protected in several countries (Mexico, Costa Rica, and Panama), such protection is impossible to enforce in the remote areas the birds inhabit. Guatemala has proclaimed this species as its national bird, but this designation has served only to increase its rate of exploitation both in Guatemala and elsewhere; and one wild-bird dealer exported 100 quetzals from Costa Rica in a single year (Kern 1968; Mountfort 1988). The species occurs in at least six Central American countries, from Mexico to Panama. Honduras probably supports one of the largest, if not the largest, remaining quetzal populations, a population scattered widely across the higher mountains of that country (Hanson 1982). Costa Rica also has a good population, one that is now better protected than it has been in the past, both in terms of critical forest habitat and protection from illegal capture and export. Quetzals are protected in three of Costa Rica's national parks (Volcán Poas, Braulio Carillo, and Chirripó), as well as in the Monteverde Forest Preserve. There is a private preserve on Guatemala's Mount Atitlán and a private quetzal sanctuary (Mario Dary Rivera) near Cobán that is operated by the University of San Carlos (LaBastille 1973, 1983). The availability of suitable nest trees in the last stages of decay may be limiting factors in the survival of these birds. Forest fires and clear-cutting are obvious major environmental hazards for the quetzal; a less important one is the felling of individual trees by natives to obtain wild honey. Artificial nest boxes have been designed for and used to some extent by the birds (LaBastille

Figure 17. Developmental stages of juvenile trogons, including mountain trogons at two weeks (A), after photographs by J. S. Rowley and A. F. Skutch; resplendent quetzal at two weeks (B), after a photograph by David Allen; and somewhat older juvenile golden-headed quetzal (C), after a Houston Zoological Gardens photograph.

1974). The preservation of moist montane forests will be vital to the preservation of this flagship trogonid species, but lower-altitude preserves will also be needed to provide protection during seasonal migration to more-tropical forests (Hernandez 1998). Stotz et al. (1996) listed the species as uncommon, with a medium degree of sensitivity to disturbance.

Obviously the resplendent quetzal is a very near relative of the crested quetzal; and perhaps the best reason for recognizing them as two separate species is for the conservation benefits gained by keeping the resplendent quetzal on the list of near-threatened species. Popular articles on the conservation of the resplendent quetzal include those by Bowes (1969), LaBastille (1975, 1976, 1985), Kern (1968, 1975),

Skutch (1982), Maslow (1986), Wille (1993), and Winter (1998).

Many quetzals have been brought into captivity, and the high prices paid by zoos and aviculturists for them have played a significant role in their decline in the wild. They have only rarely bred in captivity; but a male at the Bronx Zoo in New York survived for 17 years, and a female for 21 years. These examples represent much longer survival periods than have been reported for the smaller species of trogons.

CRESTED QUETZAL

Pharomachrus (mocinno) antisianus (d'Orbigny) 1837

OTHER VERNACULAR NAMES

Beautiful train-bearer, d'Orbigny's trogon; quetzal coliblanco (Spanish).

RANGE

Subtropical forests, from western Venezuela (Zulia, Trujillo, Táchira, and Mérida) through Colombia, Ecuador, and Peru to central Bolivia (La Paz, Cochabamba, and western Santa Cruz). See map 5.

SUBSPECIES

There are no recognized subspecies of the crested quetzal, but this form is frequently considered a race of *Pharomachrus mocinno*, as by Peters (1945), Zimmer (1949), and de Schauensee (1964). However, Wetmore (1968) argued that the two should be considered separate species, together making up a superspecies, in view of crest and tail differences among males.

MORPHOMETRICS

Measurements   Wing, males 182–195 mm (7.1–7.6 in; average of 3, 188.3 mm [7.3 in]; specimens, FMNH). Wing, one male 190.5 mm (7.5 in); one female 193 mm (7.5 in). Tail, one male 160 mm (6.2 in); one female 167.6 mm (6.5 in; Ogilvie-Grant 1892). Wing, unsexed adult 184 mm (7.2 in); tail 184 mm (7.2 in; Gould 1858–1875). Egg 30.1 × 25.8 mm (1.2 × 1.0 in; Schönwetter 1966).

Weights   One male 226 g (7.9 oz; specimen, MVZ). Three males 145–161 g (5.1–5.6 oz; average 150.6 g [5.3 oz]), two females 134 g (4.7 oz) and 138 g (4.8 oz; specimens, LSU). Estimated egg weight 11.0 g (0.4 oz; Schönwetter 1966).

DESCRIPTION (see plate 6)

Adult Male   Upperparts iridescent emerald to bronzy green; forehead crest bushy and golden green to bronzy green, not extended backwards but limited to the forehead and lores; wing-coverts and elongated tail-coverts (green upper tail-coverts extend only approximately 2.5 cm [1.0 in] beyond rectrices) also iridescent emerald-green; head more golden green; breast like back; wings black; underparts otherwise crimson red; three outermost pairs of rectrices black basally but entirely white as seen from below; more-interior rectrices black; soft-part colors as in preceding species, but bill more orange-yellow.

Adult Female   Head and chest brown, the latter glossed with iridescent green; feathers of crown and nape not elongated; wings black, except primaries

Crested quetzal, adult male. Photograph by author.

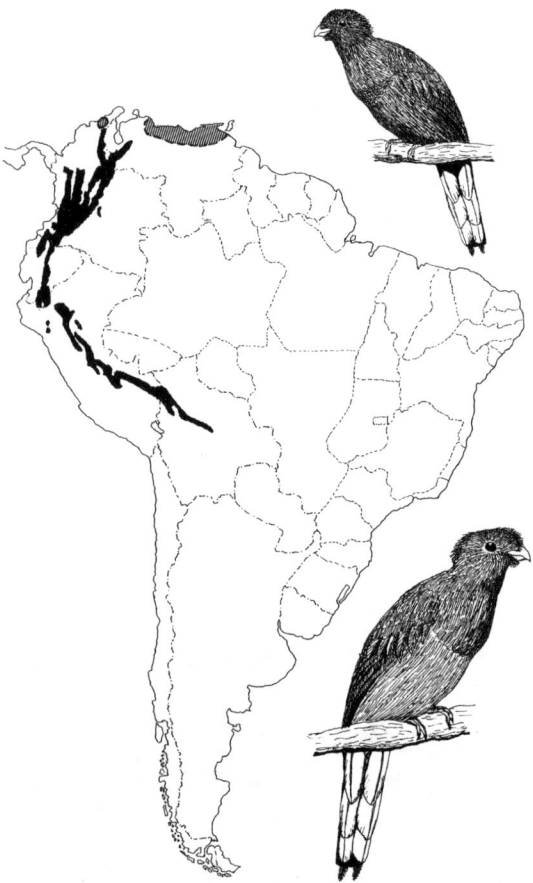

Map 5. Distribution of the crested quetzal (inked area and lower sketch) and white-tipped quetzal (hatched area and upper sketch).

brown, margined with buff; underparts red, paler posteriorly; upperparts mostly iridescent green; throat and lower underparts brownish gray, except for iridescent green breast band and pinkish on abdomen and under tail-coverts; rectrices mostly black, two or three outermost rectrices notched with white on outer vanes and narrowly tipped with white; soft-part colors as in resplendent quetzal; bill dusky.

Immature Male   Crest less developed than in adult males; head more bronzy; outer vanes of secondaries barred and notched with buff; outer vanes of primaries almost entirely buff, with some brown barring; outermost rectrices mostly white on outer vane, next two pairs notched with white on outer vanes, and both vanes tipped with white.

Immature Female   Female-like, but with rectrices more pointed, outer webs of secondaries and primaries mixed with buff, and chin and breast grayish brown.

Juvenile   Loose black plumage with buff spots and tawny buff abdomen, but with some iridescent green feathers appearing even before fledging; pale markings on outer secondaries persist for a long period (Fjeldså and Krabbe 1990).

IDENTIFICATION

In the Hand   The crested quetzal is generally quite similar to the resplendent quetzal; but males of the former have the feathers of their lores and forehead only moderately lengthened into a crest, and the crown and nape feathers are not at all lengthened. Females differ from those of the resplendent quetzal in lacking crest-like feathering, and their head plumage is brownish rather than iridescent green.

In the Field   The crested quetzal is a cloud-forest species usually occurring at moderate to fairly high elevations (1,200–3,000 m; 3,900–10,000 ft), often with the golden-headed quetzal, but sometimes at slightly lower elevations. The males have yellow bills, moderately lengthened forehead and nape feathers, wholly white outer rectrices, and iridescent green upper tail-coverts that extend only slightly beyond the longest rectrices. Females have brownish, uncrested head plumage and less-brilliant green iridescence on the breast than do males, which color becomes increasingly brown on the lower breast and abdomen. The female's outer three pairs of rectrices are strongly barred with black and white and tipped with white. Like the other quetzals, this species produces repeated loud hooting vocalizations and also utters a clear, smooth whistle. See also the vocalizations section.

GEOGRAPHIC VARIATION

None described.

ECOLOGY

Habitats   Collectively the crested quetzal ranges from 1,200 to 3,000 m (3,900–10,000 ft), in subtropical to temperate forests, especially cloud forests. The species is limited to cloud forests at 1,200–3,000 m (3,900–10,000 ft) elevation in Venezuela (de Schauensee and Phelps 1978). In Colombia it is confined to the subtropical lower mountain zone, from at least 1,400 to 2,800 m (4,600–9,200 ft; de Schauensee 1948–1952). Hilty and Brown (1986) also reported the range as 1,499–2,800 m (4,900–9,200 ft) in humid forest, forest edges, and tall secondary growth.

Foods and Foraging Ecology   De Schauensee and Phelps (1978) listed fruit, berries, insects, lizards, and frogs as foods of the crested quetzal. Hilty and Brown (1986) noted that the birds are often seen at trees bearing relatively large fruits, such as *Persea* and *Ocotea*. One stomach examined by Remsen, Hyde, and Chapman (1993) contained fruit only.

BEHAVIOR

General, Social, and Sexual Behavior   No specific information available.

Vocalizations   The most common call of the crested quetzal is a loud and deliberate *way-way-wayo*, but the alarm call is a *ha-ha-ha* . . . sequence. Vocalizations of this species were recorded in Peru by Ted Parker (Hardy, Reynard, and Coffey 1995). The song (example one) consisted of a series of distinctly double *weee-coo* notes and was higher and more squeaky than the resplendent quetzal's song; the second note dropped distinctly in pitch. The second example on the recording was more chattery and included several *waa-co'* (or *go-back*) and *waa-co-coo', co-coo'* notes, the second or final syllable louder and sharper than the first. This call was uttered faster

than the corresponding "go-back" call of the resplendent quetzal and was followed by and interspersed with what apparently were song elements. The general sound produced somewhat resembles a flicker's (*Colaptes*) chatter. Fjeldså and Krabbe (1990) described the song as a loud, rolling *way-way-wayo* and described one call as a muffled whistle *whee-coo, chuk*. The alarm call was said to be a series of *ka* notes.

BREEDING BIOLOGY

Chronology of Breeding   Crested quetzal breeding in Colombia probably occurs mainly between February and June (Fjeldså and Krabbe 1990). Hilty and Brown (1986) stated that most calling in Colombia occurs between February and April.

Nest Sites   No information available.

Eggs and Incubation   No information available.

Brood Rearing   No information available.

CONSERVATION AND EVOLUTIONARY RELATIONSHIPS

The crested quetzal is perhaps the least studied of all quetzals. It is a very close relative of the resplendent quetzal; the two forms are now ecologically separated by the Panamanian lowlands. On the basis of gene-sequencing studies, Espinosa de los Monteros (1998) considered this form and the resplendent quetzal to be sister species, along with *auriceps*. Stotz et al. (1996) listed the species as uncommon, with a medium degree of sensitivity to disturbance.

WHITE-TIPPED QUETZAL

Pharomachrus fulgidus (Gould) 1838

OTHER VERNACULAR NAMES

Bang's trogon or quetzal (*festatus*), shining train-bearer; quetzal dorado (Spanish).

RANGE

Venezuela (coastal, interior, and eastern cordilleras east to Cerro Humo) west through montane forests of northeastern Colombia (Santa Marta Mountains). See map 5.

SUBSPECIES

Pharomachrus fulgidus festatus Bangs 1899: Santa Marta Mountains, in northeastern Colombia.
Pharomachrus fulgidus fulgidus (Gould) 1838: northeastern and north-central Venezuela.

MORPHOMETRICS

Measurements   Wing, one male of *fulgidus* 172 mm (6.7 in); one male of *festatus* 183 mm (7.1 in); one female of *fulgidus* 173 mm (6.7 in); two females of *festatus* 181 and 183 mm (7.06 and 7.14 in; Peters 1929). Wing, one male of *festatus* 190 mm (7.4 in); one female 189 mm (7.4 in). Tail, one male of *festatus* 157 mm (6.1 in); one female 158.4 mm (6.2 in; Bangs 1899). The upper tail-coverts of *festatus* project farther beyond the rectrices than do those of *fulgidus* (32–39 mm [1.2–1.5 in] versus 16–19 mm [0.6–0.7 in]; Todd 1943). Bangs (1899) noted than the tail-coverts of the male *festatus* extend 48 mm (1.9 in) beyond the rectrices. No information on eggs available.

Weights   No information available.

DESCRIPTION (of *festatus*, mainly after Bangs 1899; see plate 7)

Adult Male   Head, back, rump, breast, upper tail-coverts, and wing-coverts iridescent green, appearing bronzy in some lights, especially on head, throat, and upper tail-coverts; forehead with small frontal crest of antrose feathers; abdomen and under tail-coverts scarlet-vermilion; primaries, secondaries, tertials, and greater wing-coverts black; flanks and sides black, black feathers mostly concealed; tail black, vanes of three outermost rectrices with grayish white tips; white tips of outermost rectrix 50 mm (2.0 in) long, next 59 mm (2.3 in) long, and next 32 mm (1.2 in) long; bill yellow; iris hazel; feet and toes brownish black.

Adult Female   Less brilliant than male; no frontal crest on forehead; throat and breast mixed with much drab brown; outer edges of primaries yellowish brown; breast brownish gray, bordered by green band; abdomen pink; tail black, three outermost rectrices narrowly tipped with white, terminal parts of feathers crossed with two or three black bars; outermost rectrix with three white spots on outer vane below lowest crossbar; bill yellowish brown; iris brown; feet and toes as in male.

Immature Male   Remiges edged with buff on outer edges; scapulars and greater wing-coverts intermixed with buff; tips of outer rectrices suffused with white, these feathers narrower and more pointed than in adults.

Immature Female   Wing and head plumages as in immature male; breast brown, as in adult female.

Juvenile   No information available. See species accounts of other quetzals.

IDENTIFICATION

In the Hand   Adult males of the white-tipped quetzal resemble the golden-headed and pavonine quetzals, but in the former, forehead feathers are longer and more crest-like, the head is more golden toned, and the mostly black outer rectrices have broad white tips occupying about a third of their length (thus the name white-tipped quetzal). Males differ from the closely related crested quetzal in having a smaller frontal crest, a more golden-bronze crown, and outer tail feathers that from below appear black and tipped broadly (terminal third) with white, rather than entirely white. The female white-tipped quetzal differs from the female crested quetzal in having a bronzy green crown and a greener breast, and the three outer pairs of tail feathers of the former are tipped and barred with white. White-tipped quetzals are also similar to both the golden-headed and the pavonine species but have more white on their outer rectrices, which in the latter two species are instead an all-black or nearly black appearance. In both bill size and overall size, the white-tipped quetzal is intermediate between *antisianus* and *auriceps*.

In the Field   White-tipped quetzals, like the golden-headed and crested quetzals, are cloud-forest birds: the white-tipped usually occurs at 900–1,900 m (3,000–6,200 ft) elevation in Venezuela but at 1,500–2,500 m (4,900–8,200 ft) in Colombia's Santa Marta Mountains; the crested higher at 1,200–3,000 m (3,900–9,800 ft); and the golden-headed at 2,000–3,100 m (6,700–10,000 ft). In contrast, the pavonine quetzal occurs in lowland rain forest habitats (up to 700 m [2,300 ft]). The white-tipped's vocalizations reportedly consist of "booming hoots" audible for some distance, and they are seemingly generally similar to those of the golden-headed quetzal but tend to be more three noted. See also the vocalizations section.

GEOGRAPHIC VARIATION

In addition to averaging slightly larger in wing measurements and having somewhat longer cheek plumes, *festatus* has tail-covert plumes that extend more than 25 mm (1.0 in) beyond the rectrices, whereas in *fulgidus* the plumes project only very slightly beyond the rectrices (Peters 1929).

ECOLOGY

Habitats   Collectively the white-tipped quetzal ranges from 900–2,500 m (3,000–8,200 ft), in subtropical to temperate humid forests, including cloud forests, secondary forests, forest remnants, forest edges, and moist ravines in coffee plantations. It occurs at altitudes of 900–1,900 m (3,000–6,200 ft) in Venezuela and at 1,500–2,500 m (4,900–8,200 ft) in Colombia (Hilty and Brown 1986; de Schauensee and Phelps 1978). Todd and Carriker (1922) found the species between 1,560 and 2,500 m (5,100 and 8,200 ft) in the Santa Marta Mountains of northeastern Colombia, but only in heavy forest. Breeding activities have been observed at 1,600 m (5,200 ft) and also at 2,000–2,200 m (6,700–7,200 ft).

Foods and Foraging Ecology   Like the other quetzals, the white-tipped quetzal forages on fruits and berries but probably feeds some insects and other high-protein foods to chicks. During the coffee-ripening season the birds gather at plantations to feed on the berries (Todd and Carriker 1922).

BEHAVIOR

General, Social, and Sexual Behavior   No detailed information available.

Vocalizations   Similar to the resplendent quetzal in its hoot-like calls, the white-tipped quetzal utters loud, booming *kirra* or *kirra, kip* notes and when excited produces *kierr, kip-kip, kipa-a* calls or variants (Hilty and Brown 1986). Vocalizations of this species were recorded by Ben Coffey Jr. in Colombia (Hardy, Reynard, and Coffey 1995). The recorded song, a regularly repeated *wa-coo' whew*, is at least as high pitched as the crested's but has a more plaintive quality. Another example on the recording is a sequence of 14 such triple-noted utterances, produced over a period of about 20 seconds and gradually increasing in loudness. See also the identification section above.

BREEDING BIOLOGY

Chronology of Breeding   Breeding-condition white-tipped quetzals have been collected in the Santa Marta Mountains between January and April, and pairs at nests have been seen during March and August (Hilty and Brown 1986).

Nest Sites   White-tipped quetzal nests are located in trees, sometimes in old woodpecker holes and often in dead and rotting trunks or stubs, from 4 m

(13 ft) to more than 10 m (33 ft) above ground (Hilty and Brown 1986; de Schauensee and Phelps 1978). Todd and Carriker (1922) found a nest in a cavity of a *Scapaneus* tree, about 4 m (13 ft) above ground.

Eggs and Incubation No detailed information is available on the eggs of the white-tipped quetzal, but they are probably much like those of the other quetzals.

Brood Rearing No information available.

CONSERVATION AND EVOLUTIONARY RELATIONSHIPS

The white-tipped quetzal has the smallest distribution of any quetzal, and its survival depends entirely on conservation efforts by the Colombian and Venezuelan authorities. It is probably a close relative of the crested quetzal but is geographically isolated from it as well as from the other quetzal species. Stotz et al. (1996) listed the species as uncommon, with a medium degree of sensitivity to disturbance.

GOLDEN-HEADED QUETZAL

Pharomachrus (pavoninus) auriceps (Gould) 1842

OTHER VERNACULAR NAMES

Golden-headed train-bearer; yellow-bellied trogon; viuda de la montaña (Spanish).

RANGE

Montane forests of eastern Panama (Darién), Venezuela (Sierra de Perija on the Colombian border and east into the mountains of Táchira and Mérida to northern Trujillo), Colombia, Ecuador, and eastern Peru to northern Bolivia (La Paz, Cochabamba, Santa Cruz). See map 6.

SUBSPECIES

Pharomachrus auriceps hargitti Oustalet 1891: extreme eastern Colombia and western Venezuela (Sierra de Perija and Andes from Trujillo south to Táchira), at 2,000–3,100 m (6,600–10,100 ft). This subspecies is sometimes considered a race of *Pharomachrus pavoninus.*

Pharomachrus auriceps auriceps Gould (1842): Andes (1,400–2,700 m; 4,600–8,900 ft) as indicated in species' range except for Venezuela and the Sierra de Perija (an area occupied by *hargitti*).

MORPHOMETRICS

Measurements   Wing, males from Panama 185–196 mm (7.2–7.6 in; average of 10, 190.3 mm [7.4 in]); females 178–199 mm (6.9–7.8 in; average of 9, 189.3 mm [7.4 in]). Tail, males 155–177 mm (6.0–6.9 in; average of 10, 164.8 mm [6.4 in]); females 157–172 mm (6.1–6.7 in; average of 9, 163.5 mm [6.4 in]; Wetmore 1968). Wing, males from Bolivia 187–205 mm (7.3–8.0 in; average of 3, 195.3 mm [7.6 in]); one female 210 mm (8.2 in; Bond and de Schauensee 1943). Egg 37.0 × 29.4 mm (1.4 × 1.1 in; Schönwetter 1966).

Weights   One male 222 g (7.8 oz; specimen, MVZ). Seven males 154–220 g (5.4–7.7 oz; average 176.7 g [6.2 oz]), two females 170 and 182 g (6.0 and 6.4 oz; specimens, LSU). Estimated egg weight 17.6 g (0.6 oz; Schönwetter 1966).

DESCRIPTION (mainly after Wetmore 1968; see plate 8)

Adult Male   Forehead feathers not extended forward into bushy crest, although nape feathers somewhat elongated; entire head including throat shining golden green with coppery sheen; lower hindneck, back, rump, and lesser and median wing-coverts somewhat bluish green to yellowish green; primaries, secondaries, and greater wing-coverts black;

Golden-headed quetzal, male (background) and female. Photograph by author.

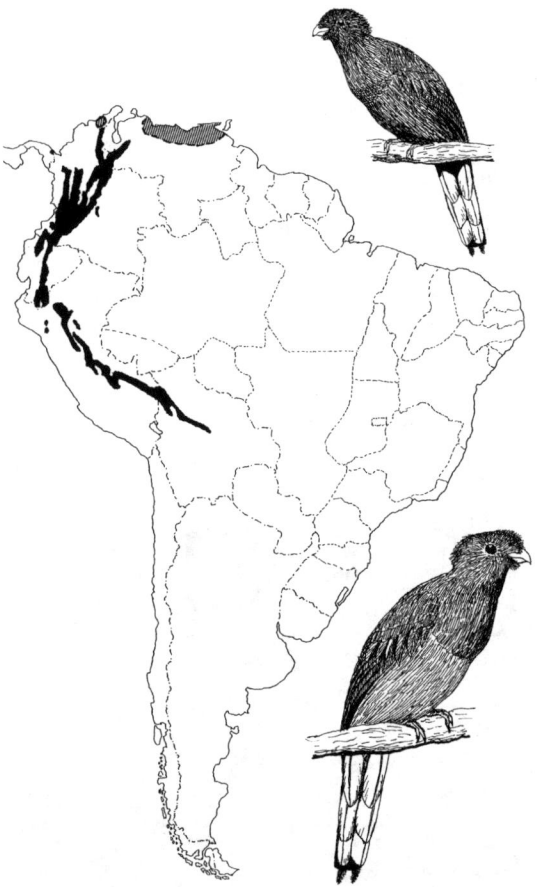

Map 6. Distribution of the golden-headed quetzal (inked area and lower sketch) and the pavonine quetzal (hatched area and upper sketch).

tail entirely dark; elongated upper tail-coverts distinctly bordered with light yellowish green; rectrices black (outer two rectrices occasionally slightly tipped with grayish white); lower foreneck, upper breast, and upper wing-coverts like back; rest of underparts brilliant carotenoid red; bill yellow; iris reddish brown to brown; feet and toes fuscous to olive-brown.

Adult Female   Crown and sides of head dull golden olive-brown; rest of upperparts like those of male, but primaries (except outermost) edged with dull brownish buff; throat brownish slate, with faint wash of golden brown; rest of foreneck rather dull light green; breast, flanks, and upper abdomen brown; lower abdomen and under tail-coverts clear red; tail black, two outermost pairs of rectrices usu-

ally entirely black but sometimes slightly barred and tipped with white; bill dull brown, paler on lower sides of mandible; iris, feet, and toes as in male.

Immature Male   Crest lacking; throat and cheeks buffy; rest of head blackish, tipped with golden bronze; primaries and secondaries edged with buff; large buff spots occur on scapulars and greater wing-coverts; rectrices black with pointed tips; tips of outer rectrices suffused with white; bill black, becoming yellowish toward the base; breast and upperparts as in immature female.

Immature Female   Wings and head as in immature male; upper breast dark brown, with paler feather tips, producing a slightly barred effect, light tones predominating on lower breast and abdomen; iridescent green feathers appearing first on scapulars and rump; pale tawny edging on outer edges of remiges, and similar pale buffy to tawny spots on tertials and most of the upper wing-coverts; sometimes also with mottling toward tips of outer rectrices; bill and toes black; iris blackish brown (description based on photos of captive bird).

Juvenile   No information available.

IDENTIFICATION

In the Hand   Male golden-headed quetzals may be recognized by their lack of a distinct forehead crest; their golden-green to bronzy head plumage; their black outer rectrices (sometimes tipped slightly with grayish white); and their yellow bill. The ornamental tail-coverts extend only slightly beyond the rectrices. The outer rectrices of females are usually entirely black but sometimes have slight barring and narrow white tips; the middle abdomen of males is red, and the head of both sexes is green or bronzy green. In females the upper tail-coverts are also shorter than in adult males and do not reach the end of the tail.

In the Field   The golden-head quetzal is associated with high foothill and montane forests, including cloud forests, and sometimes also extends into forest edges and clearings. It mostly occurs between 1,400 and 3,100 m (4,600 and 10,100 ft) but occasionally reaches 3,300 m (10,800 ft) or descends to 1,000 m

(3,300 ft). Adults of both sexes differ from other potentially confusing quetzals in having outer rectrices that are entirely or almost entirely black, and males lack the definite forehead crest that occurs in the crested quetzal. Fjeldså and Krabbe (1990) described the usual call as a melodic, sad *wa-dwyi,* similar to the last two notes of the song of an ant-pitta (*Grallaria ruficapilla*). Other calls include a plaintive *ka-kaaaur* and a rapid, whinnying *WHY-dy-dy-dy-dyyrr, hoo-whooooay,* with the first part of the introductory notes the highest in pitch. See also the vocalizations section.

GEOGRAPHIC VARIATION

No geographic variations have been described for the golden-headed quetzal. However, if this species is considered subspecies of *pavoninus,* the former averages slightly larger than *pavoninus* and smaller than *hargitti.* The feathers of the head and throat are longer in males of *auriceps* than they are in *pavoninus,* the gular area is more extensive, and the bill is yellow rather than red, although the red color may be reduced or sometimes even lacking in *pavoninus.* Females show these same relative size differences and have similar differences in head-feather lengths and the extent of gular areas. The bill is brown in both *auriceps* and *pavoninus,* but it is lighter brown in *auriceps* and sometimes reddish in *pavoninus.* The biggest differences are in the outer rectrices of females, which in *auriceps* are blackish (sometimes with paler tips) but in *pavoninus* are marked with white (or whitish) tips and white spotting for some distance basally (Zimmer 1949).

ECOLOGY

Habitats   Collectively the golden-headed quetzal ranges through subtropical and temperate foothill and montane forests, including cloud forests, sometimes reaching the elfin-forest zone just below timberline. According to Fjeldså and Krabbe (1990), this species favors midaltitude, subtropical wet and humid montane forests, forest edges, and tall secondary forest and is also sometimes found in temperate cloud forests and elfin forests. As noted in the identification section, forest clearings are sometimes also used. In Panama the species has been recorded between 1,200 and 1,500 m (3,900 and 4,900 ft; Ridgely and Gwynne 1989). It occurs there between 1,400 and 3,300 m (4,600 and 10,800 ft) elevation, but probably descends somewhat lower outside the breeding season. Hilty and Brown (1986) listed its range in Colombia as 1,400–2,700 m (4,600–8,900 ft) in humid forests, in forest edges, and occasionally in secondary growth.

Golden-headed quetzal, adult female. Photograph by author.

Foods and Foraging Ecology   The eating foraging habits of the golden-headed quetzal have not yet been studied in the wild, but captive birds at the Houston Zoological Gardens favored (in order of preference) commercial bird-of-prey diet, avocados, grapes, raisins, soaked dog and cat chow, bananas, a mixture of papayas and cantaloupe, and tomatoes. Live insects (mealworms, crickets, moths and other flying insects) were seldom eaten, as were chopped newborn "pinky" mice (Shelton n.d.). Of eight stomachs examined by Remsen, Hyde, and Chapman (1993), 87.5 percent contained fruit only, and 12.5 percent contained a mixture of fruit and arthropods.

BEHAVIOR

General, Social, and Sexual Behavior   No specific information on behavior is available for the golden-headed quetzal. The captive male at the Houston Zoological Gardens performs display flights early in the breeding cycle (Lee Schoen, pers. comm.). The birds are in a rather small flight cage, and they fly conspicuously from branch to branch, and occasionally to the nest site, without vocalizing.

Vocalizations   Vocalizations of the golden-headed quetzal are little known but are said to include mellow, reedy, and somewhat hawk-like *wheee, wu-weeeoo, we-weeoo* notes with as many as four *we-weeoo* phrases in Peru and as many as seven in Ecuador (Hilty and Brown 1986). Vocalizations of the species were recorded in Peru by Ted Parker (Hardy, Reynard, and Coffey 1995). The recorded song consists of variable numbers of repeated *wee-whew'* or *wee-a-whew'* notes uttered in unbroken succession, approximately at the rate of one per second. See also the identification section above.

BREEDING BIOLOGY

Chronology of Breeding   The breeding of the golden-headed quetzal in the wild is still unstudied, but in Colombia breeding probably occurs from April to June (Fjeldså and Krabbe 1990). At the Houston Zoological Gardens nesting has occurred at least eight times: the first hatching occurred in late October 1985, the second hatching occurred in early May 1986, and the third hatching occurred in September 1986. The fourth hatching occurred in

March 1987, which was followed, after chick removal, by a repeat effort in late April. Again following chick removal, a sixth hatch occurred in July of that year. The seventh and eighth hatches occurred in January and March 1988, the last eggs having been laid less than a month after the chicks of the previous brood had fledged (Shelton n.d.).

Nest Sites   Golden-headed quetzals at the Houston Zoological Gardens did not breed until a hollow palm log (1.8 m long) was provided for the birds, which had shown no interest in wooden nest boxes or simulated fiberglass logs. The palm log was mounted vertically and filled up to the entrance with leaves, needles, and bark; a perch and removable top were also added. The male began removing fibers around the nest entrance in early April, and the female began participating in the same activity during July. She started entering the log in mid-August and by mid-September was spending prolonged periods inside. By mid-September all of the packing materials had been removed by the birds, and when more was added most of this was also removed, so that the top of the nest material was 45.7 cm (18.0 in) below the entrance hole. It was later found that such added materials must be stamped down when inserted so that the adults can excavate their own cavity; otherwise a soft, rotten log must be provided. After each breeding the hollow base of the nest log must also be refilled (Shelton n.d.).

Eggs and Incubation   The first egg laid by the Houston female golden-headed quetzal was found on October 5, and a check 4 days later revealed a second one. The first hatched on October 25, and the second a day later, suggesting they had probably been laid a day apart. Both sexes incubated, with the male sitting most of the day and the female remaining on the nest from late afternoon until early morning (Shelton n.d.). The eggs are of an unusual color, grayish blue rather than light blue or bluish green, as in the resplendent quetzal (Wetmore 1968).

Brood Rearing   On the day the first golden-headed quetzal chick hatched at the Houston Zoological Gardens, both birds remained in the nest most of the day, with the male disappearing in the afternoon for feeding but the female remaining in all day. Incubation was judged to have lasted 16–20 days. The adults' food preferences changed somewhat at this hatching, the order of preference now being avocados, commercial dog chow, grapes, "pinky" mice, and bird-of-prey chow, the last two items consumed only occasionally. Some mealworms were also eaten. One of the two chicks died, and the other was removed for hand rearing because the parents were not feeding it. Later the hand-reared chicks were found to thrive on monkey chow; a mixture of dog and cat chow; papayas; and avocados. Parent-raised chicks were found to fledge at 24–30 days (one fledging occurred at 24 days and another at 25 days), when the chicks weigh about 90–100 g (3.2–3.5 oz). Parent-raised birds were found to survive best on a commercial softbill chow diet, with fruits, giant mealworms, and occasional anoles provided as well.

Later breedings established that fertility occurs at two years (Shelton n.d.).

CONSERVATION AND EVOLUTIONARY RELATIONSHIPS

The golden-headed quetzal seems to represent part of a superspecies with *pavoninus*, the two forms being allopatrically or parapatrically distributed and ecologically separated. Like the crested quetzal, the golden-headed is distributed in montane forests through six Andean countries from Panama to Bolivia, and its conservation status is unknown. Hilty (1985) listed it as showing signs of serious population decline in Colombia, citing deforestation as a general contributing factor. Stotz et al. (1996) listed the species as fairly common, with a medium degree of sensitivity to disturbance.

PAVONINE QUETZAL

Pharomachrus pavoninus (Spix) 1824

OTHER VERNACULAR NAMES

Peacock trogon, red-billed train-bearer; viuda pico rojo (Spanish).

RANGE

Tropical evergreen forests of Venezuela (southern Amazonas and southwestern Bolívar) south and west to western Amazonian Brazil (right bank of the Tapajós River and the Río Negro), southeastern Colombia, eastern Ecuador, eastern Peru, and northern Bolivia (Beni, Pando). See map 6.

SUBSPECIES

Pharomachrus pavoninus heliactin (Cabanis and Heine) 1862–1863 (1863): western Ecuador. This form may instead belong with *auriceps* or perhaps should be considered a separate species (Fjeldså and Krabbe 1990).

Pharomachrus pavoninus pavoninus (Spix) 1824: upper Amazonian Basin areas of Brazil, Bolivia, Peru, Venezuela, and Colombia.

Pharomachrus pavoninus viridiceps Griscom and Greenway 1937: northeastern Brazil; of doubtful validity (Zimmer 1948).

MORPHOMETRICS

Measurements Wing, nineteen males of *pavoninus* from Amazonia 179–194 mm (7.0–7.6 in; Zimmer 1949). Tail, nineteen males of *pavoninus* from Amazonia 157–168 mm (6.1–6.6 in; Zimmer 1949). Wing, males (sample size unstated) from Brazil 179–182 mm (7.0–7.1 in); tail, 158–170 mm (6.2–6.6 in; Pinto 1950). No information on eggs available.

Weights One male of *pavoninus* from Brazil 158 g (5.5 oz; Sick 1993). Nine males from Peru and Bolivia 152–180 g (5.3–6.3 oz; average 162.0 g [5.67 oz]; specimens, LSU). One male of *pavoninus* from Peru 175 g (6.1 oz; specimen, FMNH).

DESCRIPTION (see plate 9)

Adult Male Forehead feathers not extended into frontal crest, but nape feathers form bushy crest; head golden brown to golden green, shading toward green on cheeks and throat; lore feathers normal in length; chest, back, lesser and median wing-coverts, rump, and upper tail-coverts iridescent green with black bases, back feathers lengthened and edged with bronze; longest upper tail-coverts barely reaching or extending only very slightly beyond tips of rectrices; ornate greater upper wing-coverts moderately long and pointed, obscuring primaries; primaries, secondaries, greater wing-coverts, and rectrices black; breast, abdomen, and remaining underparts blood-red; thighs black; maxilla and mandible yellow toward tip, otherwise bright red; iris reddish brown; legs and toes brownish or yellow-ochre.

Adult Female Similar to male but less brilliant throughout; head distinctly brownish to grayish brown, washed with green and bronze; cheeks, throat, and breast brownish gray; primaries and secondaries with buffy edging on outer vanes; scapulars, back, and upper tail-coverts rich glossy green, longest upper tail-coverts nearly reaching tips of rectrices; tips of three outer rectrices with some barring usually present (outermost one-third white or whitish buff terminally, second about one-sixth white, and third only toothed with white); abdomen brown; thighs black; maxilla red, becoming blackish toward tip; culmen brownish; mandible red, tip black; iris brown; legs and toes brownish.

Immature Similar initially to adult female, but outer vanes of flight feathers edged with yellowish brown; bill black. Outer tail feather pattern differences should help distinguish the sexes of immatures, as described for the other quetzals.

Juvenile No information available.

IDENTIFICATION

In the Hand Adult male pavonine quetzals have forehead feathers that are not elaborated into a distinct crest, and they have entirely black outer rectrices. They are most like the golden-headed quetzal, but have uniquely red rather than yellow bills. Females have brown rather than red abdomens; the three outer rectrices are extensively barred and are variably tipped and barred with white or whitish buff (resembling those of the crested and white-tipped quetzals) rather than being entirely or nearly black (as in the golden-headed quetzal); and the bill is dusky to brownish terminally with a distinctive red base.

In the Field   In addition to having the plumage and bill-color traits listed above, the pavonine quetzal is the only species of its genus that occurs in low-altitude rain forests east of the Andes, from 150–700 m (500–2,300 ft) elevation. See also the vocalizations section.

GEOGRAPHIC VARIATION

Zimmer (1948) provided comparative measurements for *P. pavoninus*, *P. auriceps* (which he regarded as conspecific), and *P. auriceps hargitti*, showing a progressive increase in wing and tail measurements, with *hargitti* having a particularly long tail. Zimmer doubted the validity of *P. p. viridiceps* (this form supposedly having darker green head tones and slightly shorter upper tail-coverts than *pavoninus*), but he had only five specimens available for study.

ECOLOGY

Habitats   Pavonine quetzals are associated with and limited to the interiors of tall, lowland evergreen forests of the tropical zone, from near sea level to about 700 m (2,300 ft) elevation. Hilty and Brown (1986) stated that in Colombia the birds range from the lowlands to at least 500 m (1,600 ft), in humid terra firme forests.

Foods and Foraging Ecology   Of nine pavonine quetzal stomachs examined by Remsen, Hyde, and Chapman (1993), 88.9 percent contained fruit only and 11.1 percent contained a mixture of fruit and arthropods.

BEHAVIOR

General, Social, and Sexual Behavior   No information available.

Vocalizations   The usual call of the pavonine quetzal is a series of five melodious notes, *ew, ewwo-ewwo-ewwo-ewwo*, and the species also has a warning note that is a descending tremolo (Sick 1993). Hilty and Brown (1986) reported it as a loud, descending whistle, followed by an emphatic *chok* note: *heeeeear, chok!* This call has a melancholic aspect and may be repeated four to six times. Vocalizations of this little-studied species were recorded in Peru by Ted Parker (Hardy, Reynard, and Coffey 1995). The recorded song consists of a series of down-slurred *wheeeoooo* notes, each followed by a brief *chok* and uttered at the rate of about four such doublets per 10 seconds.

BREEDING BIOLOGY

Chronology of Breeding   There is a probable observation of a pavonine quetzal nest site from Brazil for February (Oriki and Willis 1983).

Nest Sites   The probable pavonine quetzal nest site mentioned by Oriki and Willis was 9 m (30 ft) above ground in a large forest tree.

Eggs and Incubation   No information available.

Brood Rearing   No information available.

CONSERVATION AND EVOLUTIONARY RELATIONSHIPS

The pavonine quetzal appears to be the lowland counterpart of the golden-headed quetzal, with the Ecuadoran race *heliactin* perhaps being somewhat transitional. This little-studied species is apparently closely tied ecologically to lowland rain forests. It occurs in six countries from Panama to Brazil. Stotz et al. (1996) listed the species as uncommon and patchily distributed, with a high degree of sensitivity to disturbance.

EARED TROGON (Genus *Euptilotis* Gould 1858)

The single, relatively large (34 cm, 13.4 in) species in the genus *Euptilotis* is similar to the species of the genus *Pharomachrus* in having unserrated bill edges (but the birds do have subterminal notches on both mandible and maxilla) and narrow oval nostrils overhung with horny opercula. The feathers of the lores point backwards, exposing most of the bill; and the median wing-coverts are rounded and of normal length. The ear-coverts are filamentous and greatly lengthened in adults of both sexes, extending back from below the eye to a narrow point behind the nape. Like typical quetzals, the eared trogon has iridescent green rather than vermiculated upper wing-coverts; but the eared trogon differs from the former species in that its longest upper tail-coverts reach only to the middle of the tail, its upper wing-coverts are not elongated and decurved, and it has no bushy forehead or occipital crest. The middle rectrices are iridescent blue and truncated and lack a terminal black band. The outer rectrices are somewhat tapered and rounded at their tips (rather than untapered and distinctly truncated as in typical trogons) and are broadly white-tipped in both sexes. On the basis of gene-sequencing studies, Espinosa de los Monteros (1998) considered *Euptilotis* to be a sister taxon to *Pharomachrus* (the quetzals) and recommended that the former genus be merged into the latter.

EARED TROGON

Euptilotis neoxenus (Gould) 1838

OTHER VERNACULAR NAMES

Eared quetzal, welcome trogon; quetzal Mexicano (Spanish).

RANGE

Upland pine, pine-oak, and pine-evergreen forests of northern and central Mexico: from northwestern Chihuahua and northeastern Sonora south through Durango, Zacatecas, and Nayarit to western Michoacán. Has been reported (since 1977) from southeastern Arizona (Huachuca and Chiricahua Mountains) only rarely; seen once in southwestern New Mexico (mountains near Animas). The first breeding recorded in the United States occurred in 1991. Classified as endangered by the IUCN (Collar, Crosby, and Stattersfield 1994). See map 7.

SUBSPECIES

None recognized.

MORPHOMETRICS

Measurements Wing, males 181–200 mm (7.1–7.8 in; average of 10, 191.6 mm [7.5 in]); females 183.5–198.5 mm (7.2–7.7 in; average of 9, 192.5 mm [7.5 in]). Tail, males 161–179.5 mm (6.3–7.0 in; average of 10, 171.2 mm [6.7 in]); females 155–177 mm (6.0–6.9 in; average of 9, 165 mm [6.4 in]; Ridgway 1911). Egg of MVZ specimen 30.5 × 24.8 mm (1.2 × 1.0 in; Carla Cicero, pers. comm.).

Weights Five males 103–126.7 g (3.6–4.4 oz, average 116.2 g [4.0 oz]), six females 115–148.5 g (4.0–5.2 oz; average 128.6 g [4.5 oz]; specimens, MVZ). Estimated egg weight 9.6 g (0.3 oz).

DESCRIPTION (adapted from Ridgway 1911; see plate 10)

Adult Male Crown dark iridescent bronze-green or greenish bronze; rest of head, including chin and throat, dull black, faintly glossed with bluish; back,

Map 7. Distribution of the eared trogon.

71

scapulars, and broad tips of upper wing-coverts bright iridescent green to greenish bronze, becoming bright bluish green on rump and upper tail-coverts; six middle rectrices and basal portion of outer rectrices dark iridescent blue; three outer pairs of rectrices broadly tipped with white, white tip occupying terminal third on outermost rectrix; remiges, greater, and primary wing-coverts slate-blackish, greater wing-coverts and secondaries more or less distinctly margined with iridescent green; outer vanes of primaries more grayish, becoming grayish white basally; median and lesser wing-coverts abruptly slate-blackish basally; chest bright iridescent bronze-green, becoming more bluish toward throat; rest of underparts bright, pure red; thighs dark slate; bill bluish gray for basal half or more, blackish terminally; eye-ring gray (not red, as sometimes depicted); gape deep orange; iris dark brown; feet and toes brown.

Adult Female   Upperparts and tail like those of adult male, but crown and remiges slate colored; chin, throat, and sides of head brownish slate-gray or mouse-gray, becoming light grayish brown on chest and upper breast; red posterior underparts less intense than in adult male; soft-part colors much as in adult male; bill dark gray, becoming pale blue-gray basally.

Immature   Head and breast brown; ear-coverts not elongated; amount of black visible at base of under-tail (involving three outer pairs of rectrices) extensive; inner rectrices tipped with white, outermost rectrix half white, second about one-third white, and third about one-fourth white; outer secondaries with traces or edgings of buff; belly and under tail-coverts pinkish; bill lead colored; other soft parts similar to adults'.

Juvenile Male   Head charcoal colored; breast a mixture of rusty and green; lower breast and belly more orange than in adults (Taylor 1994). Presumably the white spots on the upper wing-coverts persist through this stage.

Juvenile Female   Head slightly darker gray than neck, both head and neck lighter than in adult; breast pearly gray to fawn colored; orange-red area

of the belly less extensive than in adults; gape orange-red (Taylor 1994). Presumably juvenile female has white-spotted upper wing-coverts, as in fledglings. Fledglings have spotted wing-covert tips arranged in several rows; breast and belly mottled; tail blue, as in adults (Taylor 1994).

IDENTIFICATION

In the Hand   Both sexes of the unique eared trogon can be readily identified in the hand by means of the long, filamentous ear-coverts. The eared trogon is similar to the quetzals in having the nostrils narrow and overhung by horny opercula, but the former's bill is more slender and is basally flattened. Like the quetzals, the eared trogon has iridescent green upper wing-coverts; but it differs from the quetzals in having median wing-coverts that are rounded but very large rather than elongated and decurved and in having upper tail-coverts that, though elongated, reach only the middle of the tail in adult males. Females generally resemble males; but the head and neck are slate-gray rather than iridescent green, and the chest is grayish brown. Both sexes have extensive areas of white on the ends of their outer rectrices; but in young birds the black at the base of these feathers is more extensive, and the inner rectrices are white tipped.

In the Field   The eared trogon, a rare Mexican endemic of upland pine forests, more closely resembles a trogon than a quetzal in the field; the ear-like feathers are only rarely visible. Perched birds may assume a "hump-backed" posture, with the lower back and rump feathers ruffled. The wings often droop below the tail; when folded, the black tips of the middle rectrices project well below the white-tipped outer ones. Unlike those of the other, larger (and red-bellied) species of Mexican trogons, the outer three rectrices of the eared trogon are entirely white over more than half their length, without any black barring or spotting, making the bird more reminiscent of a quetzal than a trogon. Unlike the songs of both the typical trogons and the quetzals, the usual "song" of the eared trogon is a series of shrill, quavering, and reedy notes, often consisting of ten to fifteen syllables but sometimes having more than twenty-four. The individual elements may sound like *whii, whii-iir,* or *wheer-i-hi;* single-noted whistles start the series, becoming progressively louder, and are replaced by double-noted and then triple-noted elements. While perching, the birds tail-pump and simultaneously utter *wEEE-k* notes. There are also grackle-like "squeal-chucks" and cackling *kac-ka-k-kac* . . . vocalizations, the latter usually uttered when males and perhaps also females are disturbed or in flight. See also the vocalizations section.

GEOGRAPHIC VARIATION

None described.

ECOLOGY

Habitats   Little is known of preferred habitats of the eared trogon, but Zimmerman (1978) suggested that these are far ranging birds that may need large canyons or extensive stands of pine-oak and madrone (*Arbutus* spp.) woodland. In the mountains of Arizona the birds have been seen as early as late May and as late as mid-December, at elevations of about 2,100–2,500 m (6,900–8,200 ft). A nest was found at about 2,400 m (7,900 ft). Most sightings of the eared trogon have occurred during fall, after the seasonally resident elegant trogons have migrated southward. Habitats in Arizona are very similar to those in Mexico and are characterized by steep trails, rock ledges, and running water (Taylor 1994). Collar et al. (1992) indicated that the records suggest that populations in northern latitudes may live at lower altitudes than those farther south, young individuals may wander away from breeding areas, and birds may perform at least sporadic autumn and winter migrations to lower altitudes.

In Mexico eared trogons occur from 1,700 to 3,000 m (5,600 to 9,800 ft) elevation but are mainly found in pine, pine-oak, or pine-evergreen forests of the temperate zone from 2,100 to 2,800 m (6,900 to 9,200 ft). There they are especially associated with barrancas, which are river canyons with sheer rock walls rising in tier-like terraces to plateaus at the top. In Durango they have been found in pine-oak forests rich in mosses, ferns, and epiphytes, at an elevation of about 2,700 m (8,900 ft). In Chihuahua they are often seen on canyon terraces about 30 m (98 ft) above the canyon floor and at elevations of about 2,300–2,500 m (7,500–8,200 ft). They are evidently rare in Sonora (Russell and Monson 1998).

Foods and Foraging Ecology   Madrone berries may be an important food for eared trogons, especially in winter. Birds have been seen taking madrone berries from trees while in flight, hovering briefly to pluck the fruit, then returning to a branch to eat it. Feeding occurred during periods of activity lasting 5 to 10 minutes, followed by resting periods. Birds have also been observed feeding on arthropods and their larvae, including woolly caterpillars, green larvae, and katydids. Sometimes the birds spiral tightly downward along the trunk of a tree toward its base while fanning their tails, apparently thus trying to flush prey (Taylor 1994).

BEHAVIOR

General, Social, and Sexual Behavior   No detailed information on the behavior of the eared trogon is available. As many as three birds (an adult male and female, plus an incompletely colored male) have been seen and heard calling together (Zimmerman 1978). Taylor (1994) stated that the first male observed in Arizona had a known range of 5.8 km (3.6 miles), and another bird probably covered this same distance in less than 24 hours, a trip that included a flight over a ridge 300 m (1,000 ft) high. Another pair of juveniles had a lineal range of about 12.6 km (7.8 miles) along a canyon. Apparently there is some competition among trogon species and between trogons and other birds that forage on madrone berries, and on two occasions male elegant trogons have been observed chasing males of the larger eared trogon (Taylor 1994).

Vocalizations   Songs and calls of the eared trogon were recorded in Arizona by Dale A. Zimmerman (Hardy, Reynard, and Coffey 1995). The individual song elements, consisting of sharp and piercing *whiii* or *eeeep* notes, varied from six to twelve whistle-like segments and were uttered without pitch changes or pauses at the rate of nearly two per second, starting in a soft and minor key but becoming progressively louder. Zimmerman (1978) described this "song-call," audible up to 200 meters (700 ft), as raptor-like, reminiscent of the food-begging call of a young falcon. When separated, both members of a pair remained in contact by means of this song-call, and when near each other produced a "whispered" version of the same call. All of these prolonged vocalizations on the recording were seemingly single-element notes rather than the double or treble notes mentioned as typical in the identification section. There were also some short *whi-cha* or *cha-cha* squeals and chattery notes as well.

According to Zimmerman (1978), the species' most distinctive call is a loud, up-slurred squeal, somewhat like the call of a great-tailed grackle (*Cassadix mexicanus*), followed by a brief *chuck*. This call is uttered repeatedly, almost simultaneously with a quick spreading of the tail and a head-jerking movement (Sheri Williamson, pers. comm.). Females also produce the squeal call. In flight the birds (at least males) may utter a *cac-ac-ac-ac* call.

Taylor (1994) recognized four basic vocalizations. The first, termed a "squeal-chuck," is a loud, squealing *wEEE-k* uttered during perching and associated with disturbance. It is uttered repeatedly, each time with an upward flick of the tail. The second, a contact note used by pairs, is a *Weeee* that may resemble a violin note or a squeaky door hinge. The third is a cackling flight call, *kac-ka-k-kac*, consistently uttered as the bird takes flight from a perch. The fourth is the "tremolo," an extended series of disyllabic whistles that begin softly but gradually increase in volume. This vocalization is uttered from April through August and is seemingly related to territoriality or resource defense. See also the identification section above.

BREEDING BIOLOGY

Chronology of Breeding   The eared trogon apparently breeds from June to October in northwestern Mexico (Howell and Webb 1995). An egg in the Museum of Vertebrate Zoology, Berkeley, California, was collected July 10, 1959, at an elevation of about 2,400 m (7,900 ft) on Sierra del Nido, Chihuahua (Carla Cicero, pers. comm.). Baicich and Harrison (1997) described the breeding season as mid-July to mid-August, sometimes later, depending on the start of the rainy season. Taylor (1994) judged the breeding season to extend from August to October, coinciding with the period of ripening fruits and the flush of insects that appear after the rainy period of July and August. In Chihuahua a nest was found in early September 1990, when a pair was found feed-

ing its young. This same hole was used for at least 3 more years. The first Arizona nest was found at an elevation of about 2,100 m (6,900 ft) in mid-October 1991, and the pair was also feeding young at the time of its discovery (Taylor 1994).

Nest Sites   Baicich and Harrison (1997) stated that the eared trogon nest is in a natural tree cavity or old woodpecker hole approximately 10 m (33 ft) above ground, and with a cavity about 13 cm (5.1 in) in diameter and 28 cm (11.0 in) deep. Taylor (1994) described a nest he found in Chihuahua: the cavity was approximately 22 m (72 ft) above ground in a partially dead pine and was probably an old flicker hole located at the point on the trunk where dead wood replaced living wood. The dead pine was located in a forest dominated by various pines, especially ponderosa pine (*Pinus ponderosa*) but also containing Douglas fir (*Pseudotsuga menziesii*), cypress (*Cupressus arizonica*), aspen (*Populus* spp.), and madrone (*Arbutus arizonica*). The first nest reported from the United States was found about 9 m (30 ft) above ground in a flicker cavity in a bigtooth maple (*Acer grandidentata*). Shortly after its discovery there was a brief but cold storm, and the chicks died, apparently from hypothermia. On examination, the interior of this nest site was found to about 14 cm (5.5 in) in diameter and 28 cm (11.0 in) deep.

Eggs and Incubation   No detailed information on the eggs of the eared trogon is available, but a single egg from a presumably incomplete clutch is housed in the Museum of Vertebrate Zoology, Berkeley, California.

Brood Rearing   No detailed information on eared trogon brood rearing is available. Taylor (1994) stated that a chick judged to have died when at least 2 weeks old was covered with black and yellow down, but this downy appearance may have resulted from the tips of the emerging contour feathers being unusually down-like, thus producing a "fluffy" appearance (see the description of a similar phenomenon in the resplendent quetzal account).

CONSERVATION AND
EVOLUTIONARY RELATIONSHIPS

The eared trogon is the only species of trogonid that is currently classified as endangered by the IUCN

(Collar, Crosby, and Stattersfield 1994). It is believed to be very locally distributed in western Mexico, where cutting of the upland pine and pine-oak forests has greatly reduced its habitat and its access to trees with suitable nesting sites. As it is a fairly large trogon, it needs cavities that have been excavated by the larger species of woodpeckers, such as northern flickers (*Colaptes auritus*). Flickers commonly occur throughout the eared trogon's entire range, and these birds excavate nest cavities with entrances up to 10 cm (4.0 in) in diameter and interiors that often are at least 15 cm (5.9 in) in diameter and 30 cm (11.8 in) deep.

In a summary of key areas for threatened Neotropical birds, Wege and Long (1995) reported recent (since 1980) records of the eared trogon from seven such areas in Mexico. They include three unprotected areas in Chihuahua (Mesa de Huaracán, Barranca del Cobre, and Cerro Mohinoro), three areas (one protected) in Durango (Mexiquillo, La Michilía Biosphere Reserve, and Monte Oscuro), and an unprotected area that has been proposed as a reserve in Jalisco (El Carricuto del Huichol).

Collar et al. (1992) summarized this species' distribution and status in detail, listing it in eight Mexican states (Sonora, Chihuahua, Sinaloa, Durango, Nayarit, Zacatecas, Jalisco, and Michoacán). However, low densities are typical everywhere, and logging and the resulting the removal of possible nesting trees constitute the primary threats, with agricultural encroachment and development also being significant. The species is perhaps still fairly common in La Michilía Biosphere Reserve in Durango, which is virtually the only location within its Mexican range where it is receiving protection. Stotz et al. (1996) listed the species as uncommon and patchily distributed, with a medium degree of sensitivity to disturbance.

WEST INDIAN TROGONS (Genus *Priotelus* G. R. Gray 1840)

In the West Indian trogons the auricular feathers of adults are filamentous and somewhat lengthened but do not reach beyond the nape, and the eyelids are uniquely feathered. The maxillary and, to varying degrees, the mandibular edges are serrated or undulate, the nostrils are rounded and are not operculated, and the two front toes are slightly fused. Foods consist of both insects and fruits. Two small to moderate-size (25–30 cm [9.8–11.8 in], 50–70 g [1.8–2.5 oz]) species are included, each placed in its own subgenus.

Both species are woodland adapted, the Cuban forms having a broader ecological range than the upland-adapted Hispaniolan trogon. Cuba was about 87 percent covered by forests in 1800, but by 1965 forest coverage had been reduced to about 15 percent. At about the same time, Haiti was still about 7 percent wooded, and the Dominican Republic was still about 45 percent covered by forest and bushland (Heske 1973).

Subgenus *Priotelus* G. R. Gray 1840

In the single species of the subgenus *Priotelus* the rectrices have uniquely scalloped tips, the upper wing-coverts are spotted with black and white, and the anterior underparts are white. The sexes are nearly identical as adults, and neither sex has vermiculated or narrowly barred upper wing-coverts. The central rectrices are bronzy green, shading to purplish blue at their deeply incised tips.

CUBAN TROGON

Priotelus temnurus (Temminck) 1825

OTHER VERNACULAR NAMES

Isle of Pines trogon (*vescus*); tocoloro, guatini (Spanish).

RANGE

Widespread in forested areas of Cuba and the Isle of Youth. Also reported from the largest keys north of Camagüey Province. See map 8.

SUBSPECIES

Priotelus temnurus temnurus (Temminck) 1825: upland forests and woodlands of Cuba.
Priotelus temnurus vescus (Bangs and Zappey) 1905: forests and woodlands of the Isle of Youth.

MORPHOMETRICS

Measurements   Wing, males (of *temnurus*) 117–128 mm (4.6–5.0 in; average of 10, 123.6 mm [4.8 in]); females 118–125 mm (4.6–4.9 in; average of 6, 121.8 mm [4.8 in]). Tail, males (of *temnurus*) 105.5–124.5 mm (4.1–4.9 in; average of 10, 116.4 mm [4.5 in]); females 106–120 mm (4.1–4.7 in; average of 4, 114.4 mm [4.5 in]). Wing, males (of *vescus*) 116.5–119 mm (4.5–4.6 in; average of 4, 117.8 mm [4.6 in]); females 116–120 mm (4.5–4.7 in; average of 4, 1 17.6 mm [4.6 in]). Tail, males (of *vescus*) 108–122 mm (4.2–4.8 in; average of 4, 112.8 mm [4.4 in]); females 111.5–116 mm (4.3–4.5 in; average of 4, 113.6 mm [4.4 in]; Ridgway 1911). Egg 29.7 × 22.9 mm (1.2 × 0.9 in; Schönwetter 1966).

Map 8. Overall distributions of the Cuban trogon (dashed line and left sketch) and the Hispaniolan trogon (dotted line and right sketch). Presumed current ranges are also shown (inked), based largely on recent forest distributions in the West Indies.

Weights   A mixed sample of three birds 53.5–60.4 g (1.9–2.1 oz; average 58.0 g [2.03 oz]; Dunning 1993). One male 40.3 g (1.4 oz; specimen, FMNH). Estimated egg weight 8.0 g (0.3 oz; Schönwetter 1966).

DESCRIPTION (of *temnurus*, adapted from Ridgway 1911; see plate 11)

Adult male   Crown dark iridescent blue, more violaceous on nape; cheek, suborbital, and auricular regions black; back, scapulars, and rump iridescent green or slightly bronzy-green, becoming more bluish green on hindneck and upper tail-coverts; six middle rectrices glossy blue-black or dark violaceous blue, middle pair with outer vanes brighter and less violaceous blue, becoming dark violet-blue terminally, their inner vanes bronze-green or greenish bronze, passing through blue, then to dark violet-blue terminally; three outer rectrices extensively white terminally, bluish black basally; outer vanes of second and third rectrices with one to three white spots on black portion of outer vane, inner vane sometimes with one or two transverse spots of white; lesser and median wing-coverts dark iridescent green, broadly margined with much brighter green; greater secondary coverts very dark bluish green narrowly edged with brighter bluish green, outer vanes with a very large terminal squarish spot of white; alular feathers and primary coverts black, the former with three white spots on outer vanes, the latter with a subterminal spot of white; inner secondaries dark iridescent bluish green; other secondaries black, edged with iridescent bluish green and with a very large subterminal squarish spot of white on outer vanes; primaries black, terminal half of outer vanes with squarish white spots, basal portion continuously white, except two outermost, which are spotted to base; chin, throat, and malar region white, becoming clear gray on chest, breast, upper abdomen, and sides, remaining underparts pure geranium-red, the demarcation between the red and the gray forming a very sharp transverse line; thighs grayish white, feathers dark grayish basally; maxilla brownish black, except basal portion, which is dull vermilion; mandible dull vermilion; iris carmine red; feet and toes dark brownish (in dried skins).

Adult female   Exactly like adult male but averaging slightly smaller.

Immature   Immature male in the Louisiana State University collection has dull, buffy white spots and tips to the secondaries, small isolated spots on inner secondaries also buff; wing-coverts duller green than in adults, with more pronounced iridescent terminal bars; crown duller green; underparts entirely buff-gray except for reddish pink on under tail-coverts; rectrices with less pronounced terminal flaring than in adults, but otherwise perhaps somewhat wider (Steven Cardiff, pers. comm.).

Juvenile   No information available.

IDENTIFICATION

In the Hand   The Cuban trogon can be immediately distinguished from all other trogons by its scalloped rectrices. The bill is reddish, the maxilla with distinctly serrated cutting edges. The nostrils are rounded and non-operculated, the ear-coverts have elongated and outwardly curved filamentous tips, and the outer vanes of the secondaries have conspicuous white spots.

In the Field   The Cuban trogon is the only trogon on Cuba and the Isle of Youth. Both sexes have brilliant white patterning on the throat and breast, white spots on the flight feathers and upper wing-coverts, and extensive areas of white on the outer rectrices. The usual vocalization and apparent song is a soft, monotonous, and dove-like *toc-coro* or *to co-co-loro*, these interpretations being the basis for one of its vernacular names (tocoloro). The female's call is similar but much lower in pitch. Although the birds sometimes call in flight, more often a pair will sit close together, or even out of sight of each other, and call for long periods. See also the vocalizations section.

GEOGRAPHIC VARIATION

The Isle of Pines race of the Cuban trogon (*vescus*) has smaller wing, tail, and middle-toe measurements; a relatively larger and broader bill; and lighter underparts than the Cuban form.

ECOLOGY

Habitats   The Cuban trogon is apparently still fairly widespread in dense woods and along river courses but is very tame and easily shot. It has been reported as occurring in both wet and dry forests at all altitudes. Collectively, it has been reported in tropical lowland evergreen forest, tropical deciduous forest, pine forest, and secondary forest, from sea level to 2,000 m (6,600 ft).

Foods and Foraging Ecology   No specific information is available on the foraging behavior of the Cuban trogon, but "fruits of various kinds" were mentioned by Bangs and Zappey (1905). Chapman (1892) noted that the species eats insects and berries, both of which the birds obtain while in flight. Bond (1993) commented that fruits and insects are eaten by both species of West Indian trogons. Clark (1918) observed that the gizzard of a specimen he studied was filled with fruits up to 8 mm (0.3 in) long. The intestinal ceca of this species are relatively long, comprising about 18 percent of the intestinal length (Miller 1918).

BEHAVIOR

General, Social, and Sexual Behavior   No specific information available.

Vocalizations   Calls and a possible song of the Cuban trogon were recorded in Cuba by G. B. Reynard (Hardy, Reynard, and Coffey 1995). The first example on the recording consisted of four brief chuckling rattles, uttered about 2 seconds apart. A second example consisted of rapid three- to four-note rattles; and a third, of more prolonged but also multiple rattle-like notes uttered in rapid succession. The dove-like song usually described for this species (see the identification section above) was not recorded.

BREEDING BIOLOGY

Chronology of Breeding   There are nesting records for the Cuban trogon for June (one) and July (one). The relatively late nesting has been ascribed to the need for the trogons to wait until the resident woodpeckers have completed their own nesting activities. The collective reported breeding season is from May to August (Raffaele et al. 1998).

Nest Sites   Cuban trogon eggs are deposited in a deserted woodpecker's hole (Barbour 1943; Bond 1936). In one case the site was about 4.5 m (15 ft) above ground in a bottle palm.

Eggs and Incubation   The clutch size of the Cuban trogon is said to be 3–4 eggs. The incubation period is unreported.

Brood Rearing   No information available.

CONSERVATION AND EVOLUTIONARY RELATIONSHIPS

The Cuban trogon and the Hispaniolan trogon are each other's nearest relatives, but presumably they have been separated for some time. The broad ecological range of the Cuban species would seem to help its conservation potential. Furthermore, Cuba has established 73 protected areas that represent 12 percent of the country's land area, and 15–18 percent of its land area remains forested (Raffaele et al. 1998). Stotz et al. (1996) listed the species as common, with a low degree of sensitivity to disturbance.

Subgenus *Temnotrogon* Bonaparte 1854

The single species of the subgenus *Temnotrogon* differs from *Priotelus* in that the rectrices are truncated in shape, the upper wing-coverts are narrowly barred with black and white, and the anterior underparts are gray. The sexes have somewhat dichromatic plumages as adults; narrowly barred upper wing-coverts are present only in males. The central rectrices are purplish blue at their tips and outer webs; the inner webs are oil-green. This species was not studied by Espinosa de los Monteros (1998).

HISPANIOLAN TROGON

Priotelus roseigaster (Vieillot) 1817

OTHER VERNACULAR NAMES

Santo Domingo trogon; cotorrita de sierra, papagayo (Spanish).

RANGE

Montane forests and sometimes coastal mangroves of Hispaniola. See map 8.

SUBSPECIES

None recognized.

MORPHOMETRICS

Measurements   Wing, males 132.5–145 mm (5.2–5.7 in; average of 9, 138.4 mm [5.4 in]); females 133.5–141 mm (5.2–5.5 in; average of 4, 136.2 mm [5.3 in]). Tail, males 154–170.5 mm (6.1–6.6 in; average of 9, 160.7 mm [6.3 in]); females 149.5–187.5 mm (5.8–7.3 in; average of 4, 163 mm [6.4 in]; Ridgway 1911). Egg, 30 × 23.5 mm (1.2 × 0.9 in; Bond 1936).

Weights   One female 74.2 g (2.6 oz; specimen, FMNH). Estimated egg weight 9.1 g (0.3 oz).

DESCRIPTION (mainly adapted from Ridgway 1911; see plate 12)

Adult male   Crown rather dull iridescent bronze-green or greenish bronze; back, scapulars, anterior lesser wing-coverts, and upper rump bright bronze-green, becoming pure iridescent green or slightly bluish green on lower rump and upper tail-coverts; outer vanes of middle pair of rectrices dark iridescent blue or greenish blue, except for small subterminal area of grayish bronze or olive-bronze; inner vanes of middle pair of rectrices grayish bronze, bronze-gray, or bronzy olive, broadly tipped with dark iridescent blue; next two pairs of rectrices wholly dark iridescent blue or greenish blue; three outer pairs of rectrices similar but extensively white terminally, the white area occupying a much greater portion of outer vanes, which have subterminal spots of dark iridescent blue or blue-black; underside of inner vanes of three outer rectrices with an area of iridescent grayish between the basal dark iridescent blue and a subterminal spot of dark iridescent blue or blue-black; posterior lesser, median, greater, and primary wing-coverts, alular feathers, and secondaries blackish, more or less glossed with bronzed greenish and narrowly but very regularly barred with white; primaries black, outer vanes marked with squarish spots of white, except proximal portion of longer primaries, which are more or less continuously white or edged with white; lores dusky, becoming slate colored on chin, upper throat and orbital, auricular and malar regions; lower throat and chest slate-gray, more or less strongly glossed (especially on chest) with bronze-green or greenish bronze; breast, upper abdomen, and anterior portion of sides clear gray, slightly paler posteriorly; rest of underparts intense geranium-red, darker along anterior margin, where line of demarcation between red belly and gray breast is sharply defined and regularly transverse; thighs slate-gray to slate, feathers of upper tarsi darker; bill bright honey-yellow; iris light orange, with no contrasting or conspicuous eye-ring; feet and toes brownish gray, undersides of toes yellow.

Adult female   Similar to adult male, but lacking bluish cast to back; wing-coverts and secondaries without white bars, the former grayish olive margined terminally with iridescent bronze-green or bronze, the latter plain slate-gray; primaries without squarish white spots on outer vanes, which are grayish distally, white proximally; primaries also margined basally and barred on terminal part of outer vanes with white; most rectrices with subterminal oval purplish-blue spot, this rudimentary or absent on outer vanes of two outer pairs of rectrices; bill pale yellow; iris orange (as in male); feet and toes flesh colored.

Immature   As they mature, immature males increasingly resemble adults, but the terminal portions of the primaries' outer vanes may retain traces of white barring, as in adult females.

Juvenile   No information available.

IDENTIFICATION

In the Hand   The bill of the Hispaniolan trogon is relatively robust and yellow and has serrated cutting edges. The ear-coverts are lengthened and curve outwardly at their tips. The nostrils are rounded and non-operculated and are mostly hidden by the bristly feathers of the chin and lores. Females rather closely resemble males, but the former's upper wing-coverts and secondaries are narrowly barred with black and white.

In the Field   The Hispaniolan trogon is the only trogon on Hispaniola. Both sexes are similar in having gray on the throat, breast, and upper abdomen, becoming abruptly red on the lower abdomen and under tail-coverts. The usual vocalization is a loud *cock-craow*, with the second note often being repeated, especially during the breeding season. See also the vocalizations section.

GEOGRAPHIC VARIATION

None described.

ECOLOGY

Habitats   Wetmore and Swales (1931) encountered the Hispaniolan trogon in montane pine forest, wet evergreen forest (e.g., rain forest), and deciduous forest thickets. Bond (1928) observed it at sea level in mangrove swamps, but others have encountered it to at least 3,000 m (9,800 ft), mostly in upland forests, mainly pine forests and montane evergreen forests. Raffaele et al. (1998) indicated that the species occurs in pine and broadleaf forests above 300 m (1,000 ft).

Foods and Foraging Ecology   No specific information is available on the foraging behavior of the Hispaniolan trogon, but Bond (1993) indicated that fruits and insects are eaten by both species of West Indian trogons.

BEHAVIOR

General, Social, and Sexual Behavior   Wetmore and Swales (1931) described an observation of a half dozen Hispaniolan trogons gathered on low perches and engaged in apparent mating displays. Two apparent rivals perched about 1 m (3 ft) apart with heads outstretched and tails hanging down. Periodically they raised their tails to an approximate right angle with the back, then quickly lowered them. They constantly uttered rattling notes, which were interspersed with the usual cooing calls produced by the others of the group.

Vocalizations   Apparent songs and calls of the Hispaniolan trogon were recorded in the Dominican Republic by G. B. Reynard (Hardy, Reynard, and Coffey 1995). The first two examples on the recording are apparently the *cock-craow'* or *cock-craow'-craow'* utterance mentioned in the identification section, these calls being uttered at about 5- to 10-second intervals. The third example is a rapid rattling chuckle, very much like that of the Cuban trogon. The fourth is a series of repeated crowing notes similar to the second element of the call described first. The fifth example is seemingly a subdued variation of the fourth, with the last note prolonged into a wail. See also the identification section.

BREEDING BIOLOGY

Chronology of Breeding   There are three sets of Hispaniolan trogon eggs in the collection of the Western Foundation of Vertebrate Zoology collected by Bond (1928). Two were obtained in June, and one was collected in May, both at elevations of about 1,800 m (5,900 ft). Wetmore and Swales (1931) found a nest in late May. Much calling was heard during April by Wetmore and Swales (1931), suggesting that a peak in breeding activity occurs during that month. Raffaele et al. (1998) stated that the breeding season extends from March to July.

Nest Sites   The three Hispaniolan trogon egg sets in the Western Foundation of Vertebrate Zoology are accompanied by notes on two sets, which were found in holes of trees or "scrub" vegetation 3.75

and 4.7 m (12 and 15 ft) above ground. Both of the cavities were old Hispaniolan woodpecker (*Melanerpes striatus*) nest sites. Wetmore and Swales (1931) found a nest in a dead stub 4.7 m (15 ft) high but were unable to investigate its contents because of the rotted condition of the wood.

Eggs and Incubation   Bond (1928) stated that the eggs of the Hispaniolan trogon are unusual in that they are very pale green (as in the resplendent quetzal and slaty-tailed trogon) and that the clutch comprises two eggs. Early, seemingly less reliable, accounts summarized by Wetmore and Swales (1931) suggest that the clutch may number three or four and that the eggs are white, this uncertainty perhaps being the result of confusing trogon nests with those of woodpeckers.

Brood Rearing   No information available.

CONSERVATION AND EVOLUTIONARY RELATIONSHIPS

The Hispaniolan trogon has an inherently small range, and in addition it is largely limited ecologically to montane forests. Conservation activities on Hispaniola are few indeed, and with major deforestation efforts continuing the species might easily become threatened. There are no governmental agencies in the Dominican Republic concerned with conservation, and in Haiti conditions are only slightly better. However, two national parks totaling 50,000 hectares (123,550 acres) have now been established in mountainous areas of Haiti (Raffaele et al. 1998), which should provide some protection for the Hispaniolan trogon. The species has been classified as near-threatened by the IUCN (Collar, Crosby, and Stattersfield 1994), and Stotz et al. (1996) listed it as fairly common, with a medium sensitivity to disturbance, and mentioned deforestation as a contributing factor.

TOOTHED TROGONS
(Genus *Trogon* Brisson 1760)

The 17 species included in the large Neotropical genus *Trogon* may be characterized as having serrated edges on their maxillae or mandibles or both, rounded nostrils that are not mostly covered by horny, shelf-like opercula, and two anterior toes that are variably united at their bases. Bare and often colorful eye-rings (at minimum, bare eyelids) are present in adults of nearly all species. In some species the iris of adults is conspicuously yellow to ivory-white, with inconspicuous surrounding eye-rings; but in most the iris is dark brown, and the eye-ring is contrastingly colored. In most species both sexes have finely barred or vermiculated patterning on the upper wing-coverts. The rectrices of adults are variably truncated in nearly all species. The middle rectrices of adult males are bronze-green, blue, or purple, with black tips, and are chestnut, gray, or blackish in females. Patterning on the three outer pairs of rectrices is often conspicuous and species specific and is usually also sex specific. Rectrix patterns of immatures are similar to those of adult females. Sexual dichromatism among adults in these outer rectrices and in general upperpart coloration is usual. Most of the species are somewhat insectivorous.

Three subgenera are recognized here, all of which have at some time been separated as discrete genera (as by Ridgway [1911]). On the basis of gene-sequencing studies, Espinosa de los Monteros (1998) considered the species of *Trogon* that he was able to study to make up two subclades. The "violaceous subclade" contains *T. curucui, T. violaceus, T. viridis, T. comptus,* and *T. melanurus.* The "elegant subclade" contains *T. mexicanus, T. elegans, T. rufus, T. collaris,* and *T. personatus.*

Dark-Tailed Trogons (Subgenus *Curucujus* Bonaparte 1854)

The subgenus *Curucujus* consists of four fairly large (25–30 cm [9.8–11.8 in], 90–175 g [3.2–6.1 oz]) Neotropical trogon species having heavy and rather strongly serrated bills, wing-coverts vermiculated with wavy black and white markings, and (in most species) dusky to black rectrices. The two anterior toes are relatively strong and are united for more than half their length. The outer rectrices are neither highly graduated in length nor strongly truncated at their tips. The outer rectrices are entirely dusky in three species and are barred but not tipped with white in one. Immatures are conspicuously spotted with white on their wings. Nests are typically excavated in termitaria or well-enclosed chambers of rotted wood (Skutch 1983).

SLATY-TAILED TROGON

Trogon massena Gould 1838

OTHER VERNACULAR NAMES

Chapman's trogon (*australis*), Hoffmann's trogon (*hoffmanni*), Massena's trogon, Narino trogon (*australis*), southern Massena trogon (*australis*); aurora grande, trogon coliplomizo (Spanish).

RANGE

Forests and woodlands from southeastern Mexico (southern Veracruz) through Belize, eastern Guatemala, Honduras (Caribbean slope), Nicaragua (both slopes), Costa Rica (both slopes), and Panama (both slopes) to western Colombia and northwestern Ecuador. See map 9.

SUBSPECIES

Trogon massena massena (Gould) 1838: southeastern Mexico to Nicaragua.

Trogon massena hoffmanni (Cabanis and Heine) 1863: Costa Rica to Panama and extreme northwestern Colombia. Considered synonymous with *massena* by Slud (1964) but recognized as valid by Wetmore (1968).

Trogon massena australis (Chapman) 1915: western Colombia and northwestern Ecuador. Considered a distinct species by Todd (1943).

MORPHOMETRICS

Measurements Wing, males (of *massena*) 163–181 mm (6.4–7.4 in; average of 28, 173.2 mm [6.8 in]); females 165–185.5 mm (6.4–7.2 in; average of 24, 174.1 mm [6.8 in]). Tail, males (of *massena*) 161–186 mm (6.3–7.3 in; average of 28, 170.8 mm [6.7 in]); females 160.5–186 mm (6.3–7.3 in; average of 24, 171.9 mm [6.7 in]; Ridgway 1911). Wing, males (of *hoffmanni*) 160–177 mm (6.2–6.9 in); females 162–177 mm (6.3–6.9 in). Wing, males (of *massena*, sample size not indicated) 171–181 mm (6.7–7.1 in); females 175–189 mm (6.8–7.4 in; Peters 1945). Average wing measurements of *massena* only slightly greater (by ca. 7 mm [0.3 in]) than those of *hoffmanni*, but average tail measurements clearly longer in the former

(by ca. 15 mm [0.585 in]; Wetmore 1968). Eggs of *hoffmanni* average 35.35 × 27.05 mm (1.4 × 1.1 in; Wetmore 1968); also 36.3 × 28.1 mm (1.4 × 1.1 in; Skutch 1972). The egg measurements given by Schönwetter (1966) differ somewhat from those of Wetmore and Skutch and are apparently based on an erroneous species identification.

Weights Twenty-six birds of mixed sex from Panama average 141.0 g (4.9 oz; Hartman 1961). Sex and sample size unspecified 141–171 g (4.9–6.0 oz; average 150 g [5.3 oz]; Smithe 1966). Unspecified Costa Rica weight 145 g (5.1 oz; Stiles and Skutch 1989). Ten unsexed individuals average 145 g (5.1 oz; Howe 1981). Two males of *massena* from Belize 144 and 163 g (5.0 and 5.7 oz); two females 147 and 149 g (5.1 and 5.2 oz; Russell 1964). One male of *massena* from Guatemala 172 g (6.0 oz); one male of *hoffmanni* from Costa Rica 137.8 g (4.8 oz; specimens, MVZ). Estimated egg weight 14.1–15.8 g (0.5–0.6 oz).

DESCRIPTION (of *massena*, adapted from Ridgway 1911; see plate 13)

Adult Male Upperparts bright iridescent green, usually golden green on back, scapulars, and upper rump, sometimes more bluish green or even intermixed with blue or violet-blue on lower rump and upper tail-coverts; four middle rectrices iridescent green or bronzed green, broadly but not sharply tipped with black; inner vanes of second and third (from the middle) rectrices and both vanes of outer rectrices uniform black, outer vane of fourth glossed with bronze-green basally, that of outermost rectrix and edges of next two toward base usually freckled or sprinkled with pale grayish; undersurface of rectrices slate, tipped with black; wing-coverts (except anterior portion of lesser wing-coverts) and secondaries delicately vermiculated with black and white; alular feathers, primary coverts, and primaries dull black or slate-black, longer primaries edged with pale grayish or grayish white proximally; chin, throat, and cheek and orbital, auricular, and malar regions dull black; chest bright iridescent green or bronze-green; remaining underparts pure geranium-red;

Map 9. Distribution of the slaty-tailed trogon. Inset map shows locations of major protected areas (mainly biosphere reserves and national parks) in southern Mexico and Central America.

thighs sooty blackish; bill deep orange-red or orange-yellow; iris hazel or dark yellow to pale yellowish orange; naked eyelids dull red to rose-red, salmon, or pink, with bare skin above and below dull green (according to Wetmore [1968]); earlier reports have erroneously described the bare eye-ring as sky-blue); feet and toes yellowish brown to yellowish gray or fuscous.

Adult Female   Head, neck, back, scapulars, rump, upper tail-coverts, chest, and breast plain slate color, darker on chin, throat, cheek and orbital, auricular, and malar regions, paler on lower breast; rest of underparts pure geranium-red; thighs blackish; tail slate-black, sometimes faintly darker than color of other upperparts; under surface of rectrices slate, glossed with violaceous; middle rectrices more slaty but tipped with black; wings blackish slate or sooty slate, coverts and secondaries minutely freckled with paler markings, longer primaries edged with white basally; maxilla mostly gray or blackish, with extreme basal lower portion plus all of mandible

dull orange or orange-yellow; iris orange-brown to orange-red; eye-ring as in male; feet and toes light brown.

Immature Male   Similar to adult male, but outer rectrices tipped with white and narrowly barred with white on at least distal portion of outer vanes and, usually, on distal portion of inner vanes (except on third from outside); middle rectrices much duller iridescent bronze-green than in adult and without black tip; secondaries (at least distal ones) spotted or mottled with dull white or buffy white along edges; bill more brownish orange or yellowish.

Immature Female   Similar to adult female, but outer rectrices and secondaries marked with white, as in immature male; undertail pattern also as in immature male; outer rectrices of both sexes of immatures narrower and more pointed than in adults, these tipped and barred with dull white.

Juvenile   No information available.

IDENTIFICATION

In the Hand   The slaty-tailed trogon is rather large, with a bright orange bill (more dusky on culmen in females and immatures than in males), vermiculated wing-coverts and secondaries (less obviously so in females and immatures), and a uniformly slaty black color on the underside of the tail (narrowly barred and tipped with white in immatures).

In the Field   The slaty-tailed trogon is found in lowland (sea level to ca. 600 m [2,000 ft] in Mexico, ca. 1,500 m [4,900 ft] in Costa Rica and Panama) humid and wet forests, as well as in secondary forest and in openings with scattered trees. No other trogons in the range of this species have slaty black undertail coloration and an orange-red bill. The orange-red bill color of adult males separates this species from the very similar black-tailed trogon of Panama and northern South America. Females and immatures may show orange only on the lower mandible. The species' vocalizations include a series of rapidly repeated, loud, and resonant *ka, koh* or *kuk* notes, uttered at the same pitch at the rate of more than two per second. See also the vocalizations section.

GEOGRAPHIC VARIATION

Central American slaty-tailed trogon specimens show considerable variation in tail markings and in the sheen of the breast and upperparts, but these differences do not appear to be geographically correlated (Todd 1943). However, on the basis of his inspection of four specimens from coastal Colombia, Todd (1943) believed that character differences in this (allopatric) southernmost population are sufficient to warrant species-level distinction. These differences mainly involve the middle rectrices, which are bluish in *australis* rather than brassy green, as in *massena*. The northernmost populations are largest, and those from Colombia are the smallest, with intergradation mainly occurring in Nicaragua (Wetmore 1968). Wetmore (1968) and de Schauensee (1964) both treated *australis* as a subspecies of *massena*, as have other, more recent authors.

ECOLOGY

Habitats   Collectively the slaty-tailed trogon occurs in tropical lowland evergreen forest (e.g., rain forest), secondary forest, and occasionally in riparian forests, plantations, and mangroves, from sea level to 1,400 m (4,600 ft). Humid forests, with tall berry- or fruit-bearing trees, are favored. In Colombia the species ranges from lowlands to 1,100 m (3,600 ft) in wet or humid forest and in tall secondary growth and sometimes visits forest edges or scattered trees in clearings. Wetmore (1968) stated that in Panama it may also occur in shade trees growing over coffee plantations, in tall trees along streams or marshes, and even in mangroves. He recorded it mostly between 575 and 1,400 m (1,900 and 4,600 ft) in Panama. Slud (1964) also reported it in mangroves in Costa Rica but stated that it usually occurs in forests of the tropical belt, sometimes reaching the subtropical zone. Monroe (1968) stated it had not been reported above 300 m (1,000 ft) in Honduras and that it favored rather dense rain forest but sometimes extended into areas of secondary forest or open forest.

Foods and Foraging Ecology   The foods preferred by slaty-tailed trogons are not noticeably different that those of other trogons of its genus, but these birds are perhaps more fruit dependent than smaller species. According to Skutch (1972), this species mainly consumes small berries and flesh-covered, or "arillate," seeds, largely the small to medium-size palm fruits. The birds also take the seeds of trees in the nutmeg family (such as species of the genus *Virola*), seeds with spicy red arils; and they also take the berries of the mangabe tree (*Didymopanax morototoni*). The fruits are taken in flight, and the interior seeds are regurgitated. Additionally, these trogons grab caterpillars and mature insects while in flight and may even eat small lizards. Wetmore (1968) noted that the stomach of one female contained two large locusts, two sphingid (sphinx moth) caterpillars, fragments of a cerambycid (long-horned) beetle, and drupe remains. Of three stomachs examined by Remsen, Hyde, and Chapman (1993), two contained arthropods only, and one contained mixed fruit and arthropods. Fruits of the nutmeg tree (*Virola sebifera*) are sometimes dispersed by this species (Howe 1981).

BEHAVIOR

General, Social, and Sexual Behavior   Generally rather solitary, the slaty-tailed trogon sometimes occurs in groups at fruiting trees or when matings are being sought (Skutch 1972). Small groups may also gather where insects are being stirred up by the movements of mammals. Slud (1964) noted that this species is often alone but is sometimes seen in somewhat separated twos and may also occur in threes or fours. He described what is probably courtship behavior as follows: The male raises his tail while performing a jerky bow or dove-like nod and exposes his red under tail-coverts and his green rump and upper tail-coverts by fluffing them. The female also raises her tail and, while holding it thus, exposes her red under tail-coverts, without jerking or nodding her head. Neither sex spreads its tail. The two birds may approach each other and, usually keeping their backs turned, continually raise and lower their tails while uttering churring chuckles. Karr (1971) described the species as being associated with medium canopy heights in moist forest and having a dominance-overlap territorial system.

Vocalizations   Skutch (1972) described the slaty-tailed trogon's vocalization as a deep, full-throated *wuk, wuk, wuk . . .* , which is harsher than the rather mellow notes of smaller species and not accelerated into a trill. The call is typically uttered at the rate of about two notes per second. However, a similar call is more rapidly repeated and may be accelerated into a long wooden rattle or may sound like a rapid cackle. In Central America, Skutch heard the repeated *wuk* call mainly during the drier months but sometimes as late as July. Slud (1964) stated that the usual call consists of up to ten or more *poop* or *kuk* notes that may descend in pitch or may be accelerated and shortened. Some single-note *cooer* or more nasal *huh* or *anh* notes are also produced. Vocalizations of this species were recorded in Mexico by Ben Coffey Jr. (Hardy, Reynard, and Coffey 1995). The recording included loud, rapid, and resonant *koh* notes, almost having a nasal or quacking quality, that become slower and louder toward the end of the series. The series ranged from as few as 12 notes uttered in about 4 seconds, to nearly 40 notes in 15 seconds, the longest series more obviously slowing toward the end, as if the bird were "running out of gas." Davis (1972) reported the song to be uttered at the rate of about three notes per second, with the pair sometimes duetting, the female's voice raised about a halftone higher than the male's. See also the identification section.

BREEDING BIOLOGY

Chronology of Breeding   Skutch (1972) reported that in Costa Rica, slaty-tailed trogons nest from as early March until June or July and that eggs were seen in April, May, and June. In Panama nesting may be extended: Wetmore (1968) noted that females taken between February and mid-April often have badly abraded tail feathers and mentioned a nest with eggs that had been found in late June. Willis and Eisenmann (1979) reported Panamanian nests between March and July. Limited Mexican data suggest nesting occurs there from April to June. In Colombia breeding-condition birds have been collected in February (Hilty and Brown 1986).

Nest Sites   Of five slaty-tailed trogon nests observed by Skutch (1953, 1972), four were in very rotted stubs or stumps of massive trees from 2.6–5.6 m (9–18 ft) above ground. The fifth was in a termitarium 4.7 m (15 ft) high. Pairs were also seen working on excavations at heights of 12–15 m (39–49 ft). Of

the four trunk nests, two were in forests, one was only a few meters from a forest edge, and one was nearly 20 m (66 ft) from the forest, in a charred stump. Skutch observed both sexes working at one excavation, the female more assiduously than the male. The cavities in tree trunks were similar in shape and size to the cavity in the termitarium, being about 21 cm high and 16 cm wide (8.3 × 6.3 in). The cavity is reached by means of an ascending tube about 13 cm long and 9 cm in diameter (5.1 × 3.5 in), which opens to the top of the nest chamber. A ridge separates the bottom of the chamber from the bottom of the entrance tube, preventing the eggs from rolling out. The chamber is unlined, but as incubation proceeds regurgitated seeds accumulate and provide a bed for the eggs. Wetmore (1968) saw pairs entering or working on nest excavations at termitaria on "numerous" occasions.

Eggs and Incubation   Gross (1930) found a slaty-tailed trogon nest containing two eggs in a termitarium. Four nests observed by Skutch (1972) contained either three eggs each or two nestlings and an egg. In one nest studied by Skutch, the male incubated from early morning until late afternoon, and the female through the night. Of five nests studied by Skutch, four were pillaged before or shortly after the eggs hatched, and the fate of the fifth was unknown. One nest that had been built in an active termitarium was later closed over by the termites. The eggs are pale bluish white, like those of the quetzals, rather than white like those of most typical trogons.

Brood Rearing   Little is known of the brood rearing phase of the slaty-tailed trogon, but Gross (1930) noted that the chicks were fed on winged insects.

CONSERVATION AND EVOLUTIONARY RELATIONSHIPS

The slaty-tailed trogon is a lowland forest species with a range that extends through nine countries, most of them in Central America. It is fairly common over much of its range and poses no current conservation concerns. It is certainly a close relative of the black-tailed trogon and is largely but not entirely allopatric with it. Stotz et al. (1996) listed the species as fairly common, with a medium degree of sensitivity to disturbance.

BLACK-TAILED TROGON

Trogon melanurus Swainson 1838

OTHER VERNACULAR NAMES

Large-tailed trogon (*macroura*), long-tailed trogon (*macroura*), western black-tailed trogon (*eumorphus*); aurora colilarga, sorocua cola negra (Spanish).

RANGE

Evergreen forest, second-growth forest, gallery forest, and forest edge from the Canal Zone of Panama south on both slopes to central Colombia (eastern Andes, Magdalena, particularly in the Cauca and Atrato Valleys) and southeastern Colombia, Venezuela (Amazonas, Bolívar), Guyana (widespread), and Suriname (interior), south to Brazil (Amazonia, Maranhão, south to northwestern Mato Grosso), and west to eastern Peru, eastern Ecuador, and northeastern Bolivia. See map 10.

SUBSPECIES

Trogon melanurus macroura Gould 1838: eastern Panama, northern Colombia (Gould described this as a species distinct from *melanurus* and called it the "large-tailed trogon").

Trogon melanurus mesurus (Cabanis and Heine) 1863: western Ecuador, northwestern Peru.

Trogon melanurus eumorphus: Zimmer 1948: southern Colombia and Ecuador, Peru, Bolivia, and Amazonian Brazil.

Trogon melanurus occidentalis Pinto 1950: southeastern Brazil (São Paulo region).

Trogon melanurus melanurus Swainson 1838: eastern Colombia, southern Venezuela, southern Guyana, and southern Suriname, south to Bolivia and southern Brazil (Mato Grosso), and east to northeastern Brazil (Maranhão).

MORPHOMETRICS

Measurements   Wing, males (of *macroura*) 157–187 mm (6.1–7.3 in; average of 7, 162.6 mm [6.3 in]); females 155.5–175.5 mm (6.1–6.8 in; average of 3, 156.5 mm [6.1 in]). Tail, males (of *macroura*) 165.5–181.5 mm (6.5–7.1 in; average of 7, 172.6 mm

[6.7 in]); females 161.5–174.5 mm (6.3–6.8 in; average of 3, 163 mm [6.4 in]; Ridgway 1911). Wing, males (of *melanurus*) 149.5–161.5 mm (5.8–6.3 in; average of 3, 155.6 mm [6.1 in]); females 149.5–152 mm (5.8–5.9 in; average of 3, 150.6 mm [5.9 in]). Tail, males 156–171.5 mm (6.1–6.7 in; average of 3, 161.3 mm [6.3 in]); females 149–162.5 mm (5.8–6.3 in; average of 3, 155.5 mm [6.1 in]; Naumburg 1930). No available information on eggs.

Weights   An unsexed sample of five *macroura* from Panama average 119 g (4.2 oz; Hartman 1961).

Termitarium nest-site of black-tailed trogon. Photograph by A. F. Skutch.

89

Map 10. Distribution of the black-tailed trogon.

Males of *melanurus* from Suriname (unstated sample size) 100–109 g (3.5–3.8 oz), females 88–108 g (3.1–3.8 oz; Haverschmidt 1968). One male of *melanurus* 104 g (3.64 oz; specimen, MVZ). One male of *melanurus* 108 g (3.8 oz), one of *mesurus* 111.1 g (3.9 oz; specimens, FMNH). Average mass of both sexes, visually estimated from graph 110 g (3.9 oz; Remsen, Hyde, and Chapman 1993).

DESCRIPTION (of *macroura*, adapted from Ridgway 1911; see plate 14)

Adult Male Upperparts bright iridescent green, more or less golden green on back, scapulars, and upper rump, more bluish green on lower rump and upper tail-coverts; middle pair of rectrices dark iridescent bluish green or greenish blue rather broadly tipped with black, second and third pairs black with outer portion of outer vanes dark iridescent green, tipped with black; remaining rectrices black, outer vanes finely freckled or sprinkled along edge, and especially toward base, with pale grayish or dull grayish white; anterior lesser wing-coverts bright iridescent green or golden green; remaining coverts, together with secondaries, delicately vermiculated with black and white; rest of wing dull black or slate-black, longer primaries edged with white; chin, upper throat and loreal, orbital, auricular, and malar regions black; lower throat and chest bright iridescent green or golden green, margined posteriorly by a band or line of white across upper breast, this white band sometimes narrowly barred with blackish; rest of underparts pure geranium-red; thighs slate-black or blackish slate; bill honey-yellow; inside of mouth (gape) yellow; iris warm brown; eye-ring orange-red to pink; feet and toes olive-green to olive-brown.

Adult Female Upperparts plain slate colored, tail darker and more brownish slate, larger wing-coverts and secondaries usually minutely freckled or vermiculated with paler slate, primaries more brownish slate or dull blackish slate, their outer vanes broadly edged with white becoming gray terminally; sides of head, chin, and throat dull brownish slate, darker on lores and anterior margin of chin; chest slate-gray becoming paler gray on breast and side, lower portion of breast more or less distinctly barred or vermiculated; rest of underparts geranium-red, paler on under tail-coverts; thighs blackish slate; underside of tail brownish slate, rectrices narrowly tipped with darker color; maxilla mostly black, with lower basal portion and entire mandible light honey-yellow; iris light brown; eye-ring mostly black, interrupted and dotted with red; feet and toes dull brown, honey-yellow on underside of toes.

Immature Male Secondaries and greater secondary wing-coverts margined and marked on the outer vanes with white and buff; chest gray, mixed variably with green; three outer pairs of rectrices tipped and barred on terminal parts of outer vanes with white; middle rectrices blackish with slight green gloss, lacking terminal black bar.

Immature Female Outer vanes of secondaries margined with white; two outermost pairs of rectrices tipped and irregularly barred terminally with white, third pair (from outside) margined with white at tip.

Juvenile   No information available.

IDENTIFICATION

In the Hand   The black-tailed trogon, like the slaty-tailed, has uniformly blackish outer and middle rectrices, and the two species are best separated by their bill color (yellow in black-tailed, orange in slaty-tailed) and by the presence of a narrow white band separating the green chest from the red abdomen that is present in the black-tailed. Females of the two species are even more similar, but the lower mandible of the black-tailed is yellowish rather than orange tinted. Female black-tailed trogons also have obscurely vermiculated upper wing-coverts and a slightly paler grayish breast than do females of the slaty-tailed and white-eyed trogons, which have uniformly gray upper wing-coverts and darker gray breasts. As to measurements, the tail is barely longer than the wing in the black-tailed, but the wing is very slightly longer than the tail in the slaty-tailed.

In the Field   Where black-tailed and slaty-tailed trogons occur together (as in Panama, Colombia, and Ecuador), they may be separated under favorable field conditions by the traits just mentioned. Both species are associated with lowland humid forests. Where the white-eyed trogon also occurs (in Colombia and northwestern Ecuador), immatures of the three are perhaps not separable in the field, but adults of the white-eyed have a pale whitish to yellowish, rather than brown, iris coloration. The repeated *kwo* vocalizations of the black-tailed trogon closely resemble those of the slaty-tailed (and also the white-eyed) but on average are perhaps more mellow and more resonant and also are uttered more rapidly and at a higher pitch. In both of these species the usual vocalization consists of a single *kwo* note, sometimes repeated progressively more frequently until it becomes a rapid trill. There is also a *KWA-kwa* alarm note in both. All three species are limited to tropic-zone forests. See also the vocalizations section.

GEOGRAPHIC VARIATION

The black-tailed trogon race *macroura*, originally considered by Gould to be a separate species, is somewhat larger than nominate *melanurus*. The former has slightly larger wing measurements (averaging

ca. 7 mm [0.3 in] larger in males, 2 mm [0.1 in] in females) and is especially longer in its tail measurements (averaging ca. 10 mm [0.4 in] longer in males, 7 mm [0.3 in] in females). There are no plumage color differences.

ECOLOGY

Habitats   Collectively the black-tailed trogon occurs in tropical lowland evergreen forest (rain forest), gallery forest, and mangrove forest, from sea level to 1,000 m (3,300 ft) elevation, but rarely may occasionally attain higher levels. In Colombia this species ranges up to 500 m (1,600 ft) east of the Andes but may reach elevations in excess of 1,000 m (3,300 ft) west of the Andes, in humid forests, forest edges, and secondary growth. In Panama this species occurs from sea level to 550 m (1,800 ft), in the tropical zone. It is evidently very similar in its ecology to the slaty-tailed trogon (Wetmore 1968). In Venezuela the black-tailed reaches only 100 m (300 ft) north of the Orinoco River, but south of that river it reaches 1,000 m (3,300 ft; de Schauensee and Phelps 1978).

Foods and Foraging Ecology   Stomachs of black-tailed trogons studied by Wetmore (1969) contained berries, palm seeds, fragments of flower tassels of the guarumo tree (*Cecropia*), and large orthopterous insects, including a stick insect. Haverschmidt (1968) reported that berries, fruit, and insects (locusts and long-horned beetles) are eaten. Of 38 stomachs examined by Remsen, Hyde, and Chapman (1993), 21.1 percent contained arthropods only, 50.0 percent contained mixed fruit and arthropods, and 29.9 percent contained fruit only.

BEHAVIOR

General, Social, and Sexual Behavior   No information available.

Vocalizations   According to Hilty and Brown (1986), the calls of the black-tailed trogon are nearly identical to those of the slaty-tailed trogon but are possibly uttered more rapidly, and the former also produces a bubbly, purring trill. Vocalizations of this species were recorded by Ben Coffey Jr. (Hardy, Reynard, and Coffey 1995). The recording included songs of one bird, plus group calling in Peru. There were two prolonged songs of 28 and 46 individual notes of constant pitch, and the notes were produced at uniform intervals of about two per second. Others such as Wetmore (1968) have reported that the song consists of single notes that speed up and become a trill; Wetmore also stated that the species' voice is higher in pitch and the notes are more rapidly uttered than in *massena*, being somewhat sonorous when heard at moderate distances. See also the identification section.

BREEDING BIOLOGY

Chronology of Breeding   Wetmore (1968) collected a female black-tailed trogon about to lay in late March in Panama but had no other direct evidence of breeding.

Nest Sites   No information available.

Eggs and Incubation   No information available.

Brood Rearing   No information available.

CONSERVATION AND EVOLUTIONARY RELATIONSHIPS

The black-tailed trogon is clearly a very close relative of the slaty-tailed trogon but is locally sympatric with it in northwestern Colombia and thus must be considered a separate species. Although their vocalizations are very similar, bill color differences (red versus yellow) and the white breast band present in males of the black-tailed may be important species-recognition characteristics. Like the slaty-tailed trogon it is fairly common in many areas and probably occurs in ten South American countries. On the basis of gene-sequencing studies, Espinosa de los Monteros (1998) considered this species to represent a basal component of his "violaceous subclade," with *T. comptus* being a sister taxon. Stotz et al. (1996) listed the species as common, with a medium degree of sensitivity to disturbance.

LATTICE-TAILED TROGON

Trogon clathratus Salvin 1866

OTHER VERNACULAR NAMES

Aurora colirrayada, trogon ojiblanco (Spanish).

RANGE

Evergreen forests of foothills and lower slopes (to 1,350 m [4,400 ft]) on Caribbean slope of Costa Rica (north to at least Volcan Miravalles) and western Panama (Caribbean slope in Bocas del Toro, Veraguas, Coclé; locally on Pacific slope in Veraguas). See map 11.

SUBSPECIES

None recognized, but see following species.

MORPHOMETRICS

Measurements   Wing, males 152–165 mm (5.9–6.4 in; average of 3, 159 mm [6.2 in]); females 151–165 mm (5.9–6.4 in; average of 9, 158.7 mm [6.2 in]). Tail, males 148.5–160 mm (5.8–6.2 in; average of 3, 153.5 mm [6.0 in]); females 152–154 mm (5.9–6.0 in; average of 9, 153.2 mm [6.0 in]; Ridgway 1911). No information on eggs is available.

Weights   Unspecified Costa Rican weight 130 g (4.6 oz; Stiles and Skutch 1989).

DESCRIPTION (adapted from Ridgway 1911; see plate 15)

Adult Male   Upperparts nearly uniform iridescent green or golden or bronzy green, upper tail-coverts rarely more bluish, but middle pair of rectrices usually more bluish green, rather broadly but not sharply tipped with black; inner vanes of second and third pairs of rectrices (from middle) black, outer vanes iridescent green, bluish green, or greenish blue, tipped with black; remaining rectrices black, crossed on outer vanes and distal portion of inner vanes with narrow white lines, outermost rectrix with greater portion of inner vanes thus barred; anterior lesser wing-coverts iridescent green or bronze-green, remaining coverts and the secondaries delicately vermiculated with black and white; remainder of wing dull black or blackish slate, longer primaries edged with grayish white broken into minute freckling on distal portion; chin, throat, and loreal, orbital, auricular, and malar regions black; chest iridescent green or golden green; remaining underparts pure geranium-red; thighs black or dark sooty slate, the longer feathers sometimes narrowly tipped with pinkish; mandible and maxilla yellow; eye-ring inconspicuous, bare edges of eyelids black; iris cream to lemon-yellow; feet and toes brownish.

Adult Female   Crown blackish slate; remaining upperparts slate color, middle rectrices darker and more brownish slate, narrowly and rather indistinctly tipped with blackish; rest of tail as in adult male; wings dull blackish slate, wing-coverts and secondaries minutely and rather indistinctly vermiculated with pale grayish, longer primaries edged with white; chin, throat, and loreal, orbital, auricular, and malar regions dark sooty; chest slate, passing through dull slate-gray into light raw umber or wood-brown, posterior portion sometimes faintly barred; rest of underparts geranium-red; thighs blackish or sooty; culmen blackish, but remainder of maxilla and entire mandible dull yellow; iris yellowish; feet and toes as in adult male.

Immature   Secondaries and greater secondary coverts distinctly vermiculated; rectrices narrower and more pointed than in adults, three outer pairs of rectrices also more coarsely and irregularly barred with black and white.

Juvenile   Warm brown upperparts, becoming more rufous on rump and upper tail-coverts; lower breast, sides, and under tail-coverts lighter, more cinnamon-brown; middle and lesser wing-coverts white, tipped with brown and with a narrow subterminal slate bar; greater wing-coverts dusky, edged and barred with white; inner secondaries margined with white on outer vanes, with a distal bar and cinnamon-buff edging, the rest freckled with white to dull buff (Wetmore 1968).

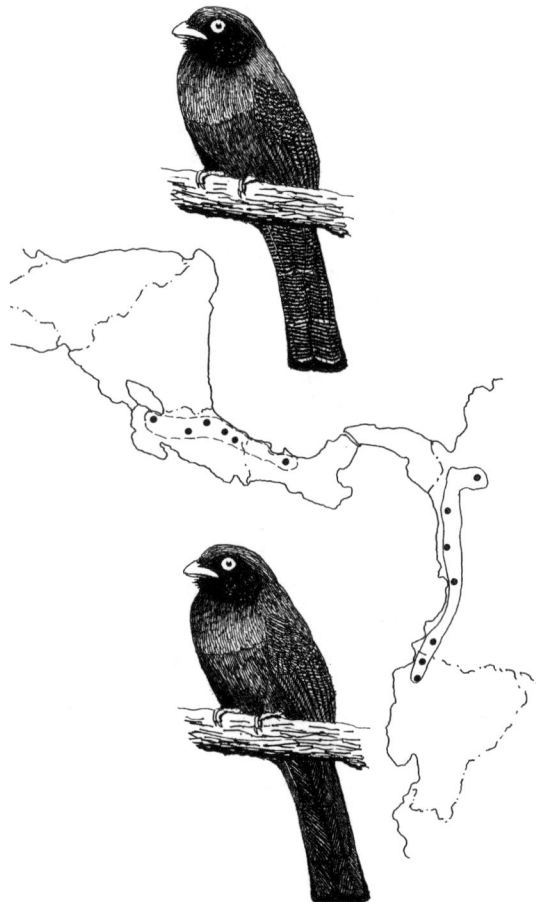

mostly dusky with narrow white barring. It is a species of lowland and foothills forest, extending up to about 750 m (2,500 ft) in Panama but reportedly to about 1,100 m (3,600 ft) in Costa Rica. Sometimes found in the same areas as the slaty-tailed trogon, the lattice-tailed has a yellow (not orange) bill, pale yellow (not brown) eyes, and a similarly blackish but obscurely barred undertail pattern. Its vocalizations are much like those of other closely related trogons, usually consisting of up to about 15 repeated *kwa* notes that tend to increase in pitch and volume in the middle of the sequence and then become faster, lower, and softer as the series ends. See also the vocalizations section.

GEOGRAPHIC VARIATION

None described.

ECOLOGY

Habitats   Collectively lattice-tailed trogons occur in tropical lowland evergreen forest (e.g., rain forest) and sometimes in forest edges or semi-open forests from 100–1,100 m (300–3,600 ft) elevation. In Costa Rica this species is almost exclusively limited to heavily wooded areas, only occasionally straying to forest edges or semi-open areas and extending from the lower tropical zone to about 1,400 m (4,600 ft), where it is widely sympatric with the more tolerant *massena* (Slud 1964). Breeding in Costa Rica occurs between 90 and 1,100 m (300 and 3,600 ft), with a movement to lower elevations after breeding and during the latter part of the rainy season (Stiles and Skutch 1989). In Panama it is relatively rare but has been reported between 150 and 1,050 m (500 and 3,200 ft), in lowland, foothill, and lower montane forests (Ridgely and Gwynne 1989).

Foods and Foraging Ecology   A lattice-tailed trogon stomach examined by Remsen, Hyde, and Chapman (1993) contained fruit only. In Costa Rica these birds are known to eat fruits of laurels (Lauraceae) and other trees, large insects, and occasionally small frogs and lizards (Stiles and Skutch 1989).

BEHAVIOR

General, Social, and Sexual Behavior   Lattice-tailed trogons inhabit the lower canopy and middle

Map 11. Distribution of the lattice-tailed trogon (dashed line and upper sketch) and the white-eyed trogon (solid line and lower sketch). Dots indicate locality records for both species (mainly after Haffer 1974).

IDENTIFICATION

In the Hand   Like most of the preceding species of its subgenus, the lattice-tailed trogon has vermiculated upper wing-coverts (indistinct in females), a yellow bill (in females the yellow is mostly confined to the lower mandible), and blackish outer (as well as middle) rectrices. The outer rectrices are narrowly banded with white, thus separating the species from the other dusky-tailed forms. Females are similar to males in their tail patterning, and adults of both sexes have pale yellow-white irises.

In the Field   Within its Central American range, the lattice-tailed trogon is the only trogon with a pale iris color and with undertail coloration that is heights of wet forests, sometimes reaching the forest's upper canopy or its open understory to forage. The birds are usually solitary or in well-separated pairs. According to Slud (1964), courtship in this species is much like that of the slaty-tailed trogon. Both sexes perform tail-raising accompanied by a liquid gurgle or a churring chuckle, the calls somewhat higher in pitch and more "yacking" in quality than those of the slaty-tailed. Presumably fluffing of the upper and lower tail-coverts and head-nodding also occur in the lattice-tailed, as in the slaty-tailed. When the male calls, his latticed outer tail feathers are spread outward from the rest or lie out of sequence, so that they are not covered by the green middle rectrices and are visible from above.

Vocalizations   Slud (1964) stated that lattice-tailed trogon's voice is higher pitched than the slaty-tailed's, that its *poop* notes are fewer in number than the *kuk* notes of *massena*, and that the first two notes of the sequence trend upward. The middle of the call sequence tends to "bulge," whereas the end tapers off. Vocalizations of this species were recorded in Costa Rica by Gary Stiles (Hardy, Reynard, and Coffey 1995). The recording includes calls of birds involved in a dispute: there were nine sequences, each with from eight to fourteen rapidly repeated *kwa* notes and lasting 4 or 5 seconds. The vocalizations often both started and ended at rather low amplitude but varied little in cadence throughout. See also the identification and general, social, and sexual behavior sections.

BREEDING BIOLOGY

Chronology of Breeding   In Costa Rica lattice-tailed trogons nest from February to May (Stiles and Skutch 1989).

Nest Sites   Lattice-tailed trogons nest in cavities of a rotting stub or snag or, occasionally, in an arboreal termitarium, at heights of 5–8 m (16–26 ft).

Eggs and Incubation   No information is available on the eggs of the lattice-tailed trogon; they are probably slightly smaller than those of the slaty-tailed trogon. Other aspects of nesting are also likely to be similar.

Brood Rearing   No information available.

CONSERVATION AND EVOLUTIONARY RELATIONSHIPS

Like the Baird's and orange-bellied trogons, the lattice-tailed trogon is limited to Costa Rica and Panama. In the former country, however, it is relatively widespread and fairly common. Its nearest relative would seem to be the slaty-tailed trogon, with which it is completely sympatric, the main differences being the lattice-tailed's yellow eyes, more yellowish bill, and slightly barred tail. These bill and eye differences might be important species-recognition signals. Vocalizations of the two are certainly extremely similar. Stotz et al. (1996) listed the species as uncommon, with a high degree of sensitivity to disturbance.

WHITE-EYED TROGON

Trogon (clathratus) comptus Zimmer 1948

OTHER VERNACULAR NAMES

Blue-tailed trogon (but also see *Harpactes reinwardti*), Zimmer's trogon; trogon ojiblanco (Spanish).

RANGE

Colombia, in tropical-zone forests from the lower Cauca Valley and the northern Atrato Valley south to Nariño; also northwestern Ecuador, south to slightly beyond Quito (Tinalandia, Santo Domingo). See map 11.

SUBSPECIES

No subspecies have been recognized, but it is possible that the white-eyed trogon should be considered a subspecies or an allospecies of *clathratus*.

MORPHOMETRICS

Measurements Wing, males 152.5–171 mm (5.9–6.7 in; average of 4, 158.2 mm [6.2 in]); females 150–165 mm (5.9–6.4 in; average of 5, 158 mm [6.2 in]). Tail, males 135–161 mm (5.3–6.3 in; average of 4, 149.9 mm [5.8 in]); females 132–161 mm (5.8–6.3 in; average of 5, 146 mm [5.7 in]; Zimmer 1949). No information on eggs available.

Weights Two males from Ecuador 100 and 108 g (3.5 and 3.8 oz; average 104 g [3.6 oz]; Dunning 1993). Adult male 126.0 g (4.4 oz), adult female 119.7 g (4.2 oz; Gary Stiles, pers. comm.).

DESCRIPTION (adapted from Zimmer 1949; see plate 16)

Adult Male Upperparts iridescent blue-green and bronzy green, becoming peacock-blue on rump, upper tail-coverts, and tail; chin, throat, lores, and sides of face deep black; breast like back; thighs black; rest of underparts red; remiges mostly black, with a white or whitish outer margin toward the base of nearly all feathers; secondaries white basally, their outer margins and exposed portions of tertials

freckled or vermiculated with grayish white and blackish; alular and primary wing-coverts black; innermost lesser wing-coverts like back; under wing-coverts plain blackish or blackish freckled with white; median rectrices dark purplish (peacock) blue with a broad black tip; next two pairs of rectrices with dark blue outer margins; remaining rectrices black with subdued light freckling on outer margins of outermost feathers toward their bases; underside of tail black; bill yellow; iris pale yellow or whitish gray; eye-ring inconspicuous, fuscous to dark grayish brown (Gary Stiles, pers. comm); feet and toes blackish.

Adult Female Generally similar to but darker than females of other species in this subgenus, such as the slaty-tailed and lattice-tailed trogons; top and sides of head, chin, throat, and chest all clouded with sooty or sooty brown; upper wing-coverts, tertials, and outer margins of secondaries uniform blackish gray rather than freckled or vermiculated; freckling or barring on outer rectrices somewhat more apparent than in adult males; soft-part colors much like those of adult male, but bill mostly blackish, with only lower mandible and base of upper mandible yellowish.

Immature and Juvenile Undescribed, but probably not significantly different from the comparable stages of preceding species, with narrow, rather pointed rectrices and buffy markings on edges of inner remiges and upper wing-coverts; tail-barring on outer two or three pairs of rectrices more conspicuous than in adults (Gary Stiles, pers. comm.).

IDENTIFICATION

In the Hand The white-eyed trogon appears to be a close relative of the slaty-tailed and lattice-tailed trogons, especially the latter. The outer rectrices of white-eyed males are uniformly black, and the bill is yellow. The upper wing-coverts and secondaries are conspicuously vermiculated, and the iris is pale yellow to whitish. Females are scarcely distinguishable from those of the slaty-tailed but are somewhat darker (sooty to sooty brown) in the slaty-tailed trogon's gray regions, and the secondaries and upper wing-coverts are uniformly dark, without obvious vermiculations.

In the Field   Within its limited Colombian and Ecuadoran range the white-eyed trogon overlaps the similar slaty-tailed trogon (but has pale eyes) and perhaps also the lattice-tailed trogon (which also has pale eyes but barred outer rectrices). The white-eyed differs from the similar black-tailed trogon, which it barely meets at the northern edge of its range in northern Colombia and again at the southern end of its range in northwestern Ecuador, in having a yellow (not orange) bill and pale (not brown) eyes. In the white-eyed trogon the outer two to three rectrices have inconspicuous fine gray barring or freckling; in the lattice-tailed this barring is rather more contrasting and occurs over both webs of these feathers, but even in this latter species it is difficult discern the barring clearly under field conditions. In the white-eyed trogon the barring is especially inconspicuous in adult males and most noticeable in young birds, but even in the latter it is virtually impossible to see in the field at any distance (Gary Stiles, pers. comm.). See also the vocalizations section.

GEOGRAPHIC VARIATION

None described.

ECOLOGY

Habitats   The white-eyed trogon has been reported in Colombia in tropical and lower subtropical forests, at elevations of 900 m (3,000 ft, Río Anchicayá), 1,500 m (4,900 ft, La Frijolera), and 1,800 m (5,900 ft, La Selva; de Schauensee 1948–1952). There it occurs in humid and wet forests and forest borders, especially in the broken and hilly terrain of the foothills (Hilty and Brown 1986). Ecuadoran localities are tropical forest sites situated between about 200 and 650 meters (700 and 2,100 ft). Collectively the range extremes are from about 200 to 1,800 m (700 and 5,900 ft), mostly in forests of the tropical zone.

Foods and Foraging Ecology   Of seven white-eyed trogon stomachs examined by Remsen, Hyde, and Chapman (1993), 42.9 percent contained arthropods only, 42.9 percent contained mixed fruit and arthropods, and 14.3 percent contained fruit only.

BEHAVIOR

General, Social, and Sexual Behavior   No detailed information about white-eyed trogon behavior is available. Males may sometimes gather in loose singing assemblages (Haffer 1975).

Vocalizations   Haffer (1975) stated that the white-eyed trogon's vocalizations resemble those of the slaty-tailed and black-tailed trogons but are slower and lower in pitch. Songs and calls of this species were recorded in Ecuador by Murray Gell-Man (Hardy, Reynard, and Coffey 1995). Most of the song examples consisted of four- and five-note *kwa* sequences, each rapidly uttered and lasting less than 2 seconds. There are also some rapid chatter-like calls, some very quick trill-like utterances, and some low-pitched, almost grunting notes. Although Hardy, Reynard, and Coffey suggested that the white-eyed's vocalizations are specifically distinct from potentially related forms such as *massena*, they clearly are quite similar to those of *clathratus*.

BREEDING BIOLOGY

Chronology of Breeding   Breeding condition white-eyed trogons have been obtained in Colombia in early March.

Nest Sites   No information available.

Eggs and Incubation   No information available.

Brood Rearing   No information available.

CONSERVATION AND EVOLUTIONARY RELATIONSHIPS

The white-eyed trogon, the most recently described species of trogon, is limited to western Colombia and northwestern Ecuador and is generally considered uncommon. It seems likely on geographical grounds alone that the species' nearest relative is the slightly larger black-tailed trogon, with which it is essentially parapatric. The males of the two species differ mainly in that white-eyed trogons have whitish eyes, are more bluish above and anteriorly, and have finer shoulder freckling and a darker blue and distinctly shorter tail. Female white-eyed trogons are generally darker throughout than female slaty-tailed and black-tailed trogons but are very similar to both. Zimmer (1948), who first described the species, also evidently considered the black-tailed trogon its nearest relative. On the basis of gene-sequencing studies, Espinosa de los Monteros (1998) considered the black-tailed trogon to be this species' sister taxon. Stotz et al. (1996) listed the species as uncommon, with a medium degree of sensitivity to disturbance, and they saw a need for monitoring, owing to deforestation.

Dark-Winged Trogons (Subgenus *Trogon* Brisson 1760)

The subgenus *Trogon* consists of four small to medium-size (25–28 cm [11.4–11.0 in], 50–100 g [1.8–3.5 oz]) species, which, when compared to the species of the subgenus *Curucujus,* have more graduated tails, middle rectrices with more distinctly truncated tips, and underparts that are usually bright yellow (reddish orange in one species). In three of the four species the upper wing-coverts of adults are uniformly slate colored, and in the fourth they are very narrowly barred or vermiculated with white on a dark background. The outer rectrices are extensively patterned with white in both sexes. All species have relatively short tarsi, and their front toes are only slightly fused basally. Species in this genus excavate sites either in termitaria or in well-rotted wood, making an ascending entrance tube leading to a chamber that wholly encloses the incubating bird (Skutch 1983). Males often gather in singing groups during the courting season and have a rattling song phrase of a "bouncing ball" type. Their outer tail feathers are typically mostly white or at least broadly tipped with white (Davis 1972).

BAIRD'S TROGON

Trogon (viridis) bairdii Lawrence 1868

OTHER VERNACULAR NAMES

Vermilion-breasted trogon; aurora de Baird, trogon vientribermejo (Spanish).

RANGE

Lowland and foothill evergreen forests of Costa Rica (southern Pacific slope, north regularly to Caraça, sometimes to Ortina), south to western Panama (western Chiraquí). See map 12.

SUBSPECIES

No subspecies have been recognized, but *bairdii* is sometimes considered conspecific with *viridis*.

MORPHOMETRICS

Measurements Wing, males 142–153 mm (5.5–6.0 in; average of 18, 146.5 mm [5.7 in]); females 143–153.5 mm (5.6–6.0 in; average of 13, 146 mm [5.7 in]). Tail, males 147–172.5 mm (5.7–6.7 in; average of 18, 155.9 mm [6.1 in]); females 148–170 mm (5.8–6.6 in; average of 13, 157.8 mm [6.2 in]; Ridgway 1911). No information on eggs available.

Weights Unspecified Costa Rica weight 95 g (3.3 oz; Stiles and Skutch 1989); male 93.7 g (3.3 oz; Wetmore 1968).

DESCRIPTION (adapted from Ridgway 1911; see plate 17)

Adult Male Crown and hindneck black, usually mixed with violet-blue; back, scapulars, anterior lesser wing-coverts, and upper rump bright iridescent green to bluish green; becoming violet-blue or bluish violet on lower rump and upper tail-coverts; middle rectrices dark iridescent bluish green to violet-blue, tipped with black; inner vanes of second and third rectrices black; outer rectrices extensively white terminally, the white on the third from outside covering from one-third to half the vane; outermost rectrix with all the exposed vane white, black base concealed by reddish under tail-coverts; wings black, longer primaries edged on anterior vanes with white; sides of head, chin, throat, and upper breast glossed with violet-blue; remaining underparts bright reddish orange to orange-red; thighs black; bill bluish white, darker on mandible and at base of maxilla; iris dark brown; eye-ring pale blue to whitish; feet and toes lead-gray.

Map 12. Distribution of Baird's trogon.

Adult Female   Upperparts plain slate or dark slate-gray, four middle rectrices darker, narrowly tipped with black; wings black, wing-coverts and secondaries very narrowly barred with white; longer primaries edged basally with white and usually dotted or specked with white terminally; outer rectrices black, narrowly tipped (2–9 mm, 0.1–0.4 in) with white; remaining rectrices narrowly barred with white (white bars much narrower than black) distally and on most of outer vanes; lores and orbital region blackish; sides of head and neck, chin, throat, chest, breast, sides, and flanks slate-gray or brownish slate; underparts somewhat paler posteriorly, where sometimes barred with paler grayish and tipped with orange-pinkish; abdomen and under tail-coverts reddish orange, flanks more or less tinged with the same; thighs sooty slate; bill mostly bluish, becoming black to dark grayish basally; iris brown; eye-ring, feet, and toes all probably as in adult male.

Immature   Older immatures increasingly resemble adults of their respective sex, but the rectrices of the former are narrower, more pointed, and more coarsely and irregularly barred than those of the latter.

Juvenile   Nestlings somewhat resemble the adult female, but the abdomen, flanks, and sides of the former are grayish sooty, becoming pale fulvous or buffy around vent; under tail-coverts are white to pale buff. In older juveniles the wings and tail are patterned as in the adult female, but the white tips of the outer rectrices of the former are somewhat larger than those of the latter.

IDENTIFICATION

In the Hand   Baird's trogon is the only member of the subgenus having orange-red rather than yellow underparts, although the hue is not so red as in many other trogons and quetzals. Like the white-tailed trogon, males of Baird's trogon show no black on their outer rectrices as seen from below; but females are more heavily barred black and white on these feathers, and the feathers are distinctly orange-tinted rather than yellow below. Both sexes of Baird's trogon have pale blue eye-rings and bluish to bluish-white bills; in the somewhat similar orange-bellied trogon, the bill is yellow and the eye-ring (apparent in females only) is white.

In the Field   Baird's trogon is associated with moist lowland and foothill forests, forest edges, and sometimes open woodlands. The only other trogon occurring in the same region and exhibiting orange underparts is the orange-bellied trogon: males of the orange-bellied have outer rectrices narrowly barred with black and white, whereas females are mostly white on their outer rectrices, with only faint black markings (the tails Baird's have exactly the opposite patterning). Baird's trogon occurs in lowlands, foothills, and low mountains to about 1,200 m (3,900 ft). Its vocalizations include a prolonged series of "barking" notes that often accelerate and then slow down abruptly. Repeated soft and melodious notes have also been described. See also the vocalizations section.

GEOGRAPHIC VARIATION

None described.

ECOLOGY

Habitats   Collectively Baird's trogon has been observed in forests of the tropical and lower subtropical zones, from a few hundred meters elevation to 1,200 m (3,900 ft). In Costa Rica this species is "virtually confined" to the humid tropical zone of the Pacific southwest but extends slightly into the zone of dry tropical deciduous forests and altitudinally into the lower subtropical belt of that region (Slud 1964). Locally it may reach 1,200 m (3,900 ft) in Costa Rica, but it is primarily a lowland and foothills species (Stiles and Skutch 1989). In Panama it is similarly distributed but rare, with records at least as high as 1,200 m (3,900 ft; Wetmore 1968; Ridgely and Gwynne 1989).

Foods and Foraging Ecology   No detailed information on the foraging behavior of Baird's trogon is available; but fruiting trees and shrubs are visited, and insects are plucked from foliage or the air.

BEHAVIOR

General, Social, and Sexual Behavior   Baird's trogon is found in the canopy zone of tall rain forest, sometimes descending to the upper understory or nearby tall secondary forest or semi-open forest (Stiles and

Skutch 1989); the species is only rarely seen at eye level. Unlike many other trogons, this species can often be found in small (perhaps family) groups rather than singly or in pairs (Slud 1964). Skutch (1983) stated that pairs are formed early in the year, at which time groups of five or six birds may be seen. When alarmed, males of this species rapidly spread and fold the tail, producing flashes of white to observers behind. Females may do the same, but the strong visual effect produced by the male is absent.

Vocalizations   The voice of Baird's trogon is somewhat like that of the slaty-tailed, the individual *chowp* notes similarly starting slowly but soon speeding up and "bouncing" along with increasing strength until near the end, when the notes slow and become weaker. There are also rapid chicken-like cackles, repeated dove-like notes, and repeated *chuck* notes accompanied by nervous flashes of the tail (Slud 1964). Vocalizations of this species were recorded in Costa Rica by J. W. Hardy (Hardy, Reynard, and Coffey 1995). The song examples consist of increasingly rapid *whew* notes that remain at the same pitch but quickly speed up to become nearly a trill, with each sequence lasting about 5 seconds or less. A slowing down of the sequence toward the end, as mentioned in the field identification section, was not present in the recorded examples. See also the field identification section.

BREEDING BIOLOGY

Chronology of Breeding   In Costa Rica Baird's trogons breed between April and August (Stiles and Skutch 1989). Skutch (1983) stated that his earliest egg date was April 1, and latest was in August.

Nest Sites   Skutch (1983) reported finding seven occupied Baird's trogon nests in Costa Rica. These nest openings were from 1.8 to 10.6 m (6–35 ft) high, in the decaying trunks of massive trees. Four nests were made in successive years in a single tree stub standing at the edge of a pasture, about 30 m (100 ft) from the forest edge. Each year the stub was shorter, so that the birds nested progressively closer to the ground (entrance holes starting at 5.5 m [18 ft] and dropping to 2 m [7 ft]). Finding a tree with the proper degree of interior hardness is evidently difficult; one trogon pair started six different holes but

completed none because the wood was too rotten. The birds bite and tear at the wood rather than chisel it, as might woodpeckers. The male chooses the site and then tries to convince a female to accept both him and the site. Males alternate between excavating a nest site and singing from the top of a nearby tree. Once mated, the pair work on the excavation together, one excavating as the other perches nearby. The entrance hole is dug in an ascending direction. When it is deep enough, one bird may work inside while the other digs or waits at the entrance. Should a female reject a hole that has been started, the male may try to entice her back but will eventually abandon his effort. Completed nest chambers have an entrance tube about 8.3 cm in diameter and 18 cm long (3.2 × 7.1 cm), with the end of the tube entering the top of the nesting chamber. A ridge of wood serves to help keep the eggs from rolling out of the chamber, as in the slaty-tailed trogon.

Eggs and Incubation   Skutch (1983) reported that of the five Baird's trogon nests he studied four had two eggs and one had three. Eggs are laid at intervals of 2 to 3 days. In one nest there was a 2-week interval between the start of the nest and the laying of the first egg, while in a late nesting there was a month-long interval. Like most other trogons, the male Baird's trogon incubated from early morning until late afternoon. In 3 of the 4 years in which the birds nested in the same tree the nestings were failures, but in 1 year two of the three eggs hatched, after 16 days. In another nest, the incubation period was also judged to be about 16 days, the female apparently not starting to incubate until the clutch was completed.

Brood Rearing   Skutch (1983) found that young Baird's trogons hatch completely naked, with their eyes remaining closed until they were 9–10 days old. Their first feathers remained in long sheaths until the birds were about 2 weeks old, at which time the coverings quickly sloughed off to reveal a dark juvenal plumage. Of five nests studied, the eleven eggs produced four chicks, only one of which survived to fledging. This chick did not fledge until it was 25 days old, probably because it was heavily infested with dipteran parasites. Both parents initially brooded the young, and the female continued to brood through the night until the chicks were 11 days old. Large, green-winged insects (probably orthopterans) were the bulk of the chicks' food; but green caterpillars were also fed to the chicks, and a stick insect was brought to the nest.

CONSERVATION AND EVOLUTIONARY RELATIONSHIPS

Skutch (1983) noted that recent deforestation in the very restricted range of Baird's trogon has placed it in danger of extinction. It and the white-tailed trogon constitute a very closely related pair of allospecies; Peters (1945) considered them conspecific. Wetmore (1968) suggested that the abdominal color differences (orange-yellow in *bairdii* vs. yellow in *viridis*) and the tail patterning differences warrant species separation. He observed that *Trogon viridis chionurus* is now separated from *bairdii* by half the length of the Panamanian isthmus. Since both are lowland forms, there presently is no barrier separating the two, and presumably the splitting of the gene pools must have occurred at some period when the isthmus was submerged. The species was classified as near threatened by the IUCN (Collar, Crosby, and Stattersfield 1994). Stotz et al. (1996) listed the species as uncommon, with a medium degree of sensitivity to disturbance, and declining as a result of deforestation.

WHITE-TAILED TROGON

Trogon viridis L. 1766

The white-tailed trogon was identified as *T. strigulatus* L. by Gould and many later authors, including Ridgway (1911) and Peters (1945). However, Linnaeus used *strigulatus* for the female of this species and *viridis* for the male, with the latter name having taxonomic priority (Zimmer 1949).

OTHER VERNACULAR NAMES

Greater yellow-bellied trogon, green-backed trogon (*viridis*), snow-tailed trogon; sorucua cola blanca, trogon coliblanca (Spanish).

RANGE

Evergreen forest, forest edges, and second-growth woodlands from western Panama (Caribbean slope, central Bocas del Toro east though the Canal Zone; also recorded from eastern Panama province on Pacific slope to Colombia (tropics east of eastern Andes, also south along the Pacific coast; also occurs in north-central Colombia in the Santa Marta Mountains and in the Magdalena Valley) and east to Venezuela (widely distributed), Guyana (widespread), Suriname (widespread), French Guiana, and Trinidad Island (common); also Brazil (Amazonas, south and east to western Minas Gerais, northern São Paulo, southern Mato Grosso), western Ecuador, eastern Peru, and northern Bolivia (Pando, Beni). See map 13.

SUBSPECIES

Trogon viridis chionurus Sclater and Salvin 1870: eastern Panama, western Colombia, and western Ecuador.

Trogon viridis viridis L. 1766: Colombia to Venezuela, Guyana, Suriname, French Guiana, Trinidad Island, western Brazil east to Maranhão, and eastern Peru.

Trogon viridis melanopterus Swainson 1838 (1837): southeastern Brazil.

MORPHOMETRICS

Measurements Wing, males (of *viridis*) 134–148 mm (5.2–5.8 in; average of 7, 141 mm [5.5 in]); females

135–144 mm (5.3–5.6 in; average of 4, 140.2 mm [5.5 in]; ffrench 1991). Wing, males (of *viridis*, sample size unstated) 145–148 mm (5.7–5.8 in); females 140–143 mm (5.5–5.6 in; Haverschmidt 1968). Tail, males (of *chironurus*) 136–144 mm (5.3–5.6 in; average of 10, 139.1 mm [5.4 in]); females 134–143 mm (5.2–5.6 in; average of 10, 139.3 mm [5.4 in]; Wetmore 1968). Egg of *viridis* 33.1 × 23.9 mm (1.3 × 0.9 in), of *chionurus* 28.1 × 23.9 mm (1.1 × 0.9 in), of *melanopterus* 28.5 × 22.5 mm (1.1 × 0.9 in; Schönwetter 1966). Haverschmidt (1968) provides similar egg measurements for *viridis*.

Weights Five birds from varied locations 77–99 g (2.7–3.5 oz; average 87.6 g [3.1 oz]; Burton 1973;

White-tailed trogon, adult male. Photograph by author.

Dick, McGillivray, and Brooks 1984; ffrench 1991). Four males from Trinidad 77–99 g (2.7–3.5 oz; average 89.2 g [3.1 oz]; ffrench 1991). One male from Brazil 93 g (3.3 oz; Sick 1993). Unstated number of males from Suriname 78–95 g (2.7–3.3 oz), females 79–86 g (2.8–3.0 oz; Haverschmidt 1968). One male from Suriname 92 g (3.2 oz), one female 79 g (2.8 oz); two females from Peru 90 and 107 g (3.2 and 3.7 oz; specimens, MVZ). Two males of *chionurus* 87.1 and 94.5 g (3.0 and 3.3 oz), two females 53.5 and 93.9 g (1.9 and 3.3 oz); three males of *viridis* 88–99.4 g (3.1–3.5 oz; average 95.0 g [3.3 oz]), seven females 78–98 g (2.7–3.4 oz; average 89.7 g [3.1 oz]; specimens, FMNH). Weight of fresh eggs 10.5 g (0.4 oz; Haverschmidt 1968). Estimated egg weight of *chionurus* 8.5 g (0.3 oz), of *viridis* 9.7 g (0.3 oz), of *melanopterus* 7.7 g (0.3 oz; Schönwetter 1966).

Map 13. Distribution of the white-tailed trogon.

DESCRIPTION (of *chionurus*, adapted from Ridgway 1911; see plate 18)

Adult Male   Head, neck, and upper chest black glossed with violet; lower chest and sides of breast dark iridescent bluish violet or violet-blue; back, scapulars, and upper rump brilliant iridescent green, becoming violet-blue on upper margin of back, lower rump, and upper tail-coverts; middle pair of rectrices iridescent bluish green or greenish blue, abruptly tipped with black; next two pairs or rectrices similar on outer vanes, but inner vanes wholly black; fourth rectrix (from inside) black, with about terminal half of outer vane white, inner vane with much less white, with base of the area forming an acute angle next to shaft; fifth rectrix with the white occupying approximately terminal two-thirds of outer vane and terminal half of inner vane; outermost (sixth) rectrix with only the extreme base of outer vane and no more than basal third of inner vane black, sometimes with black reduced to a small spot; wings black, primaries edged basally with white; median portion of breast, abdomen, flanks, and under tail-coverts rich orange-yellow; sides dark slaty or blackish slate; thighs slate-black; bill bluish white to pale greenish; iris dark reddish brown; eye-ring light blue to pale grayish blue or even nearly white; feet and toes dark gray to fuscous.

Adult Female   Upperparts uniform blackish slate, rectrices without black tips; wings rather darker,

coverts and secondaries very narrowly barred with white, primaries edged with white basally and with small spots of white along edge for terminal half or more; three outer rectrices with much less white than in adult male, this broken with broad bars confluent with the slate-black basal area; side of head and neck, chin, throat, chest, and breast plain blackish slate, slightly paler posteriorly and on sides and on outer portion of flanks; rest of underparts orange-yellow; thighs blackish slate; bill grayish blue at base of maxilla and over entire mandible, grading to black toward tip of maxilla; iris warm brown; eye-ring light blue; feet and toes neutral gray to blackish.

Immature Male   Similar to adult male, but outer rectrices much as in adult female (the white more extended, however, and the broad black bars not reaching to shaft on inner vane); wing-coverts and proximal secondaries very narrowly barred with light gray; distal secondaries with small white spots along edge of outer vanes; anterior underparts (in younger specimens) intermixed with dull blackish slate.

Immature Female Immature females probably much like adult females, except for differences in shape and patterning of outer rectrices and barring on secondaries and wing-coverts.

Juvenile No information available.

IDENTIFICATION

In the Hand Male white-tailed trogons are easily identified by means of the combination of bright yellow underparts and outer rectrices that are white for all or most of their exposed areas. Females have blackish-slate rather than glossy violet breasts, and the outer rectrices are not pure white but have regular black barring over most of their exposed length, and unmarked white tips (see the geographic variation section).

In the Field The white-tailed trogon is associated with moist lowland and foothill forests, forest edges, and clearings, as well as with second-growth woodlands and plantations. In Panama the males of this species seem to have entirely white outer rectrices, but in Venezuela some black markings are apparent toward the base of the tail. Both sexes have pale blue eye-rings (versus yellow in the violaceous trogon) and bluish-gray bills (versus greenish yellow in the black-throated). In females the outer rectrices may be barred or blotched with black, especially near the base and on the inner vanes. The birds are found in lowland forests, up to about 550 m (1,800 ft), in Panama. Vocalizations include a series of dove-like *coo* notes, which gradually accelerate and may end with some notes that are slower or lower or both. See also the vocalizations section.

GEOGRAPHIC VARIATION

Wetmore (1968) noted that white-tailed trogons from northwestern Colombia are grayer dorsally than those from Panama and perhaps should be recognized as distinct. The form *chionurus* has more-purplish upperparts than nominate *viridis*, which is more greenish in that area (Todd 1943). Males of *chionurus* have black only at the base of their outer-

most rectrices, but in *viridis* and *melanopterus* black occupies about two-thirds of the feather; in females of *chionurus* the outer third of the outermost rectrices is white, whereas in *viridis* and *melanopterus* these feathers are only tipped and toothed with white.

ECOLOGY

Habitats The white-tailed trogon is a widespread species with a rather broad ecological range. Collectively it ranges from a few hundred meters to 1,200 m (3,900 ft), in moist tropical and subtropical forests, forest edges, secondary forests (especially younger forests), and gallery forests. In Panama it occurs in the tropical and lower subtropical zones, mostly in lowland evergreen forest below 1,000 m (3,300 ft). In Colombia it ranges to 1,000 m (3,300 ft) west of the Andes and to 1,200 m (3,900 ft) east of the Andes, as well as extending to humid lowlands. In most areas it is one of the commonest of trogons and also one of the tamest.

Foods and Foraging Ecology White-tailed trogons forage in the usual trogon manner, on drupes and various insects. Wetmore (1969) reported finding large ants and small stick insects (phasmids) in one stomach, and Haverschmidt (1968) noted fruits, berries, larval lepidopterans, orthopterans (locusts), and small lizards (Teidae). Fruits in the families Melastomaceae and Sapotaceae have been mentioned as foods (Ruschi 1979). Of 29 stomachs examined by Remsen, Hyde, and Chapman (1993), 12 contained arthropods only, 6 contained mixed fruit and arthropods, and 11 contained fruit only.

BEHAVIOR

General, Social, and Sexual Behavior The white-tailed trogon is a fairly tame and relatively confiding species. It is most often found in pairs, but sometimes several males may gather in calling groups, presumably as part of a courting assemblage.

Vocalizations The typical vocalization of the white-tailed trogon is a sequence of rapidly uttered *chook*, *kyoh*, or *cow* notes that often accelerate and become louder toward the end of the sequence. The sequence may also end with a few slower *cow* notes, and low purring sounds or individual *chucks* may be uttered too. Hilty and Brown (1986) stated that up to 16 or so brisk *kyoh* or *cow* notes may be uttered in uniform or accelerating sequence, the sequence becoming louder toward the end and sometimes ending with a few slow *cow* notes. Vocalizations of this species were recorded in Peru by Ben Coffey Jr. (Hardy, Reynard, and Coffey 1995). The song examples on the recording consist of an extended series of *woo* (or *coo*) notes having a cuckoo-like quality, rapidly uttered at about the same pitch and sometimes becoming louder toward the end. The notes are produced at a rate of nearly three per second, and examples on the recording vary in length from 7 to 19 notes. See also the identification section above.

BREEDING BIOLOGY

Chronology of Breeding   In Suriname white-tailed trogon eggs have been recorded from December to July, although most records are from February to June. (Haverschmidt 1968). In Trinidad there are records from April to July (Herklots 1961), and in Colombia birds in breeding condition have been reported from January to June (Hilty and Brown 1986). In Panama breeding has been reported from March to July (Willis and Eisenmann 1979).

Nest Sites   At least three white-tailed trogon nests in arboreal termitaria have been observed in Panama (Wetmore 1968); and in Suriname, Haverschmidt (1968) also found nests in arboreal termitaria. According to Haverschmidt, both sexes excavate the nesting hole, and both sexes incubate. Nests have also been found in cavities of dead palms. Ruschi (1979) mentioned a nest 13 m (43 ft) high in a termitarium and stated that two to four eggs may be present, but the egg measurements he gives seem too large for this species.

Eggs and Incubation   White-tailed trogon clutch sizes usually vary from two to three eggs. Sets in the Panard collection from Suriname include two clutches of three, four clutches of two, and three single-egg clutches (Hellebrekers 1942). The incubation period has been estimated at 18 days (Ruschi 1979).

Brood Rearing   No detailed information is available on the brood rearing of white-tailed trogons. A fledging period of 25 days has been reported (Ruschi 1979).

CONSERVATION AND EVOLUTIONARY RELATIONSHIPS

As noted above, the white-tailed trogon is an allopatric replacement form of Baird's trogon but is much more widespread, occurring in 11 countries from Panama to Bolivia. It is also one of the most common trogons in lowland forests of the Amazon Basin and poses no conservation problems. On the basis of gene-sequencing studies, Espinosa de los Monteros (1998) considered this species' closest relatives (among those ten species of *Trogon* he was able to study) to be *T. violaceus* and *T. curucui*. Stotz et al. (1996) listed the species as common, with a medium degree of sensitivity to disturbance.

CITREOLINE TROGON

Trogon citreolus Gould 1835

OTHER VERNACULAR NAMES

Lemon-breasted trogon; trogon citrino (Spanish).

RANGE

Arid woodlands of Mexico, from northern Sinaloa to western Chiapas. See map 14.

SUBSPECIES

The races of *melanocephalus* are sometimes also included in *citreolus*, as by Peters (1945).
Trogon citreolus citreolus Gould 1835: western Mexico.
Trogon citreolus sumichrasti Brodkorb 1942: southern Mexico.

MORPHOMETRICS

Measurements   Wing, males (both races) 130.5–145.5 mm (5.1–5.7 in; average of 10, 137 mm [5.3 in]); females 130.5–141 mm (5.1–5.5 in; average of 9, 136.6 mm [5.3 in]). Tail, males (both races) 139–158 mm (5.4–6.2 in; average of 10, 147.8 mm [5.8 in]);

females 140.5–160 mm (5.5–6.2 in; average of 9, 148 mm [5.8 in]; Ridgway 1911). Wing, males (of *citreolus*) 133–141 mm (5.2–5.5 in; average of 9, 137.6 mm [5.4 in]); females 134–138 mm (5.2–5.4 in; average of 7, 135.7 mm [5.3 in]). Wing, males (of *sumichrasti*) 141–150 mm (5.5–5.9 in; average of 7, 147 mm [5.7 in]); females 141–144 mm (5.5–5.6 in; average of 5, 143 mm [5.6 in]; Brodkorb 1942). Average of ten eggs 29.6 × 23.4 mm (1.2 × 0.9 in; Rowley 1984).

Weights   Seven *sumichrasti* (both sexes) from Yucatan 72.6–91.0 g (2.5–3.2 oz; average 83.1 g [2.9 oz]; Paynter 1955). Birds of unspecified sex and sample size 66–95 g (2.3–3.3 oz; average 80 g [2.8 oz]; Smithe 1966). Two *sumichrasti* males from Belize 48.2 and 71.5 g (1.7 and 2.5 oz; average 59.9 g [2.1 oz]), one female 74.7 g (2.6 oz; Russell 1964). Eight males of both subspecies 69.8–79.9 g (2.4–2.8 oz; average 73.6 g [2.6 oz]), one female of *citreolus* 71.7 g (2.5 oz; specimens, MVZ). One male of *sumichrasti* 89.7 g (3.1 oz), two females 88.6 and 90.2 g (3.1 and 3.2 oz; specimens, FMNH). Estimated egg weight 13.1 g (0.5 oz).

Map 14. Distribution of the citreoline trogon.

DESCRIPTION (mainly adapted from Ridgway 1911; see plate 19)

Adult Male Crown, hindneck, and sides of head and neck blackish slate, grading into slate on chin, throat, and chest; back and scapulars iridescent golden green, becoming more bluish on rump and upper tail-coverts; middle pair of rectrices iridescent green to bronzy green, tipped narrowly with black; outer vanes of next two pairs of rectrices similar, but inner vanes entirely black; three outer pairs black basally, extensively white terminally, the white on outermost rectrix covering about half the inner vane and nearly all the outer vane; the white on third from outside pair of rectrices covering about a third of outer and a fourth of inner vane; wing-coverts and secondaries pale slate, lesser coverts tipped with green; primaries blackish slate, nearly all broadly edged with white; upper breast white or yellowish white, becoming yellow on rest of underparts; thighs slate; bill bluish gray; iris yellow; eye-ring dark blue; feet and toes bluish gray to ash-gray.

Adult Female Similar to male, but outer vanes of greater wing-coverts and secondaries conspicuously edged with white; tertials similarly mottled; middle rectrices lighter, more brownish, and lacking black tips; underparts yellowish white; soft-part colors similar to male.

Immature Usually more extensively black than adults on undertail surface; with black bars present basally on outer vanes; iris color of immatures may remain dark brown until early winter (Howell and Webb 1995). In young males the belly is whitish, and the middle rectrices are grayish black, glossed with bronze-green. In young females the outer vanes of the secondaries are blotched or margined with buff, and the outer vanes of the three outer rectrices are extensively toothed with white.

Juvenile No information available.

IDENTIFICATION

In the Hand The yellow iris (not the eye-ring) immediately sets the citreoline trogon apart from other Mexican trogons. The very similar black-headed and violaceous trogons also have yellow underparts, but both have dark-colored irises and more black visible on the three outer pairs of rectrices as seen from below.

In the Field The citreoline trogon is associated with open woodlands, deciduous forests, plantations, and scrub, mainly in the tropical zone. Essentially limited to western Mexico, this is the only trogon of that region having bright yellow underparts, yellow eyes (with an inconspicuous dark blue eye-ring), and unmarked white outer rectrices (as seen from below). The very similar (but probably not conspecific) black-headed trogon has dark brown eyes with a conspicuous pale blue eye-ring and occurs on the Atlantic slope (sympatry is possible in the isthmus region but is still unproved), and the violaceous trogon has strongly barred outer rectrices. The citreoline occurs as high as about 1,000 m (3,300 ft) but is mostly found in lowland arid to semi-arid forests. The species' vocalizations include a series of *cow* notes that start slowly but typically accelerate toward the end. See also the vocalizations section.

GEOGRAPHIC VARIATION

The race *sumichrasti* averages distinctly larger than the nominate race (Brodkorb 1942). The former also has more white present on the outer rectrices than does the nominate form, which distinction Binford (1989) has suggested may represent selection for species differences from *melanocephalus* in those areas where these two forms might have come into secondary contact.

ECOLOGY

Habitats Collectively the citreoline trogon has been reported from tropical deciduous forest, gallery forest, and secondary forest, from sea level to 1,000 m (3,300 ft). In Mexico these birds occur from sea level to 1,000 m (3,300 ft) and are found in woodlands, thorn forests, plantations, and mangroves (Howell and Webb 1995). Skutch (1948) encountered the species in low, thorny scrub and cacti of western Mexico.

Foods and Foraging Ecology Eguarte and del Rio (1985) studied the foraging of the citreoline trogon during the dry season in Jalapa, Mexico. Four species of fruits were consumed during the study period (April), all of which were plucked from the trees by perching birds. During the 3-week observation period, only two individuals were seen catching insects, which the birds snatched from the undersides of leaves while hovering. Fruits taken in the largest quantities (30 each) were figs (*Ficus pertusa*, Moraceae) and drupes of *Reccia mexicana* (Simaroubaceae), a small tree. Taken in slightly smaller quantities (25) were the berries of a vine (*Trichostigma octandrum*, Phytolaceae). Seven drupes of *Comocladia engleriana* (Anacardiaceae) were also eaten. The figs not only represent fairly substantial food items, weighing nearly a gram each, but also are available throughout the year and in large quantities. On average, the birds required about 90 seconds to find, handle, and swallow each fig. An average of nearly four birds arrived per 5-minute interval, the birds apparently foraging in loose flocks. Little intraspecific aggression was observed during this period, which was about 2 or 3 months prior to the breeding season.

BEHAVIOR

General, Social, and Sexual Behavior   No specific information on behavior is available. Like the blackheaded trogon, male citreoline trogons sometimes form singing assemblages that resemble leks (Howell and Webb 1995).

Vocalizations   The calls of the citreoline trogon are not obviously different from those of the blackheaded trogon; Skutch (1948) did not distinguish between them. Both species utter an accelerating series of nasal or wooden hoots that become a chatter. They also utter dry or clucking *kek* notes in alarm. Vocalizations of this species were recorded in Mexico by Richard Bradley (Hardy, Reynard, and Coffey 1995). The song on the recording consists of very rapid, almost chattering, and somewhat wooden *cow* notes, the sequences being of varied lengths but often lasting about 5 seconds, occasionally with brief hesitations or breaks, especially near the beginning. See also the field identification section.

BREEDING BIOLOGY

Chronology of Breeding   In the collection of the Western Foundation of Vertebrate Zoology, there are twelve sets of citreoline trogon eggs from Oaxaca; and Rowley (1984) has summarized breeding records for these sets. The dates range from May 6 to May 19, with fresh eggs being found as late as May 15. In Chiapas in mid-July, Skutch (1948) found a nest containing two fresh eggs.

Nest Sites   The 12 sets of eggs in the Western Foundation of Vertebrate Zoology mentioned in the previous section were collected at altitudes of about 125–250 m (400–800 ft), and all were in termite nests at heights of 2–4.7 m (7–15 ft). Rowley (1984) found nine of these nests and observed only the female working on the excavation process. The nest found by Skutch (1948) was in a termitarium about 5.6 m (18 ft) above ground.

Eggs and Incubation   As noted in the chronology of breeding section, Rowley (1984) listed twelve sets of citreoline trogon eggs from Oaxaca. One incomplete set contained one egg, and there were five sets of two eggs, four sets of three, and two sets of four.

Brood Rearing   No specific information is available, but the brood rearing behavior of the citreoline trogon is probably comparable to that described for the black-headed trogon.

CONSERVATION AND EVOLUTIONARY RELATIONSHIPS

The citreoline trogon is limited to the western slope of Mexico and is generally regarded as fairly common over much of its range. It is allopatric or perhaps parapatric with the black-headed trogon of Mexico's eastern slope, and the two species have previously often been considered conspecific. Like the eared trogon, the citreoline is confined entirely to Mexico but has a broader ecological tolerance and is not in the same threatened state. Stotz et al. (1996) listed the species as common, with a low degree of sensitivity to disturbance.

BLACK-HEADED TROGON

Trogon (citreolus) melanocephalus Gould 1836

The black-headed trogon has often previously been considered a subspecies of *citreolus*, as by Peters (1945), but is now generally given full specific status and considered as part of a superspecies with *citreolus*.

OTHER VERNACULAR NAMES

Black-headed citreoline trogon, slaty-headed trogon (*illaetabilis*); aurora de pecho gris, trogon cabecinegro (Spanish).

RANGE

Forest edges, plantations, and woodlands of Central America from southern Veracruz in Mexico south through the Yucatán Peninsula (including islands off Quintana Roo), eastern Guatemala, Belize, Honduras (both coasts), El Salvador (mostly coastal), and Nicaragua to northwestern and northeastern Costa Rica. See map 15.

SUBSPECIES

The black-headed trogon is sometimes included within *citreolus*; possible previous or current sympatry between these two forms is still unproved.

Trogon melanocephalus melanocephalus Gould 1836: Mexico to northeastern Costa Rica.
Trogon melanocephalus illaetabilis Bangs 1901: western Costa Rica.

MORPHOMETRICS

Measurements Wing, males (of *melanocephalus*) 130–146.5 mm (5.1–5.7 in; average of 29, 137.9 mm [5.4 in]); females 133–141.5 mm (5.2–5.5 in; average of 18, 136.4 mm [5.3 in]). Tail, males (of *melanocephalus*) 136–165 mm (5.3–6.4 in; average of 29, 148.2 mm [5.8 in]); females 138–155 mm (5.4–6.0 in; average of 18, 147 mm [5.7 in]). Corresponding measurements of *illaetabilis* average slightly larger (3–7 mm [0.1–0.3 in]; Ridgway 1911). Egg of

Map 15. Distribution of the black-headed trogon.

111

melanocephalus 28.6 × 22.5 mm (1.1 × 0.9 in; Schön-wetter 1966).

Weights   Unspecified Costa Rica weight 90 g (3.2 oz; Stiles and Skutch 1989). An unidentified sample from Guatemala 66–95 g (2.3–3.3 oz; average 80 g [2.8 oz]; Smithe 1966). One male of *melanocephalus* 74 g (2.6 oz), one female 85 g (3.0 oz); five males of *illaetabilis* 84–94.3 g (2.9–3.3 oz; average 89.4 g [3.1 oz]), six females 78–88.8 g (2.7–3.1 oz; average 84.0 g [2.9 oz]; specimens, MVZ). Three males of *melanocephalus* 81–91.1 g (2.8–3.2 oz; average 86.0 g [3.0 oz]), one female 82.7 g (2.9 oz; specimens, FMNH). Estimated egg weight 8.1 g (0.3 oz; Schön-wetter 1966).

DESCRIPTION (of *melanocephalus*, mainly adapted from Ridgway 1911; see plate 20)

Adult Male   Head, neck, and chest uniform black or slate-black, nape and hindneck sometimes glossed with bluish; back, scapulars, anterior lesser wing-coverts, and upper rump bright iridescent bluish green to golden green, usually more bluish next to black of hindneck and sometimes mixed with violet-blue; lower rump and upper tail-coverts iri-descent blue, violet-blue, or bluish violet; four mid-dle rectrices bronze-green to bluish green, tipped with black; inner vanes of second and third rectrices (from middle) black; three outer pairs of rectrices black, tipped broadly (15–30 mm, 0.6–1.2 in) with white; wings slate-black, longer primaries edged basally with white; sides and flanks blackish slate, flanks mixed with orange-yellow; remaining under-parts orange-yellow, becoming yellowish white ante-riorly, where a whitish band is formed behind black chest; thighs blackish; bill pale whitish yellow to greenish glaucous or pale light blue; iris brown; eye-ring sky (delft) blue to light azure to even whitish; feet and toes gray to dusky gray or slaty horn. In El Salvador the annual molt of adults occurs between August and late October (Dickey and van Rossem 1938).

Adult Female   Similar to adult male, but iridescent coloring of upperparts replaced by slate or blackish slate; maxilla black, except for yellowish lower basal portion; mandible light bluish horn; eye-ring delft-blue; feet and toes bluish horn.

Immature Male   Similar to adult male, but middle rectrices mostly dull black; outer rectrices narrower at tip and with two or three white spots on subter-minal portions of outer vanes, and outer vanes of secondaries spotted and edged with white, as are upper wing-coverts; maxilla blackish along culmen; other soft parts as in adult.

Immature Female   Similar to adult female, but middle rectrices lacking black tips; outer rectrices narrower with outer vanes more or less spotted, toothed, or barred with white; secondaries edged with white.

Juvenile   Wing-coverts strongly spotted with buff (these spots persisting in immature birds); under-parts blotched with white and buffy; upperparts grayish brown; body plumage lost by September or October, but rectrices (and probably wing-coverts) carried for a year.

IDENTIFICATION

In the Hand   Very similar to the citreoline trogon, the black-headed trogon likewise has yellow under-parts and outer rectrices that are predominantly white on their exposed underside areas. However, the latter is distinguished by its dark brown iris color, pale blue eye-ring, darker (more blackish) head, generally darker yellow abdomen, darker breast and upper wing coloration, and the greater amount of black visible at the base of the outermost rectrices as seen from below. The wings average darker, the upper surface of the tail is more bluish, and the amount of white visible on the breast is usually more extensive. Among females of the black-headed trogon the undertail pattern of the folded tail is one of alternating black and white barring, whereas in the citreoline trogon the outer rectrices appear to be all white except for their largely hidden blackish bases.

In the Field   The black-headed trogon is mostly associated with open woodlands, partially cleared forests, plantations, and scrubby or bushy areas, of-ten in more humid areas than those inhabited by the citreoline trogon. As noted above, differences in iris (and eye-ring) color provide the most obvious field distinctions between the citreoline and black-headed trogons, and males of the latter are also somewhat more blackish on the head. Both sexes of the black-headed show more black at the base of the outer-most pair of rectrices than do the citreoline trogons. Western Chiapas (in the general vicinity of Oco-zocuatla) represents the point of nearest possible contact (sympatry) between these two forms. The black-headed trogon's songs have been described as a series of short, rapid notes similar to those of some antbirds and as a rapid, rattling series of clear bark-ing notes that accelerates into a chuckling trill that falls in pitch. The voice of this species is said to be harsher and more nasal than that of Baird's trogon and to include whining and barking notes as well (Stiles and Skutch 1989). See also the vocalizations section.

GEOGRAPHIC VARIATION

No geographic variation has been described, al-though as noted above the black-headed trogon has often been considered as representing an extreme variant of the citreoline trogon rather than a sepa-rate species. Although the two species are believed not to show any intergradation between the two types, Schaldach, Escalante, and Winker (1997) col-lected one possibly intermediate bird from Sarabia, on the Atlantic slope of Oaxaca. The bird, a male with a bright yellow iris and a slate-blue (not pale blue) eye-ring, was collected in an area where only *melanocephalus* phenotypes were expected and where 19 other typical adults were collected.

ECOLOGY

Habitats   Collectively the black-headed trogon has been reported from tropical lowland evergreen for-est (rain forest), tropical deciduous forest, gallery forest, secondary forest, plantations, pastures, and mangroves, from sea level to 1,000 m (3,300 ft) but usually below 600 m (2,000 ft). In Belize, where this is the most common species of trogon, the birds are most commonly found in tall to medium secondary forest and forest edges but extend into pinelands and areas of scattered trees. Only rarely do the birds occur in heavy forest (Russell 1964). In Guatemala, where they are also common and occur from sea level to 500 m (1,600 ft), they occupy humid forests, woodlands, plantations, and secondary forest (Land

1970). Skutch (1948, 1983) found birds of this form barely reaching 600 m (2,000 ft), in pastures with scattered trees, in light secondary forest woodland, along the edges of banana plantations, and in fringe gallery forests but not in dense rain forests. Mon-roe (1968) did not record the species above 600 m (2,000 ft) in Honduras, considering it most common in monsoon forest and arid scrub. In El Salvador it reaches about 450 m (1,500 ft), preferring humid mangroves, marshes, and swampy areas (Dickey and van Rossem 1938).

Foods and Foraging Ecology   Stomachs from speci-mens of black-headed trogons taken in El Salvador contained fruit pulp (three stomachs), berries (three), caterpillars (three), and fruit pulp plus a large cater-pillar (one); see Dickey and van Rossem (1938). Skutch (1983) reported that this species consumes both fruits and insects, the latter including dragon-flies, mantids, grasshoppers, other orthopterans, caterpillars (hairy and smooth), and many smaller insects. Fruits eaten include the orange pulp of the Central American rubber tree (*Castilla*), green fruit-ing spikes of the guaramo tree (*Cecropia*), and various kinds of berries. Like other trogons, black-headed trogons obtain much of their food while in flight.

BEHAVIOR

General, Social, and Sexual Behavior   Skutch (1983) referred to the black-headed trogons as "dignified, gentlemanly birds," never having observed them engaged in actual fighting, even when competing for mates. During the mating season several trogons of both sexes may gather, perching closely together and calling at intervals or chasing one another from perch to perch. Lek-like assemblages of up to ten birds (of both sexes) have also been reported (How-ell and Webb 1995).

Vocalizations   Skutch (1983) stated that the call of the black-headed trogon is an accelerated rattle, much like that of Baird's trogon but harsher and more nasal. The call of the former is even more like that of the citreoline trogon, with both species utter-ing throaty *cuck* notes that may remain at a constant cadence or, more typically, become accelerated. Slud (1964) stated that the notes consist of accelerating throaty chucks, which may change to a higher or lower pitch in the middle, may rise at the end, or may descend somewhat and finish as a harsh cackle. Either sex may call singly, or the two may call back and forth to each other. There are various short notes, including a single *kop* that is accompanied by tail jerking. Davis (1972) reported that the usual song phrase is a "bouncing-ball" type of rattle, which first accelerates and then slows down, producing a cuckoo-like effect. Vocalizations of this species were recorded in Mexico by Ben Coffey Jr. (Hardy, Reynard, and Coffey 1995). The recorded songs are distinctly higher pitched and more rapidly uttered than the songs of the citreoline trogon, the utterances sounding almost like those of a passerine in terms of speed of notes and pitch. The song tends to speed up and perhaps becomes louder toward the end of each sequence. The sequences represented on the recording are variable in length, each usually lasting about 3 to 5 seconds, and several birds are singing simultaneously or in rapid succession. See also the identification section.

BREEDING BIOLOGY

Chronology of Breeding   In Costa Rica black-headed trogons breed from March to July (Stiles and Skutch 1989). Nests found by Skutch in Guatemala ranged in time from April to July.

Nest Sites   Skutch (1948, 1983) found three nests of the black-headed trogon in Guatemala; all were in termitaria. One was on a wooden post only about 1 m (3 ft) high, another was in a fallen willow tree less than 1 m (3 ft) above ground, and a third (a replacement nest for the pair that built the second) was about 7 m (23 ft) away on a post about 1.5 m (5 ft) above ground. The entrance hole was below the nest and led sharply upward, for about 18 cm (7.1 in). In one case the excavation was done in 6 days, the two sexes working alternately and the male doing most of the work. Work began late in the morning and continued over much of the afternoon.

Eggs and Incubation   Each of the three black-headed trogon nests studied by Skutch (1983) contained three eggs, which in one case had been laid on alternate days. One replacement clutch was laid about a month after the first effort failed. During incubation the male tended the nest from early morning until the latter part of the afternoon, when replaced by the female, who then remained until the next morning. The incubation period (19 days) was established for one egg.

Brood Rearing   At hatching black-headed trogon chicks are completely naked, but their pinfeathers are already sprouting (Skutch 1983). Slight peeping sounds are uttered by two days. The young were fed small mantids, green caterpillars, dragonflies, other small insects, and rarely berries. By 11 days the eyes were open, and at 2 weeks these pinfeathers were ready to expand, exposing the juvenal plumage. At this stage the nestlings were quieter, making hissing sounds while their parents fed them. By then the parents were feeding the young three to five times per hour. Fledging occurred at 16 or 17 days, or only a few days after the feather sheaths had been sloughed off.

CONSERVATION AND EVOLUTIONARY RELATIONSHIPS

Although earlier taxonomic literature (e.g., Peters 1945) considered the black-headed trogon as a subspecies of *citreolus*, most more recent authors have accorded it species-level distinction. These include Slud (1964), Skutch (1983), Stiles and Skutch (1989), Binford (1989), Howell and Webb (1995), Sibley and Monroe (1990), and the AOU (1998). Since no proven secondary contact exists (but see Schaldach, Escalante, and Winker [1997]), classification is difficult, but perhaps DNA studies will help to resolve the issue. The species is fairly widespread, occurring in seven Central American countries, from Mexico to Costa Rica. Stotz et al. (1996) listed the species as common, with low degree of sensitivity to disturbance.

Vermiculated Trogons (Subgenus *Trogonurus* Bonaparte 1854)

The subgenus *Trogonurus* includes a group of nine small to moderate-size (23–30 cm [9.0–11.8 in], 40–90 g [1.4–3.2 oz]) species having serrated maxillae and anterior toes that are partially fused basally. Adults have upper wing-coverts that are vermiculated or finely barred and middle rectrices that are black tipped. The outer rectrices are white-tipped with truncated tips, and the underparts are red, orange, or yellow. Nests are usually shallow niches in decaying wood that barely hide the sitting bird; but the nest of the violaceous trogon may be excavated in a papery wasp's nest or in a cavity made among the dense roots of an epiphyte, and the elegant trogon most often takes over an old woodpecker hole (Skutch 1983). Males of this group have song phrases that comprise well-separated elements that remind one of the songs of cuckoos (Davis 1972). On the basis of gene-sequencing studies, Espinosa de los Monteros (1998) grouped five of these species (*T. mexicanus, T. elegans, T. collaris, T. personatus,* and *T. rufus*) within a single subclade but regarded *T. violaceus* and *T. curucui* as part of a different subclade.

MOUNTAIN TROGON

Trogon mexicanus Swainson 1827

OTHER VERNACULAR NAMES

Mexican trogon; trogon Mexicano (Spanish).

RANGE

Upland forests in Mexico from eastern Sinaloa, southern Chihuahua, and southern Tamaulipas south discontinuously through Durango, Zacatecas, and San Luis Potosí to the mountains of southern Mexico, Guatemala (western and central uplands), El Salvador (rare), and Honduras (interior forests). See map 16.

SUBSPECIES

Trogon mexicanus clarus Griscom 1932: northwestern Mexico, from Sinaloa south to Durango.

Trogon mexicanus mexicanus Swainson 1827: central Mexico from Nayarit and Jalisco to western Guatemala.

Trogon mexicanus lutescens Griscom 1932: El Salvador and Honduras. Probably an invalid race (Monroe 1968).

MORPHOMETRICS

Measurements Wing, males (of *mexicanus*) 138.5–152 mm (5.4–5.9 in; average of 16, 145.6 mm [5.7 in]); females 137–157 mm (5.3–6.1 in; average of 14, 142.8 mm [5.6 in]). Tail, males (of *mexicanus*) 165.5–190 mm (6.5–7.4 in; average of 16, 179.9 mm [7.0 in]); females 175–200 mm (6.8–7.8 in; average of 14, 184.5 mm [7.2 in]; Ridgway 1911). Egg of *mexicanus* 30.2 × 23.7 mm (1.2 × 0.9 in; Schönwetter 1966).

Weights One male from Mexico 69.3 g (2.4 oz; Davis 1945). Eleven males of various subspecies 61.5–85 g (2.2–3.0 oz; average 71.5 g [2.5 oz]); five females 63.7–77.5 g (2.2–2.7 oz; average 70.3 g [2.5 oz]; specimens, MVZ). Three males of *mexicanus* 68–72.7 g (2.4–2.5 oz; average 70.9 g [2.5 oz]; specimens, FMNH). Estimated egg weight 6.3 g (0.2 oz; Schönwetter 1966).

DESCRIPTION (adapted from Ridgway 1911; see plate 21)

Adult Male Upperparts bright iridescent green to bronzy; middle pair of rectrices green to bluish

Map 16. Distribution of the mountain trogon. Inset map shows approximate distribution of temperate and montane dry (mostly pine and oak) forests in the species' range.

green, tipped with black; wings dark slaty to blackish, he wing-coverts and secondaries finely vermiculated with grayish white; interior rectrices (except middle pair) black; three outer pairs of rectrices broadly tipped with white; face, chin, and throat dull black to slaty black; chest iridescent green, bordered posteriorly by a broad white pectoral band across upper breast; remaining underparts pure red; thighs blackish; bill yellow; iris brown; bare eye-ring narrow, reddish; feet and toes cinnamon-brown.

Adult Female   Upperparts brown, middle pair of rectrices nearly russet, tipped with black, and sometimes with paler subterminal band; wing-coverts and secondaries vermiculated with dusky and pale brown or buffy; primaries broadly edged with white; rectrices (except middle pair) dull black, three outer pairs broadly tipped with white, and terminal half of outer vanes white, broadly barred with blackish, the terminal white of inner vanes forming a narrow wedge along inner side, next to shaft, edge of inner vanes also notched with white; orbital area of white feathers in front of and behind eye; sides of head more dusky than rest of upperparts; chin, throat, and chest brown, like upperparts; breast paler brown, with a more or less distinct breast band of white across upper portion, brown of middle breast gradually merging with deep red or pink of lower underparts; maxilla yellowish brown to brownish; mandible blackish, sometimes paler at tip; iris brown, bare eye-ring inconspicuous; feet and toes yellow.

Immature Male   Similar to adult male, but white tail markings as in adult female but less regularly marked, and middle pair of rectrices green basally and tipped with pale buffy brown or cinnamon; wing coverts browner than adult, with buff spotting; often with brown mottling on upper abdomen; breast immediately below the white band and under tail-coverts mixed with brownish olive and buff.

Immature Female   Similar to adult female, but middle pair of rectrices narrowly tipped with whitish or buff; scapulars, wing-coverts, and tertials spotted with pale buff.

Juvenile   Head, chest, and upperparts tawny brown to umber; rump and upper tail-coverts lighter and more rufescent, these feathers narrowly tipped with

Arboreal nest-site of mountain trogon. Photograph by A. F. Skutch.

dusky; underparts brownish buff to clay, barred or spotted with brown; under tail-coverts tawny; wing-coverts and tertials with very large, roundish spots of pale buff margined with blackish; these coverts and outer webs of secondaries otherwise narrowly barred buff and blackish; feathered white ring circles eye; primaries and rectrices as in immature female; mandible and maxilla yellowish; iris brown; bare eye-ring yellow.

IDENTIFICATION

In the Hand   The mountain trogon is a red-bellied, yellow-billed trogon, and adult males have a distinctive undertail pattern, in which three large white rectangular patches (rectrix tips) appear to be separated by unbroken areas of black (rectrix bases). The pattern of females (and immature males) is more complex: there are large black markings at the middle of the tail, while the white edges are interrupted by black bars of the same width as the white bars separating them. Both sexes have drooping white "necklaces" below the green (males) or brown (females) chests; females have a brownish breast band below this line, rather than grading gradually to red posteriorly, as in the collared trogon. A similar but less distinct brown breast band occurs in females of the elegant trogon, but females of that species also exhibit a distinctive white ear patch.

In the Field   The mountain trogon is a midlevel montane species (1,200–3,500 m; 3,900–11,500 ft) sympatric with both the collared and elegant trogons, but it extends higher on mountains than either of these. All three species are similar in appearance, with green heads and breasts and bright red underparts, the latter two areas separated by a white pectoral band (or drooping "necklace"). The strongly patterned underside of the tail in adult male mountain trogons (see description above) is diagnostic. Females of the three species are even more similar; but in the female mountain trogon the white necklace is bordered below by a brown band similar in color to the upper breast, and the underside of the tail is strongly marked with black and white (see description). The mountain trogon has a presumably territorial "song" that consists of repetitive downslurred plaintive whistles, these whistles usually uttered as triplets or in four-note sequences but sometimes prolonged in an extended *kyow-kyow-kyow . . .* series. This sequence reportedly often is followed by a bouncing-ball series of notes. The usual call has been described as similar to that of a ferruginous pygmy-owl (*Glaucidium brasilianum*), and the song as a series of *coo* notes similar to the call of a greater roadrunner (*Geococcyx californianus*). See also the vocalizations section.

GEOGRAPHIC VARIATION

The reputed southernmost race (*lutescens*) of the mountain trogon averages somewhat larger in size than nominate *mexicanus*, but variation within this species is quite limited. Monroe (1968) considered the Honduran population inseparable from the nominate form.

ECOLOGY

Habitats   Collectively the mountain trogon occurs in drier montane forest from Mexico to Honduras, particularly in pine-oak forest, pine forest, and montane evergreen forest; and the species typically ranges from 800 to 3,500 m (2,600–11,500 ft). In Mexico it occurs at altitudes of 1,200–3,500 m (3,900–11,500 ft), mostly in pine-oak and pine-evergreen forests (Howell and Webb 1995). At least in Sonora the birds appear to be migratory and are present from May to September (Russell and Monson 1998). In Guatemala the species ranges from 800–2,900 m (2,600–9,500 ft), in cloud forest, pine and oak woodland, and secondary forest (Land 1970). In Honduras it is mostly associated with pine and pine-oak forests above 1,200 meters (3,900 ft), but it occasionally descends to as low as 600 m (2,000 ft; Monroe 1968).

Foods and Foraging Ecology   No detailed information available.

BEHAVIOR

General, Social, and Sexual Behavior   No information available.

Vocalizations   According to Davis (1972), the song of the mountain trogon consists of 8–20 *cuc* notes delivered at the rate of two per second and a more hurried and "whispered" phrase of similar but lower-pitched *cu* notes. Vocalizations of this species were recorded in Mexico by Ben Coffey Jr. (Hardy, Reynard, and Coffey 1995). The recorded sequence consists of some individual down-slurred *kyow* notes with interspersed brief *wa-chew'* elements, as well as two sets of four-note down-slurred *kyow* sequences of uniform cadence and loudness, each lasting about 3 seconds. See also the identification section.

BREEDING BIOLOGY

Chronology of Breeding   Three mountain trogon egg sets from Mexico in the collection of the Museum of Vertebrate Zoology are from May 14 to

June 12. Binford (1989) reported that L. L. Short found a nest with two eggs on April 3, in Oaxaca. Robins and Head (1951) found a nest in Tamaulipas on May 27. Skutch (1944, 1983) found three nests in Guatemala between March 21 and early April.

Nest Sites   The four Mexican mountain trogon egg sets mentioned in the previous section were all in tree cavities from 1.25 to 12.5 m (4–41 ft) high, in pines or oaks. Rowley (1966) found a nest with two juveniles in late May in Oaxaca, about 3 m (10 ft) up in an old pine stump The nest found by Robins and Head had an entrance measuring 10 × 6.4 cm (3.9 × 2.5 in), and the cavity bottom was only about 2.5 cm (1.0 in) below the base of the entrance hole. The three Guatemalan nests found by Skutch were all in low rotting stumps or in the broken stumps of branches, at heights of 84–124 cm (33.0–48.7 in). Other uncompleted nest sites were found, one as high as 3.6 m (12 ft), these efforts apparently having been abandoned because the wood was either too soft or too hard. Several apparently completed but unused cavities were also found. At the altitudes these birds nested, no wasp or termite nests were available as potential nest sites.

Eggs and Incubation   Each of the four Mexican sets of mountain trogon eggs mentioned above had two eggs, and the two nests with juveniles also had two young. The three Guatemalan nests observed by Skutch each had two eggs. As with other trogons studied by Skutch, the male incubated during daytime hours and the female at night. Skutch estimated that in a period of more than 12 hours the female sat twice, over periods totaling 428 minutes; and the male four times, totaling 213 minutes. The eggs were covered during 89 percent of the entire observation period. In one nest studied the incubation period was estimated at 19 days after the laying of the second egg, and at another the incubation period was estimated to be 18 or 19 days. In one nest the eggs hatched a day apart.

Brood Rearing   The nestings of the mountain trogon are hatched naked, with closed eyes. Skutch reported that in the nest he watched closely, the

nestlings were fed white and green larvae (caterpillars), adult moths, and other insects. The female was a much more efficient food provider than the male, who at least initially often neglected to pass on the food items he brought to the nest. The female seemed to do the majority of the brooding. At about 1 week of age the eyes of the nestlings began to open, and their juvenal feathers began to break free of their sheaths. Soon thereafter they began to hiss and crouch at Skutch's approach. Hissing noises were also made when the adults approached with food, and the nestlings produced low and repeated *cup* calls when hungry. Skutch judged that the relatively early appearance of the juvenal plumage in this montane species is an adaptation to the cool climate. Late in the nestling period the female stood in the entrance opening at night, apparently trying to keep out the cold night air. Fledging occurred at 15 and 16 days, and the chick that hatched a day after the first fledged the same day as its sibling. Juveniles remain with their parents for some time and may be fed by them even as they are molting out of their juvenal plumage into their first adult-like plumage, about 2 months after hatching. The white spots on their wing coverts are lost about 2 months later, after which time the young look identical in plumage to their parents. Skutch was uncertain whether pair bonds persist through the nonbreeding period but found no evidence of pairs roosting together.

CONSERVATION AND EVOLUTIONARY RELATIONSHIPS

The mountain trogon extends through the wooded highlands of Mexico, Guatemala, Honduras, and extreme northern El Salvador. It is fairly common in pine-oak forests, cloud forests, and even second-growth forests over this broad area and should not be in any direct foreseeable conservation danger as long as some of these forest types remain. It is obviously a close relative of both the collared and elegant trogons, differing from them mainly in undertail patterning. It is broadly sympatric with both species but occurs at higher elevations than either. Stotz et al. (1996) listed the species as common, with a medium degree of sensitivity to disturbance.

ELEGANT TROGON

Trogon elegans Gould 1834

OTHER VERNACULAR NAMES

Coppery-tailed trogon, doubtful trogon (*ambiguus*), Goldman's trogon (*goldmani*), graceful trogon (but also see black-throated trogon); trogon elegante (Spanish).

RANGE

Open woodlands and gallery forests from Arizona, New Mexico, and (rarely) Texas south through Mexico (Sonora to Oaxaca, Nuevo León to Oaxaca, also Tres Marías Islands) and from Guatemala (upper Motagua Valley, drier Pacific lowlands) though Honduras (Pacific lowlands and interior), El Salvador (widespread), western Nicaragua, and Costa Rica (Pacific lowlands south to Gulf of Nicoya). See map 17.

SUBSPECIES

Trogon elegans canescens van Rossem 1934: southern Arizona (Atascosa, Chiricahua, Huachuca, and Santa Rita Mountains (recent reports of breeding in Brown Canyon, Santa Cruz County); northwestern Mexico (Sonora, northern Sinaloa, west-ern Chihuahua). Also reported from New Mexico (Peloncillo and Animas Mountains, especially Guadeloupe Canyon; one breeding record in 1992 from Skeleton Canyon). Migratory in the United States and also in the northernmost Mexican portion of its range (north of lat. 28°30' N); usually present in breeding range from April to early November (Kunzmann, Hall, and Johnson 1998). Considered synonymous with *goldmani* by Webster (1984).

Trogon elegans ambiguus Gould 1835: southern Texas (Starr, Hidalgo, Cameron, and Brewster Counties, where casual; no known breeding records); also eastern and central Mexico from southern Sinaloa southeast to Veracruz. Considered a separate species by Gould.

Trogon elegans goldmani Nelson 1898: Tres Marías Islands (María Madre and María Magdalena Islands), off western Mexico (present status unknown).

Trogon elegans elegans Gould 1834: southeastern Guatemala.

Trogon elegans lubricus Peters 1945: Honduras, Nicaragua, and Costa Rica. This form was previously known as *Trogon elegans australis* Griscom 1930, an epithet that is preoccupied.

Map 17. Distribution of the elegant trogon. Hatching indicates migratory population.

MORPHOMETRICS

Measurements   Wing, males (of *ambiguus*) 127–136 mm (5.0–5.3 in; average of 13, 130.8 mm [5.1 in]); females 124–137 mm (4.8–5.3 in; average of 13, 132 mm [5.1 in]). Tail, males (of *ambiguus*) 153.5–171 mm (6.0–6.7 in; average of 13, 165.6 mm [6.5 in]); females 165–185 mm (6.4–7.2 in; average of 13, 176.5 mm [6.9 in]). Wing measurements of male *elegans* average very slightly smaller than those of *ambiguus*, but tail measurements show no clear differences. Wing and tail measurements in *goldmani* also tend to be slightly smaller than those in *ambiguus* (Ridgway 1911). Egg of *ambiguus* 28.3 × 23.4 mm (1.1 × 0.9 in; Schönwetter 1966). Additional egg measurements for *ambiguus* and *canescens* were provided by Kunzmann, Hall, and Johnson (1998), and they found that the latter's eggs average very slightly larger than those of the former.

Weights   Ten birds of both sexes from Mexico 60–78.6 g (2.1–2.8 oz; average 67.3 g [2.4 oz]; Dunning 1993), including three males 60–70 g (2.1–2.5 oz; Martin, Robins, and Head 1954). Fourteen males from Mexico (various subspecies) 53–70.8 g (1.9–2.5 oz; average 65.4 g [2.3 oz]), seven females 61–76.7 g (2.1–2.7 oz; average 68.3 g [2.4 oz]; specimens, MVZ). Four males of *elegans* 72.2–83 g (2.5–2.9 oz; average 77.0 g [2.7 oz]), one female 82 g (2.9 oz; specimens, FMNH). Four adult males from Arizona average 65.8 g (2.3 oz), two adult females average 73.8 g (2.6 oz; Kunzmann, Hall, and Johnson 1998). Estimated egg weight 8.3 g (0.3 oz; Schönwetter 1966).

DESCRIPTION (of *ambiguus*, adapted from Ridgway 1911; see plate 22)

Adult Male   Forehead back, nape, hindneck, and most other upperparts bright iridescent green, becoming green to bronze-green on upper tail-coverts; middle pair of rectrices bronzy, broadly tipped with black; next pair of rectrices similar on outer vanes, but inner vanes darker and more purplish, becoming black basally; remaining rectrices very broadly tipped with white, subterminal part and most of outer vanes white, vermiculated or flecked with blackish, basal part of inner vanes blackish; wing-coverts vermiculated with black and white; primaries slate-black, outer vanes mostly grayish white; face, chin, and throat uniform black to slate-black; chest bright iridescent green or bronzy, followed posteriorly by a crescentic band of white; remaining underparts pure red; thighs slaty, longer feathers tipped with pink; bill cadmium-yellow to dull ochre; iris dark brown; eye-ring red to yellowish orange; feet and toes brownish to dusky.

It should be noted that Oberholser (1974) described a postjuvenal plumage sequence of (his terms) first winter, first nuptial, second winter, and subsequent "adult" nuptial and winter plumages in both sexes, suggesting that there is an annual complete postnuptial molt, producing the winter (or in current American terminology, definitive basic) plumage and an annual incomplete prenuptial molt, resulting in the nuptial (definitive alternate) plumage in all birds after their first winter. The adult winter plumage of males was described as being "similar" to the nuptial plumage of males by Oberholser, but no distinguishing features were identified.

Adult Female   Forehead and anterior crown slate to brownish slate, becoming plain brown on nape, hindneck, and upperparts; rump and upper tail-coverts paler; wing-coverts similar to rump in color but minutely vermiculated; outer vanes of secondaries similar but paler than in male; primaries slaty, outer edges edged with white or grayish white; middle pair of rectrices brown, broadly tipped with black; next pair of rectrices blackish brown to blackish, outer vanes more or less broadly edged with lighter brown; remaining rectrices extensively white terminally and on greater part of outer vanes, a broad terminal area immaculate white, rest of white portion more or less barred with blackish or dark slaty; auricular area crossed with a broad bar of buff and tipped with oblique black bar; a brownish white feathered eye-ring; chin and throat brownish slate to grayish brown, becoming brown on chest; breast much paler brown, becoming brownish white posteriorly and crossed anteriorly by a crescentic band of brownish white or pale brownish buff; lower abdomen and remaining underparts light red; thighs slaty, longer feathers tipped with whitish; bill dull yellow (duller than in male), maxilla sometimes tinged with brownish; eye-ring orange to orange-red; other soft parts as in male.

As mentioned above, Oberholser (1974) posited two molts annually in both sexes, the winter plumage of females (produced by a complete postnuptial molt) being described as "similar" to the adult nuptial plumage (reportedly produced by an incomplete prenuptial molt). Dickey and van Rossem (1938) stated that adults have an annual complete molt in August and a body molt in spring. The postjuvenal molt occurs in late summer to early fall and is completed by about November. No spring molt was indicated in the report by Kunzmann, Hall, and Johnson (1998).

Immature Male   Similar to adult male, but tail mostly like female's, except middle pair of rectrices bronzy basally, becoming light brown to cinnamon at tip, or edged with black. Younger males have breast mostly pale buffy grayish, narrowly barred or vermiculated, becoming buffy white posteriorly; secondaries each with a terminal white spot; most wing-coverts widely tipped with white and edged with black. Bill brownish orange; eye-ring dull orange or orange-yellow. Oberholser (1974) called this stage the "first winter" plumage, which is followed by a partial prenuptial molt into the first nuptial plumage, the differences between the two plumages most conspicuously involving the three outer pairs of rectrices.

Immature Female   Similar to adult female, but middle rectrices tipped with blackish; three outer pairs of rectrices more broadly barred and little if at all vermiculated with black; wing-coverts and secondaries more or less spotted with pale buff or whitish; red underparts replaced by white; undertail patterns of females more coarsely barred than those of males, and no iridescent feathers are present on the upperparts.

Juvenile   Upperparts and chest brown, becoming rufous on rump; white pectoral zone present; breast and belly buff-white and brown; young females like males, but middle rectrices of females not marked with black toward tip. In juveniles of both sexes the underparts are mostly mottled brown and white, and the wing-coverts are heavily spotted with white. According to Oberholser (1974), juvenile females resemble males but are duller and lighter above and below, and the purplish-black bars on the tail are broader and less numerous.

IDENTIFICATION

In the Hand   Of the three red-bellied, yellow-billed, white-necklaced trogons of Central America, the elegant trogon has the palest undertail patterning, and females are unique in having a white patch extending from the rear of the eye to the lower ear region. The upper surface of the tail is rather coppery toned in males from Mexico (more golden green in southern races). Undertail patterns also differ regionally: adult males of northern birds have mostly white undertails with dusky vermiculations toward the base, and females have dusky barring on the underside of the tail but are predominantly pale toned. Among birds from Guatemala and southward both sexes have more definite blackish barring and broader black tips on the outer rectrices.

In the Field   The elegant trogon is a lowland to lower montane species (reaching 1,300–2,300 m [4,300–7,500 ft] in Mexico, 1,900 m [6,200 ft] in Guatemala, and 750 m [2,500 ft] in Costa Rica) and is often found in relatively arid woodlands. It also occurs in pine-oak woodland or open forest, in second-growth forest, and occasionally in more humid woods. Field marks of the female include the distinctive white ear patch and a broad, indefinite white breast band, or "necklace," immediately below the brown chest. Males of the Mexican population have a copper-tinted uppertail color, a drooping white breast band, or "necklace," between the green chest and the red abdomen, and an undertail pattern that appears to be black-tipped (a visual effect produced by the black middle rectrices extending beyond the shorter outer rectrices) but is otherwise predominantly white, with faint (in Mexico) to definite (in more southern races) but narrow black barring. Vocalizations include an extended series of coarse and wooden or turkey-like *kowm* notes, often increasing progressively in loudness, or more distinctly disyllabic *krow'h* notes that may be rapidly repeated up to about six times. See also the vocalizations section.

GEOGRAPHIC VARIATION

The major geographic variations in the elegant trogon concern *ambiguus*, which along with *goldmani* and *canescens* has often been taxonomically separated specifically. Males of the northern forms have golden bronze to golden green upperparts, coppery bronze middle rectrices, vermiculated (and white-tipped) rather than narrowly barred (but also white-tipped) outer rectrices, and distinctly shorter tails than southern forms. The southern birds are more greenish or golden green dorsally and on the upper tail surface, rather than copper toned. Females of the northern forms have their three outer pairs of rectrices irregularly barred or mottled with gray to brown or dusky markings, whereas in the southern races these rectrices are regularly barred with grayish black. The race *goldmani* on the Islas Tres Marías off western Mexico resembles the southern populations in the color of its central rectrices but is otherwise associated with the northern assemblage. The race *canescens* has slightly longer average wing and tail lengths than does *ambiguus* (perhaps related to its migratory tendencies) and tends to be somewhat paler.

ECOLOGY

Habitats   Collectively the elegant trogon occurs in tropical deciduous forest, pine-oak forest, and broadleaf riparian woodland near pine-oak woodland, from sea level to 2,500 m (8,200 ft). Pines and live oaks, scrubby deciduous woodlands, plantations, thorn forests, shady canyons, and riparian woodlands rich in sycamores are all used by this species. Nesting in Arizona preferentially occurs in riparian canyons of the pine-oak zone, at elevations of 1,515–2,120 m (5,000–7,000 ft); and the overall range of regular usage there is about 1,350–2,300 m (4,400–7,500 ft; Hall 1996; Hall and Karubian 1996; Taylor 1978–1983). In Mexico the species ranges generally from near sea level to 1,800 m (5,900 ft; Howell and Webb 1995), specifically in Sonora from about 200 to 2,000 m (700 to 6,600 ft), and from pine-oak woodland to tropical deciduous forest and tropical thornscrub (Russell and Monson 1998). In Guatemala it ranges from near sea level to 1,900 m (6,200 ft; Land 1970). In El Salvador it extends through the tropical zone to about 1,100 m (3,600 ft), mainly in low trees and scrub (Dickey and van Rossem 1938). In Honduras it mainly occurs below 1,500 m (4,900 ft), especially in drier areas such as regions of seasonally moist, or "monsoon," forest (Monroe 1968). In Costa Rica these birds range to about 750 m (2,500 ft).

They occupy hillside or ravine forests, evergreen forests, and relatively taller and drier forests than those used by the black-headed trogon (Stiles and Skutch 1989).

Foods and Foraging Ecology   Bent (1940) has summarized early information on the foods consumed by the elegant trogon. Vegetable foods include wild grapes, cultivated cherries, fruits of cut-leaved cissus (*Cissus*), and red peppers. Adults and larvae of lepidopterans were found in one stomach, and another contained 68 percent insect remains, including nymphs and eggs of grasshoppers, three mantids, three stink bugs, other heteropterans, a leaf beetle, a larval hawkmoth (sphingid), and larvae of sawflies and unidentified lepidopterans. Dickey and van Rossem (1938) listed stomach contents as including berry pits and pulp (six stomachs); fruit pulp and caterpillars (four); caterpillars and grasshoppers (two); small beetles, grasshoppers, and fruit pulp (one); and fruit pulp and grasshoppers (one). The birds usually occur well up in the forest canopy but move lower to take insects, berries, and fruits from the foliage (Stiles and Skutch 1989). Taylor (1994) reported that oak trees and fruiting trees are favored foraging sites, and Hall (1996) further noted that both sexes, but especially males, favor oaks for their foraging, with dead or dying trees used more often than would be expected by chance. However, acorn consumption has evidently never been reported for any trogon, and the birds were probably feeding on insects associated with these trees. A stomach examined by Remsen, Hyde, and Chapman (1993) contained arthropods only.

BEHAVIOR

General, Social, and Sexual Behavior   Elegant trogons seem to be fairly solitary, like most trogons. Male courtship behaviors observed by Hall and Karubian (1996) included perching near females and "tail-flicking" (tail-pumping), expanding the crimson chest while facing the female, and following females from perch to perch while calling in low-pitched voices. A major component of courtship was nest-advertisement calling by the male, either from a perch or at a nest cavity. Copulation was observed repeatedly and in three cases was seen at least twice in a single hour. According to Sheri Williamson (pers. comm.), copulation is preceded by some quiet conversational notes by the two birds as they perch together but is not marked by obvious precopulatory or postcopulatory postural displays. Territoriality is well established in this species. In Arizona, males begin advertising their locations and courting females soon after they arrive in the spring. They may call to one another over fairly long distances of at least 50 m (200 ft), and playbacks of male vocalizations have been used as a censusing tool. Territories are evidently rather linear, averaging about a kilometer in length; although one territory was estimated at 1.6 kilometers (1.0 mile), and in another case two nests were located only about 183 m (600 ft) apart. Evidently the entire home range is defended (Taylor 1994). In addition to engaging in territorial encounters with other elegant trogons, elegant trogons defend their nesting cavities against eared trogons, northern flickers, sulfur-bellied flycatchers (*Myiodynastes luteoventris*), and whiskered screech-owls (*Otus trichopsis*); and parent birds may also attack gray squirrels (*Sciurus arizonensis*; Kunzmann, Hall, and Johnson 1998). Mobbing calls consist of seven to ten staccato chucking notes, with the male's notes higher pitched than those of females (Cully 1986).

Vocalizations   Hall and Karubian (1996) found that male elegant trogons call much more frequently than do females (at least during the breeding season, which was the period under study) and, when disturbed, produce rattling or cackling notes, combined with lateral tail flaring or vertical tail pumping. The species' most common vocalization is a rather hoarse, somewhat disyllabic and down-slurred *koink, kwa'h* or *kow-r*, repeated five to ten times with no apparent change in loudness or rate of calling. This vocalization is the male's primary territorial vocalization, or "song, " and is commonly heard during the breeding season, sometimes even during moonlit nights (Elliott 1983). Vocalizations of this species were recorded in Mexico by Ben Coffey Jr. (Hardy, Reynard, and Coffey 1995). The recording contains five apparent song sequences, each consisting of five or six individual notes. All notes are produced at the same pitch and loudness and at a uniform rate of about two notes per second, with brief intervals between each sequence. A sonogram of this vocalization appears in the report by Kunzmann, Hall, and

Johnson (1998), although in this example the song is produced at a faster rate (four notes per second) and includes thirteen notes. A similar *koa* or *coa* call, more monosyllabic and sometimes extended into a series of up to 50 notes, is uttered by birds of either sex when they are disturbed by intruding trogons, other birds, or mammals. A softer *kow* or *kuh* or both is uttered in various situations, such as before copulation, by males when luring females to prospective nest sites, or by a pair of birds visually separated from each other. The tonal quality of all these calls is distinctly wooden, reminiscent of sounds made with an artificial turkey call (sound is produced by rubbing a scraper over a hollow wooden box). The similarity of this species' voice to that of a hen turkey has been mentioned by others (Bent 1940). A male may also utter a sharp *ha!* note when chasing intruders from his territory. A rather different and more metallic *w-k-k* is used as a flight call, typically uttered on takeoff from a perch and repeated at half-second intervals while the bird is in flight. A prolonged and more extreme version of this apparent alarm note, a *w-kkkk*, may also occur when the trogon is attacking another species. Then the call may become accelerated into a slurred series of notes lasting up to a minute or more, finally trailing off into a series of hoarse *cluck* notes When the alarm call is uttered from a perch the tail is sharply raised, then slowly lowered. Finally, nestlings utter long series of clear, bell-like *tu-u* notes that may continue until the birds are fed (Kunzmann, Hall, and Johnson 1998; Taylor 1994). See also the identification section for possible other variations.

BREEDING BIOLOGY

Chronology of Breeding   In an Arizona study of elegant trogons, six nests were started in May, eight in June, and six in July (Hall and Karubian 1996). Eggs in the collection of the Foundation of Vertebrate Zoology from Mexico range from April 27 to June 16, with one April record, seventeen May records, and six June records. The second or third week of May seems to be the peak Mexican breeding period; Russell and Monson (1998) stated that in Sonora nesting coincides with the start of the summer rainy season. Bent (1940) reported that 23 nest records from Mexico ranged from March 29 to June 30, with 12 occurring between April 15 and

May 13, making the peak nesting period somewhat earlier than that described for the eggs in the MVZ collection. Breeding in El Salvador extends from May 1 to August 1 (Dickey and van Rossem 1938). In Costa Rica the nesting season is from April to July, with two broods frequently occurring (Stiles and Skutch 1989).

Nest Sites   Of 46 elegant trogon nest sites in Arizona, 80 percent were in pre-existing cavities in sycamore trees (*Plantanus wrightii*), 4 percent Arizona white oaks (*Quercus arizonica*), 4 percent in Gooding's willows (*Salix goodingii*), and the remainder were in a variety of other oaks, a pine snag, and a black walnut (*Juglans major*). Most of these sites were in areas with no nearby running water, but intermittent standing water was present in most nesting areas (Hall and Karubian 1996). However, trogons have seldom been observed drinking water, and there are few observations of bathing in water (young birds only). Taylor (1978–1983, 1994) found that 83 percent of 59 nest sites in Arizona were in woodpecker holes, mostly those of flickers. Most of the nests found in southeastern Arizona have been within 300 m (1,000 ft) of permanent water, with sycamores being a favored nesting choice, presumably because this rather soft hardwood tree is also a favored choice of woodpeckers. Of 58 sites observed by Hall (1996), 91 percent were in riparian areas, and the rest were in upland locations. One large sycamore was used for nesting during 4 consecutive years, and about one-quarter of all nests are reused at least once, although not necessarily in consecutive years. The nest used for 4 years was unusual in being very deep (over 100 cm [39.3 in]) compared to the usual size of sycamore nests (average 50 cm [19.7 in]) and pine and oak nests (average less than 30 cm [11.8 in]) (Taylor 1978–1983, 1994).

Seventeen sets of eggs in the collection of the Western Foundation of Vertebrate Zoology were from tree nests ranging in height from 3 to 9 m (10 to 30 ft), the median height being about 6 m (20 ft). The nest cavity is about 13 cm (5.1 in) in diameter and may be about 30 cm (11.8 in, in oaks or pines) to 48 cm (18.9 in, in sycamores) deep. Ten sets mentioned by Bent (1940) were all in tree cavities at heights ranging from 3 to 12.5 m (10 to 41 ft) above ground. One Mexican nest in an oak was located about 3 m (10 ft) high and had a cavity 40 cm (15.7 in)

deep (Martin, Robins, and Head 1954), presumably made by a woodpecker. In Costa Rica the sites are 2–6 m (7–20 ft) high, the birds perhaps occupying or enlarging old woodpecker holes (Stiles and Skutch 1989). In a study of nest sites, Hall (1996) found that both the height of vegetation south of the nest and the diameter of the nesting tree correlated with nesting success rates (higher vegetation and larger diameters correlating with higher nesting success), but other measured variables failed to show significant correlations with nesting success.

Eggs and Incubation   Sets of elegant trogon eggs in the collection of the Western Foundation of Vertebrate Zoology range from two to four eggs, with seven sets of two, six sets of three, and nine sets of four. Fourteen clutches of *ambiguus* had forty-seven total eggs (average clutch 3.4 eggs); seven clutches of *canescens* had fifteen eggs (average clutch 2.1 eggs; Kunzmann, Hall, and Johnson 1998). The usual clutch in Costa Rica is three eggs (Stiles and Skutch 1989). Incubation is performed about equally by the two sexes, as are brooding and feeding of the chicks. The incubation period averaged 19 days, with a range of 17–21 days (Hall and Karubian 1996). Taylor (1994) estimated a 17-day incubation period.

Brood Rearing   Elegant trogon chicks are hatched naked, but by 10–12 days they are covered by downlike feathers everywhere but the abdomen. Fledging occurs at 15–17 days (Hall and Karubian 1996). By 20–23 days the rectrices are about one-third adult length, and the remaining growth of these feathers occurs during the month following fledging (Kunzmann, Hall, and Johnson 1998). Hall and Karubian (1996) reported that males and females fed young at roughly equal rates, and the identified foods included eleven orthopterans, ten lepidopterans, six coleopterans, two each of Homoptera, Hymenoptera, and Odonota, and a single mantid. No fruit was observed among the foods seen by Hall and Karubian, but Taylor (1994) did observe berries being fed to chicks; it is possible that the absence of fruit in the observed diet simply resulted from its scarcity during dry years. The duration of brooding was estimated at 17 days by Hall and Karubian (1996) and at 20–23 days by Taylor (1994). The chicks may remain with their parents for a month or two and continue to be fed by them for nearly a month. The two parents may divide the brood and care for them separately at this time (Kunzmann, Hall, and Johnson 1998). Although elegant trogons generally have only one brood (at least in Arizona), failed clutches may result in a second nesting effort at a new location (Kunzmann, Hall, and Johnson 1998). Of 47 nesting efforts in Arizona, 68 percent were successful (Hall 1996). This success rate is close to the estimated annual fledging success rate of 50–78 percent for 36 nests that were studied over several years by Taylor (1978–1983).

CONSERVATION AND EVOLUTIONARY RELATIONSHIPS

The elegant trogon occurs from the southwestern United States to Costa Rica, its range encompasses seven countries, and it is relatively common over much of its range. Its nearest relative is perhaps the collared trogon, with which it is widely sympatric, but it is clearly also related to the mountain trogon. Stotz et al. (1996) listed the species as fairly common, with a medium degree of sensitivity to disturbance.

In Arizona this species has always been rare; Marshall (1957) estimated densities of only one to two pairs per 1.6 km (1.0 miles) of riparian vegetation in known nesting areas of Arizona and adjacent Mexico. Taylor (1994) estimated a total Arizona population of about 100 adult elegant trogons. A few of these birds were found in the Atascosa Mountains, about 8–12 birds were found in Madera Canyon in the Santa Rita Mountains, and a similar number occurred in the Huachuca Mountains. The rest were in the Chiricahua Mountains, where the South fork of the Cave Creek watershed may have supported six to ten nesting pairs. Numbers in Arizona have seemingly remained quite stable from the time of Taylor's early surveys in the late 1970s and early 1980s (averaging 66.8 trogons) to the early 1990s (averaging 76.0 trogons). More recently, between 1993 and 1995, annual trogon numbers per canyon have averaged 12.5 in the Atascosas, 5.7 in the Chiricahuas, 12.9 in the Huachucas, and 8.4 in the Santa Ritas. Along 5 km (3 miles) of road in southeastern Sonora 75 birds were found (15 birds/km), suggesting the population may be quite high in favorable areas of the core range (Kunzmann, Hall, and Johnson 1998).

The elegant trogon is a candidate species for Arizona's list of threatened native wildlife. There have as yet been no reports of the birds using nest boxes erected for trogons. Gradual loss and destruction of sycamore nesting habitat and accidental or intentional disturbance of breeding birds by humans and vehicles are all regarded as potential threats to the species' continued existence in Arizona (Kunzmann, Hall, and Johnson 1998).

COLLARED TROGON

Trogon collaris Vieillot 1817

OTHER VERNACULAR NAMES

Bar-tailed trogon (*puella*), Colombian collared trogon (*subtropicalis*), Jalapa trogon (*puella*), Pallatinga trogon (*virginalis*), rayed-tailed trogon, red-bellied trogon, Tabatinga trogon (*castaneus*), Venezuelan collared trogon (*exoptatus*); sorocua acollarado, trogon collarejo (Spanish).

RANGE

Humid evergreen and mixed forests from east-central Mexico (southwestern San Luis Potosí) south through Guatemala, Belize, Honduras, El Salvador (rarely), Nicaragua, Costa Rica, Panama, Colombia, Venezuela, Trinidad Island (northern range), Tobago (formerly common), Guyana (local), Suriname (local), French Guiana, Brazil (southeast to Rio de Janeiro and south to Mato Grosso), west to northwestern Ecuador, eastern Peru, and northern Bolivia. See map 18.

SUBSPECIES

Trogon collaris puella Gould 1845: central Mexico (San Luis Potosí and Guerrero) to western Panama. Considered a separate species by Gould as well as by others, such as Ridgway (1911). Includes *xalopensis* Phillips (1966), which reputedly occurs on the Caribbean slope of Mexico and Guatemala.

Trogon collaris extimus Griscom 1929: eastern Panama.

Trogon collaris heothinus Wetmore 1967: Darién, Panama.

Trogon collaris virginalis Cabanis and Heine 1862–63 (1863): western Colombia, western Ecuador, and northwestern Peru.

Trogon collaris subtropicalis Zimmer 1948: central Colombia.

Trogon collaris exoptatus Cabanis and Heine 1862–63 (1863): northern Colombia, northern Venezuela, Trinidad and Tobago Islands.

Trogon collaris collaris Vieillot 1817: Colombia to Bolivia, Venezuela, Guyana, Suriname, and French Guiana.

Trogon collaris castaneus Spix 1824: southeastern Colombia, northwestern Brazil, eastern Peru, and northern Bolivia.

Trogon collaris eytoni Fraser 1856: eastern Brazil.

MORPHOMETRICS

Measurements Wing, males (of *collaris*) 119–123 mm (4.6–4.8 in; average of 5, 121.7 mm [4.7 in]); females 117–125 mm (4.6–4.9 in; average of 5, 120.4 mm; ffrench 1991). Wing, males (of *puella*) 115–138 mm (4.5–5.4 in; average of 35, 126.6 mm [4.9 in]); females 120–136.5 mm (4.7–5.3 in; average of 17, 127.1 mm [5.0 in]; Ridgway 1911). Tail, males (of *puella*) 132.6–136.6 mm (5.2–5.3 in; average of 10, 133.4 mm [5.2 in]); females 131.3–139.6 mm (5.1–5.4 in; average of 10, 134.8 mm [5.3 in]; Wetmore 1968). Wing, six males (of *eytoni*) 113–117 mm (4.4–4.6 in); two females 110 and 111 mm (4.29 and 4.33 in); tail, six males 124–145 mm (4.8–5.7 in); two females 134 and 145 mm (5.2 and 5.7 in; Pinto 1950). Wings and tails of *extimus* average smaller than those of other Panamanian races and those of *virginalis* from South America (Wetmore 1968). Egg of *puella* 28.3 × 22.6 mm (1.1 × 0.9 in), of *virginalis* 28.3 × 22.6 mm (1.1 × 0.9 in), of *collaris* 28.0 × 24.1 mm (1.1 × 0.9 in; Schönwetter 1966). Egg of *heothinus* 28.65 × 22.55 mm (1.1 × 0.8 in; Wetmore 1968). Average of extremes of five eggs of *puella* 29.5 × 23.2 mm (1.2 × 0.9 in; Rowley 1966).

Weights Twenty-nine males from Panama average 63.4 g (2.2 oz), eighteen females average 65.4 g (2.3 oz; Hartman 1961). One male of *exoptatus* from Trinidad 61 g (2.1 oz), five females 51–62 g (1.8–2.2 oz; average 55.4 g [1.9 oz]; ffrench 1991). Three males of *subtropicalis* 54.5–63.6 g (1.9–2.2 oz; average 60.5 g [2.15 oz]), two females 59.0 and 59.6 g (2.07 and 2.09 oz; Miller 1963). One male of *puella* from Belize 63.1 g (2.2 oz; Russell 1964). Sex and sample unspecified 41–67 g (1.4–2.3 oz; average 60 g [2.1 oz]; Smithe 1966). Three males of *collaris* 48.5–58 g (1.7–2.0 oz; average 53.3 g [1.9 oz]), one female 55 g (1.9 oz); one male of *castaneus* 61.5 g (2.2 oz), four males of *puella* 66.6–78.6 g (2.3–2.8 oz; average

Map 18. Distribution of the collared trogon.

72.9 g [2.6 oz]), two females 68.9 and 73.2 g (2.4 and 2.6 oz); one male of *virginalis* 54.1 g (1.9 oz), two females 52.3 and 56.8 g (1.8 and 2.0 oz; specimens, FMNH). Seven males of various subspecies 54.5–69 g (1.9–2.4 oz; average 65.6 g [2.3 oz]), four females 53.1–63.9 g (1.9–2.2 oz; average 59.2 g [2.1 oz]; specimens, MVZ). Estimated egg weight of *puella* 7.6 g (0.3 oz), of *virginalis* 7.7 g (0.3 oz), of *collaris* 8.5 g (0.3 oz; Schönwetter 1966).

DESCRIPTION (of *puella*, adapted from Ridgway 1911; see plate 23)

Adult Male   Upperparts except wings iridescent golden green, sometimes grass-green or golden bronze; middle rectrices tipped with black; second and third pairs of rectrices iridescent green on outer vanes, inner vanes black; three outer pairs of rectrices black, tipped rather narrowly with white (tips 1.5–9 mm [0.1–0.4 in] wide), remaining rectrices black, narrowly barred with white (white bars about twice as wide as black), but white bars lacking on inner vanes of third rectrix (from outside); wings mostly black, lesser wing-coverts narrowly barred with white, greater wing-coverts and secondaries finely vermiculated with white; longer primaries edged with white; chin, throat and loreal, orbital, auricular, and malar regions black; chest iridescent green like upperparts; a broad white breast band crosses upper breast; rest of underparts pure deep red; thighs slaty; bill yellow; iris brown; bare eye-ring inconspicuous brown, with pinkish-buff eyelids; feet and toes brown or blackish.

Adult Female   Upperparts plain brown, darker on crown, more tawny on lower rump and upper tail-coverts; middle pair of rectrices chestnut, tipped with black; outer vanes of next two rectrices chestnut, tipped black, inner vanes brownish black; three outer pairs of rectrices mostly black on upper surfaces, grayish below, narrowly tipped with white, and each crossed by a subterminal narrow black bar, area next to which is vermiculated; outer vanes of rectrices grayish white, with or without minute freckles or vermiculations, becoming darker basally; secondaries similarly patterned; primaries slate-black, edged with white; a white eye-ring of feathers interrupted on eyelids and most evident before and behind eye; face otherwise dusky to blackish slate, chin and throat somewhat lighter; chest uniform brown, like back; rest of underparts as in adult male; maxilla dull yellowish or yellowish green, grading to black on culmen; mandible olive-yellow; iris umber-brown; eye-ring inconspicuous, bare eyelids dusky to brown; feet and toes neutral gray to light brown.

Immature Male   Head and body largely brown, with scattered green feathers on breast and upperparts; white spot in center of breast; upper wing-coverts and tertials brown, finely vermiculated with blackish, and with triangular buff spots near their tips, and blackish edges; secondaries with outer margins vermiculated brown; middle rectrices green basally and rufous toward tips, sometimes with dusky tips; next two pairs of rectrices like middle pair, but inner vanes brownish black; three outer pairs of rectrices as in adult females, but sometimes with complete barring as in adult male, but white tips longer, feathers more slender and their tips more rounded (Zimmer 1949).

Immature Female   Similar to adult female, but coarsely freckled with blackish and white on inner vanes of undertail surface (Howell and Webb 1995).

Juvenile   Upperparts of both sexes mostly rich brown, with medium brown breast, deep buff abdomen, and rufous under tail-coverts; wing-coverts spotted with buff. Juveniles and immatures also have narrower vanes and more pointed outer rectrices than adults.

IDENTIFICATION

In the Hand   The broad range of the collared trogon brings it into contact with several other similar red-bellied, yellow-billed trogons having white breast bands separating their green (males) or brown (females) chests from their red bellies. Males in Mexico and northern Central America are recognizable simply by their uniformly narrowly striped black-and-white undertail pattern, but in South America (Colombia to Peru) the masked trogon has a very similar, but more finely barred, undertail pattern. Females of the collared trogon in both Central and South America have mostly grayish to brownish undertail colors with freckling but little distinct patterning, whereas the masked trogon female is more distinctly barred on the undertail.

In the Field   The collared trogon is a species of the humid to wet evergreen and semideciduous forests of lowlands and foothills (reaching 1,600 m [5,200 ft] in Mexico, 2,000 m (6,700 ft] in Guatemala, 700–2500 m [2,300–8,200 ft] in Costa Rica and Panama, and 250–2,300 m [800–7,500 ft] in Venezuela). In its broad range it encounters several similar species, especially the mountain and elegant trogons to the north and the masked and black-tailed trogons to the south. All have yellow bills (less apparent in females), red bellies, and white breast bands separating the green or brown breasts from the red abdominal colors. Undertail patterns are the best means of separating these species, as described above. The collared trogon's vocalizations include a territorial

or courtship "song" of multiple-note *caow-caow* or *kyow-kyow-kyow* elements that may extend into a series of five to eight such sequences. See also the vocalizations section.

GEOGRAPHIC VARIATION

Although nine subspecies of the collared trogon are generally recognized, all are very similar in appearance, differing mainly in such traits as the amount of white present on the tips of the outer rectrices and the width of the white barring between the intervening black bars. There are also minor differences in tail lengths among the races. Although *puella* has at times been considered a separate species, its plumage differences primarily involve a somewhat darker color, especially on the wings and head, and reduced amounts of white relative to black on the barred vanes and white tips of the outer rectrices. Some geographic variations in the Andean forms of *collaris* have been described by Zimmer (1948).

ECOLOGY

Habitats   Collectively the collared trogon occurs in lowland tropical evergreen forest, montane forest, cloud forest, rain forest, gallery forest, and tropical deciduous forest, at elevations from sea level to 2,500 m (8,200 ft). In the northern parts of its range, such as Mexico, Guatemala, and Honduras, the species is generally found between 600 and 2,000 m (2,000 and 6,600 ft). In Honduras it is most common between 600–1,000 m (2,000–3,300 ft), in low montane rain forest, but it extends into lowland rain forest and into cloud forest above 1,200 m (3,900 ft; Monroe 1968). In Costa Rica it is most common between 1,050 and about 2,000 m (3,400 and 6,600 ft), with an upper limit of 2,440 m (8,000 ft), which is also its upper limit in Venezuela (Skutch 1983). In Costa Rica, as generally elsewhere, these birds favor montane forests, often cloud forests or rain forests, but may extend out into forest edges, older secondary forest, and coffee plantations. Following breeding the birds may migrate down to the foothills (150–600 m [500–2,000 ft]; Stiles and Skutch 1989). In Panama the species extends from 690–2,400 m (2,300–7,900 ft), mostly occurring between 1,200 and 2,400 m (3,900 and 7,900 ft; Ridgely and Gwynne 1989). In Colombia it ranges mostly from 400–2,000 m

(1,300–6,600 ft) but may have been reported as high as 3,100 m (10,200 ft), in humid and wet forests, forest edges, and secondary growth, including varzea forests east of the Andes (Hilty and Brown 1986).

Foods and Foraging Ecology   Foods of the collared trogon in Brazil include fruits of Melastomaceae, Palmae, Sapotaceae, Rubiaceae, and other tropical plant families, as well as insects and insect larvae, especially caterpillars (Ruschi 1979). Stiles and Skutch (1989) stated that the species is relatively insectivorous, taking beetles, homopterans, smooth and spiny caterpillars, and crickets. Of 25 stomachs examined by Remsen, Hyde, and Chapman (1993), 56 percent contained arthropods only, 16 percent contained mixed fruit and arthropods, 12 percent contained fruit only, and 16 percent contained unidentified materials.

BEHAVIOR

General, Social, and Sexual Behavior   Skutch (1983) described the collared trogon as "quiet, retiring, dignified." Except during the nesting season the birds are solitary, perching fairly high in forest trees. He noted that, when uttering its rattling or churring alarm call, the bird initially slightly spreads and then quickly closes its tail, thus briefly exposing the barred outer tail feathers; and then slowly elevates the tail in a deliberate way that contrasts with the rapid earlier movement.

Vocalizations   Most descriptions of the collared trogon's call indicate that it consists of up to about eight rather hoarse to more mellow, coo-like notes, uttered either at the same pitch or with each note downwardly deflected. Churring notes, usually accompanied by tail movements, are probably alarm calls, and a descending trill has also been heard. Vocalizations of this species were recorded in Mexico and Colombia by Ben Coffey Jr., J. W. Hardy, and Ralph J. Raitt (Hardy, Reynard, and Coffey 1995). The recording contains songs, calls, and trills; several song sequences consist of series of from three to five *kyow* notes on nearly the same pitch and of uniform volume and duration (lasting up to about 2 seconds). This utterance is considerably higher pitched and less wooden in tone than the mountain trogon's. A second vocal example is a rather harsh down-slurred *churr* call with interspersed chipping notes, presumably an alarm call. See also the identification section.

BREEDING BIOLOGY

Chronology of Breeding   Nine sets of collared trogon eggs from Mexican nests in the collection of the Western Foundation of Vertebrate Zoology range from May 1 to June 11, with six May and three June records. An additional clutch from Ecuador was obtained in mid-June. Breeding in Trinidad and Tobago occurs from March or April to July, and there is a Belize record for April. Costa Rican records date from January to April (Stiles and Skutch 1989), and there is a Panama egg record for February (Wetmore 1968). In Colombia breeding-condition birds have been obtained from January to May (Hilty and Brown 1986; Miller 1963). The breeding season in northern Venezuela extends from April to June (Schäfer and Phelps 1954), and in Brazil breeding occurs between October and January (Ruschi 1979).

Nest Sites   Most of the collared trogon eggs in the collection of the Western Foundation of Vertebrate Zoology were obtained in Oaxaca by J. S. Rowley, who found the nests in dead stubs or dead limbs of trees, at heights of 1.25–3.75 m (4–12 ft; Rowley 1966, 1984). The nests were found at elevations of 250–1,900 m (800–6,200 ft), mostly in cloud-forest vegetation. One set of eggs from Ecuador was found in a partly rotted stump about 3.5 m (11 ft) above ground, in a nest with an entrance hole measuring 15 × 9 cm (5.9 × 3.5 in) and a depth of about 5 cm (2.0 in). In Trinidad and Tobago the birds may nest either in tree holes or in termite nests (ffrench 1991; Herklots 1961). Stiles and Skutch (1989) characterized the nest as a shallow, unlined niche that leaves much of the sitting bird exposed, at heights of 1.2–5 m (4–16 ft), usually in a slender and decaying stump. One nest found by Skutch was 3.6 m (12 ft) above ground in a soft-wooded tree that stood about 23 m (75 ft) from heavy forest. The aperture was pear shaped, and the nest cavity extended only a short distance below the entrance, so that the sitting bird could be seen easily from the front. Another nest was in a decaying stump 2.1 m (7 ft) high, with an entrance 1.5 m (5 ft) above ground. The entrance of this nest was also pear shaped, 15 × 7.3 cm in height and maximum width (5.9 × 2.9 in). The cavity was 11.4 cm deep and 10 cm wide (4.5 × 3.9 in) and extended 6.4 cm (2.5 in) below the lower edge of the entrance.

Eggs and Incubation   Nine clutches of collared trogon eggs from Mexican nests in the collection of the Western Foundation of Vertebrate Zoology have either two (seven sets) or three (two sets) eggs. One nest studied by Skutch (1983) was incubated by the male from 8:30 or 9:00 A.M. to 4:00 or 5:00 P.M. and by the female from early evening until the next morning. The incubation period could not be established.

Brood Rearing   The collared trogon eggs in the nest studied by Skutch (1956, 1983) hatched in early February, with both chicks apparently hatching on the same day. The eggshells were never removed by the parents. The young were fed crickets, grasshoppers, hairy caterpillars, and some unidentified green insects, probably also grasshoppers. Pinfeathers became visible at 5 days. When 9 days of age the young began to slough the sheaths of their pinfeathers, so that by 11 days they were clad in their juvenal plumages. One of the chicks died at 13 days, the other at 16 days. Ruschi (1979) stated that both sexes care for the young, which fledge in 25–28 days.

CONSERVATION AND EVOLUTIONARY RELATIONSHIPS

The collared trogon has one of the broadest ranges of all trogons, occurring in 18 countries between Mexico and Brazil. It is also generally common over nearly all of its range, making it one of the most abundant of all trogon species. Stotz et al. (1996) listed the species as common, with a medium degree of sensitivity to disturbance.

ORANGE-BELLIED TROGON

Trogon (collaris) aurantiventris Gould 1856

OTHER VERNACULAR NAMES

Underwood's trogon (*underwoodi*); trogon anaran-
jado, trogon vientrianaranjado (Spanish).

RANGE

Evergreen forests of Costa Rica (Guanacaste, Tila-
ran, Central, and Talamanca cordilleras) and Panama
(east to western Panama province, mainly on Pacific
slope, but local on Caribbean slope in Bocas del
Toro). See map 19.

SUBSPECIES

The orange-bellied trogon should very possibly be
considered as only a color morph of *collaris*, as suggested by Stiles and Skutch 1989.

Trogon aurantiventris underwoodi Bangs 1908: north-
    western Costa Rica (Guanacaste, Alajuela, and at
    least previously in the uplands of the Nicoya
    peninsula).
Trogon aurantiventris aurantiventris Gould 1856:
    central Costa Rica to western Panama; includes

T. a. flavidior (Griscom) 1925; not recognized by
Wetmore (1968).

MORPHOMETRICS

Measurements   Wing, males (of *aurantiventris*)
122.5–128 mm (4.8–5.0 in; average of 10, 125.6 mm
[4.9 in]); females 120–124 mm (4.7–4.8 in; average of
3, 122.7 mm [4.8 in]). Tail, males (of *aurantiventris*)
132.5–157.5 mm (5.2–6.1 in; average of 10, 142.4 mm
[5.6 in]); females 135–144 mm (5.3–5.6 in; average of
3, 138.2 mm [5.4 in]; Ridgway 1911).Wing and tail
measurements of *underwoodi* (and *flavidior*) show no
clear distinguishing differences from those listed in
this section. Egg of *aurantiventris* 26.9 × 22.5 mm
(1.0 × 0.9 in; Blake 1956).

Weights   Three birds of both sexes from Panama
61.5–68.5 g (2.2–2.4 oz; average 65.9 g [2.3 oz];
Hartman 1961; Strauch 1977). Three males from
Panama 66–72 g (2.3–2.5 oz; average 69.3 g [2.4 oz];
specimens, LSU). Unspecified Costa Rica weight 70 g
(2.5 oz; Stiles and Skutch 1989). Estimated egg weight
7.5 g (0.3 oz).

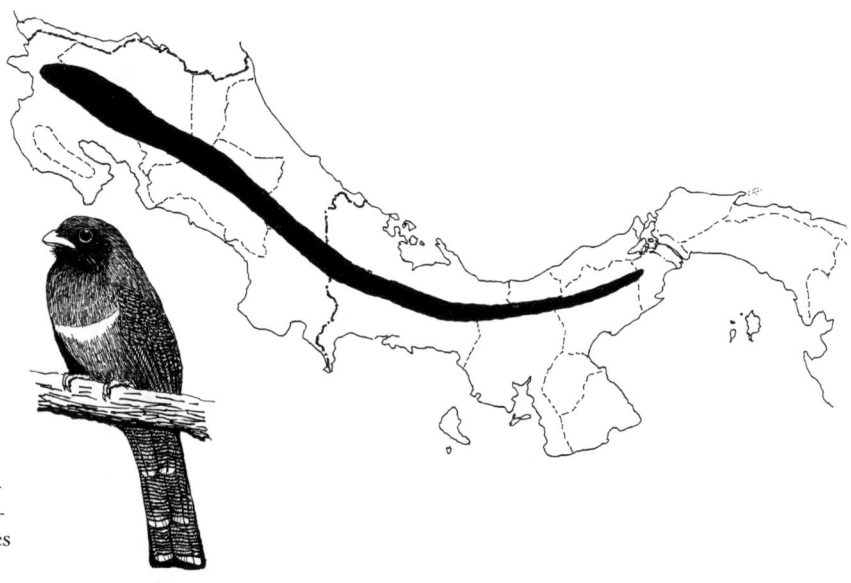

Map 19. Distribution of
the orange-bellied trogon.
Current status in the
Nicoya Peninsula is uncertain; deforestation has perhaps eliminated the species
from that region.

Plate 1. Resplendent quetzal, male and female. Acrylic painting by John O'Neill.

Plate 2. Narina trogon, male and female. Hand-colored lithograph by J. Gould, H. C. Richter, and W. Hart (Gould 1858–1875). Courtesy Kenneth Spencer Research Library.

Plate 3. Bare-cheeked trogon, male and female.
Watercolor by James D. McClelland.

Plate 4. Bar-tailed trogon, male, female, and nestling.
Watercolor by Daniel Lane.

Plate 5. Resplendent quetzal, male and female. Hand-colored lithograph by J. Gould, H. C. Richter, and W. Hart (Gould 1858–1875). Courtesy Kenneth Spencer Research Library.

Plate 6. Crested quetzal, male and female. Hand-colored lithograph by J. Gould, H. C. Richter, and W. Hart (Gould 1858–1875). Courtesy Kenneth Spencer Research Library.

Plate 7. White-tipped quetzal, male and female. Hand-colored lithograph by J. Gould, H. C. Richter, and W. Hart (Gould 1858–1875). Courtesy Kenneth Spencer Research Library.

Plate 8. Golden-headed quetzal, male and female. Hand-colored lithograph by J. Gould, H. C. Richter, and W. Hart (Gould 1858–1875). Courtesy Kenneth Spencer Research Library.

Plate 9. Pavonine quetzal, male and female. Hand-colored lithograph by J. Gould, H. C. Richter, and W. Hart (Gould 1858–1875). Courtesy Kenneth Spencer Research Library.

Plate 10. Eared trogon, male and female. Hand-colored lithograph by J. Gould, H. C. Richter, and W. Hart (Gould 1858–1875). Courtesy Kenneth Spencer Research Library.

Plate 11. Cuban trogon, male and female. Hand-colored lithograph by J. Gould, H. C. Richter, and W. Hart (Gould 1858–1875). Courtesy Kenneth Spencer Research Library.

Plate 12. Hispaniolan trogon, male and female. Hand-colored lithograph by J. Gould, H. C. Richter, and W. Hart (Gould 1858–1875). Courtesy Kenneth Spencer Research Library.

Plate 13. Slaty-tailed trogon, male and female. Hand-colored lithograph by J. Gould, H. C. Richter, and W. Hart (Gould 1858–1875). Courtesy Kenneth Spencer Research Library.

Plate 14. Black-tailed trogon, male and female. Hand-colored lithograph by J. Gould, H. C. Richter, and W. Hart (Gould 1858–1875). Courtesy Kenneth Spencer Research Library.

Plate 15. Lattice-tailed trogon, male and female. Hand-colored lithograph by J. Gould, H. C. Richter, and W. Hart (Gould 1858–1875). Courtesy Kenneth Spencer Research Library.

Plate 16. White-eyed trogon, male and female. Watercolor by Daniel Lane.

Plate 17. Baird's trogon, male and female. Hand-colored lithograph by J. Gould, H. C. Richter, and W. Hart (Gould 1858–1875). Courtesy Kenneth Spencer Research Library.

Plate 18. White-tailed trogon, male and female. Hand-colored lithograph by J. Gould, H. C. Richter, and W. Hart (Gould 1858–1875). Courtesy Kenneth Spencer Research Library.

Plate 19. Citreoline trogon, male and female. Hand-colored lithograph by J. Gould, H. C. Richter, and W. Hart (Gould 1858–1875). Courtesy Kenneth Spencer Research Library.

Plate 20. Black-headed trogon, male and female. Hand-colored lithograph by J. Gould, H. C. Richter, and W. Hart (Gould 1858–1875). Courtesy Kenneth Spencer Research Library.

Plate 21. Mountain trogon, male and female. Hand-colored lithograph by J. Gould, H. C. Richter, and W. Hart (Gould 1858–1875). Courtesy Kenneth Spencer Research Library.

Plate 22. Elegant trogon, male and female. Hand-colored lithograph by J. Gould, H. C. Richter, and W. Hart (Gould 1858–1875). Courtesy Kenneth Spencer Research Library.

Plate 23. Collared trogon, male and female. Hand-colored lithograph by J. Gould, H. C. Richter, and W. Hart (Gould 1858–1875). Courtesy Kenneth Spencer Research Library.

Plate 24. Orange-bellied trogon, male and female. Hand-colored lithograph by J. Gould, H. C. Richter, and W. Hart (Gould 1858–1875). Courtesy Kenneth Spencer Research Library.

Plate 25. Masked trogon, males and female. Hand-colored lithograph by J. Gould, H. C. Richter, and W. Hart (Gould 1858–1875). Courtesy Kenneth Spencer Research Library.

Plate 26. Black-throated trogon, male and female. Hand-colored lithograph by J. Gould, H. C. Richter, and W. Hart (Gould 1858–1875). Courtesy Kenneth Spencer Research Library.

Plate 27. Surucua trogon, male and female. Hand-colored lithograph by J. Gould, H. C. Richter, and W. Hart (Gould 1858–1875). Courtesy Kenneth Spencer Research Library.

Plate 28. Blue-crowned trogon, males and female. Hand-colored lithograph by J. Gould, H. C. Richter, and W. Hart (Gould 1858–1875). Courtesy Kenneth Spencer Research Library.

Plate 29. Violaceous trogon, male and female. Hand-colored lithograph by J. Gould, H. C. Richter, and W. Hart (Gould 1858–1875). Courtesy Kenneth Spencer Research Library.

Plate 30. Blue-tailed trogon, male and immature. Hand-colored lithograph by J. Gould, H. C. Richter, and W. Hart (Gould 1858–1875). Courtesy Kenneth Spencer Research Library.

Plate 31. Malabar trogon, male and female. Hand-colored lithograph by J. Gould, H. C. Richter, and W. Hart (Gould 1858–1875). Courtesy Kenneth Spencer Research Library.

Plate 32. Red-naped trogon, male and female. Hand-colored lithograph by J. Gould, H. C. Richter, and W. Hart (Gould 1858–1875). Courtesy Kenneth Spencer Research Library.

Plate 33. Diard's trogon, male. Watercolor by Dana Gardner.

Plate 34. Philippine trogon, male and female. Hand-colored lithograph by J. Gould, H. C. Richter, and W. Hart (Gould 1858–1875). Courtesy Kenneth Spencer Research Library.

Plate 35. Whitehead's trogon, male and female. Hand-colored lithograph by J. Kuelemann (Sharpe 1888).

Plate 36. Cinnamon-rumped trogon, male and female. Hand-colored lithograph by J. Gould, H. C. Richter, and W. Hart (Gould 1858–1875). Courtesy Kenneth Spencer Research Library.

Plate 37. Scarlet-rumped trogon, male and female. Hand-colored lithograph by J. Gould, H. C. Richter, and W. Hart (Gould 1858–1875). Courtesy Kenneth Spencer Research Library.

Plate 38. Orange-breasted trogon, male and female. Hand-colored lithograph by J. Gould, H. C. Richter, and W. Hart (Gould 1858–1875). Courtesy Kenneth Spencer Research Library.

Plate 39. Red-headed trogon, male and female. Hand-colored lithograph by J. Gould, H. C. Richter, and W. Hart (Gould 1858–1875). Courtesy Kenneth Spencer Research Library.

Plate 40. Ward's trogon, male and female. Painting by David Reiser.

DESCRIPTION (of *aurantiventris*,
adapted from Ridgway 1911; see plate 24)

Adult Male   Upperparts bright iridescent green to golden green, including middle rectrices, which are tipped with black; second and third pairs of rectrices with inner vanes wholly black; outer three pairs of rectrices black, narrowly tipped (2–5 mm, 0.1–0.2 in) with white, rest of vanes barred with black and white (white bars several times narrower than black), on inner vane of third rectrix (from outside) bars confined to subterminal portion, except along edge; on second rectrix bars limited to about terminal half and edge; anterior lesser wing-coverts bright iridescent green, like upperparts, other coverts narrowly barred with white; greater wing-coverts and secondaries finely vermiculated with white; primaries black, edged anteriorly with white; face, chin, and throat black; chest bright iridescent green, margined posteriorly with white breast band; rest of underparts pure orange, becoming paler on under tail-coverts; thighs black; bill yellow; iris brown; eye-ring dull black and inconspicuous; feet and toes dull light olive-green.

Adult Female   Plain brown above, darker on crown, paler on lower rump and upper tail-coverts; middle pair of rectrices chestnut, tipped with black; inner vanes of second and third pairs of rectrices (from middle) dark brown; anterior lesser wing-coverts brown, like crown; remaining coverts and secondaries finely vermiculated or freckled with blackish; primaries dull slate, mostly edged with white; loreal and postorbital region dusky; eye-ring of pale whitish feathers, broader behind eye and interrupted on eyelids; face, chin, and throat dusky brown; chest tawny brown, like upperparts; underparts as in adult male but averaging paler; bill mostly yellow, with some blackish on maxilla; iris brown; eye-ring black and inconspicuous, as in male; feet and toes as in adult male.

Immature   Seemingly indistinguishable from those of *collaris*.

Juvenile   Seemingly indistinguishable from those of *collaris*.

IDENTIFICATION

In the Hand   Both sexes of the orange-bellied trogon are virtually identical to the collared trogon but have orange rather than red underparts. This single difference visually separates them. Both species evidently occur in some areas of Panama and Costa Rica, so the orange-bellied cannot be identified by range differences alone. It is especially similar to *collaris puella*; but females of *aurantiventris* are much paler orange below, and males usually differ noticeably in their underpart coloration.

In the Field   As noted above, the orange to (locally) orange-red underpart color is the only feature separating the orange-bellied trogon from the extremely similar (but red-bellied) collared trogon. The former may occur at somewhat lower elevations (400 m to at least 1,900 m [1,300–6,200] versus 700–2,500 m [2,300–8,200] for the collared). A comparison of the two species' vocalizations is needed for identification, and possible acoustic differences are noted in the vocalizations section; detailed studies might establish whether the forms warrant species-level distinction.

GEOGRAPHIC VARIATION

The *underwoodi* race of the orange-bellied trogon differs from the nominate form in having orange-red rather than orange posterior underparts. (In *underwoodi* the underparts range individually from salmon to nearly scarlet but are usually not as reddish as those of *collaris puella*.) This intermediate color type casts additional doubt on the validity of the orange-bellied trogon as a distinct species.

ECOLOGY

Habitats   Collectively orange-bellied trogons occur in montane forest and secondary forest from 400 to 1,900 m elevation (1,300–6,100 ft). The habitats of the orange-bellied and collared trogons are very similar but not identical. In Panama the former occur from foothills (400 m; 1,300 ft) to at least 1,800 m (5,900 ft) but is common between 600 and 1,800 m (2,000 and 5,900 ft) in open forest and stands of large secondary forest (Wetmore 1968) and in humid forest and forest borders (Ridgely and Gwynne 1989). In Costa Rica the birds range from 750–1850 m (2,500–6,100 ft) but are rarer at high altitudes. Although the Pacific coastal race *underwoodi* is evidently not sympatric with the collared trogon, the nominate form is "everywhere sympatric" with it (Slud 1964). Although the mountains of the Nicoya Peninsula once supported a population of the orange-bellied trogon, massive deforestation during the 1950s and 1960s may have eliminated it from the region (F. G. Stiles, pers. comm.).

Foods and Foraging Ecology   The foods and foraging behaviors of the orange-bellied trogon are apparently virtually the same as those of the collared trogon. Foods taken include fruits of at least ten plant families, such as *Hasseltia* (Flacourtiaceae), figs (Moraceae), and one or more species of such groups as Lauraceae, Solanaceae, Ericaceae, Myrtaceae, Rubiaceae, and Symplocaceae (Wheelwright et al. 1984).

BEHAVIOR

General, Social, and Sexual Behavior   No specific information available.

Vocalizations   According to Stiles and Skutch (1989) and Slud (1964), the orange-bellied trogon's vocalizations are identical to those of the collared trogon. Vocalizations of the former were recorded in Costa Rica by Gary Stiles (Hardy, Reynard, and Coffey 1995). The examples on the recording consist of songs and trill calls: the song example provided is almost exactly like that of the collared trogon but consists of single or double *kyow*-note sequences rather than three- to five-note sequences. The descending *churr* call is also seemingly identical.

BREEDING BIOLOGY

Chronology of Breeding   Little is known about the breeding biology of the orange-bellied trogon.

Wetmore (1968) collected females in breeding condition or with abraded tail feathers during March in Panama. Apparently only a single nest has been reported, which was found early in July (Blake 1956).

Nest Sites   The orange-bellied trogon nest described by Blake (1956) was located in the rotted cavity of a large tree, the opening 2 m (7 ft) above ground, and contained a single egg.

Eggs and Incubation   No information available.

Brood Rearing   No information available.

CONSERVATION AND EVOLUTIONARY RELATIONSHIPS

Only tradition (a notoriously poor guide) prevents taxonomists from merging *aurantiventris* with *collaris*. The questionable species status of the orange-bellied trogon actually goes back at least to comments by Ogilvie-Grant (1892), who considered it "doubtful" that species recognition for *aurantiventris* was warranted. Wetmore (1968) thought the question to be still unsettled and urged that further detailed studies be undertaken, especially in view of an early report that no mixed pairing was observed in an area (Boquete) where both red-bellied and yellow-bellied phenotypes were present. Assortative mating between birds of the same color morph is well known among species such as the white and "blue" forms of the snow goose (*Chen caerulescens*) and does not of itself prove species-level separation. Stiles and Skutch (1989) "strongly suspect" the two types to be color morphs of a single species. Stotz et al. (1996) listed the species as fairly common, with a medium degree of sensitivity to disturbance.

MASKED TROGON

Trogon personatus Gould 1842

OTHER VERNACULAR NAMES

Andean masked trogon (*assimilis*), black-faced trogon, highland trogon (*temperatus*); Santa Marta masked trogon (*sanctaemartae*); sorocua enmascarado (Spanish).

RANGE

Evergreen subtropical forests of Venezuela (Zulia to Amazonas and Bolívar), southwestern Guyana, and Brazil (extreme north) and from Colombia (Santa Marta Mountains and western, central, and eastern Andes) south through Ecuador (both slopes) and Peru (eastern Andes, local in western Andes in Piura) to central Bolivia (La Paz, Cochabamba, western Santa Cruz). See map 20.

SUBSPECIES

Trogon personatus sanctaemartae Zimmer 1948: northern Colombia (Santa Marta Mountains).

Trogon personatus personatus Gould 1842: western Venezuela, eastern Colombia, eastern Ecuador, and eastern Peru, at elevations of 1,250–2,800 m (4,100–9,200 ft).

Trogon personatus duidae Chapman 1923: Venezuela (Mount Duida).

Trogon personatus roraimae (Chapman) 1929: southeastern Venezuela (Mounts Roraima and Auyán-epuí), Guyana.

Trogon personatus ptaritepui Zimmer and Phelps 1946: southeastern Venezuela (Pantepui).

Trogon personatus assimilis Gould 1846: Colombia (western Andes), western Ecuador, and northwestern Peru.

Trogon personatus temperatus (Chapman) 1923: central Colombia, Ecuador, and Peru, at elevations of 3,000–3,850 m (9,800–12,600 ft). Sometimes considered a separate species (the so-called "highland trogon"), as sympatry with nominate *personatus* has been suggested (e.g., by Todd [1943] and de Schauensee [1949, 1964]).

Trogon personatus heliothrix Tshudi 1844: Peru, at elevations of 1,800–2,700 m (5,900–8,900 ft).

Trogon personatus submontanus Todd 1943: Bolivia, Peru, at elevations of 675–2,700 m (2,200–8,900 ft).

MORPHOMETRICS

Measurements Wing, males (of *ptaritepui*) 118–124 mm (4.6–4.8 in; average of 5, 121 mm [4.7 in]); females 114–124 mm (4.4–4.8 in; average of 4, 118.5 mm [4.6 in]). Tail, males (of *ptaritepui*) 128–134 mm (5.0–5.2 in; average of 5, 132 mm [5.1 in]); females 120–126 mm (4.7–4.9 in; average of 4, 122 mm [4.8 in]; Zimmer and Phelps 1946). Egg of *submontanus* 28.1 × 21.9 mm (1.1 × 0.9 in; Schönwetter 1966).

Weights An unsexed bird from Peru 60.3 g (2.1 oz; Weske 1972). One male of *assimilis* 68.3 g (2.4 oz; Miller 1963). One male of *assimilis* 68.3 g (2.4 oz), two males of *personatus* 63 and 66 g (2.2 and 2.3 oz; specimens, MVZ). One male of *duidae* 65 g (2.3 oz), one of *submontanus* 63.5 g (2.2 oz; specimens, FMNH). Average mass (both sexes), visually estimated from graph, 60 g (2.1 oz; Remsen, Hyde, and Chapman 1993). Estimated egg weight 7.1 g (0.2 oz; Schönwetter 1966).

Masked trogon, adult female. Photograph by author.

Map 20. Distribution of the masked trogon.

DESCRIPTION (of *sanctaemartae*,
partly after Zimmer 1949; see plate 25)

Adult Male   Upperparts mostly iridescent green,
bluer on head and rump; forehead, sides of head,
chin, and throat black; breast like top of head, bordered posteriorly by a white breast band; rest of
underparts scarlet- to rose-red becoming lighter
posteriorly; thighs black; remiges blackish, outer
margins of second to sixth primaries from outside
narrowly and sharply white toward base but more
freckled distally and black not reaching tips; fifth
primary (from outside) to inner secondaries have
white patch across both vanes at extreme base; secondaries and tertials broadly vermiculated with
white; alular and primary coverts blackish, upper
wing-coverts marked like inner remiges; under wing-
coverts blackish; axillars white; middle pair of rectrices, most of outer vanes of next pair, and outer
margins of third pair iridescent green to citrine, but

with broad black tip; inner vanes of second and
third pairs (from middle) of rectrices also blackish;
outer three pairs of rectrices with broad white tips
(width 10–15 mm [0.4–0.6 in]), remainder of these
outer rectrices blackish, crossed by narrow whitish
bars, except for large unbarred area on inner vanes
and to a more limited extent on outer vanes, the
black becoming progressively more extensive from
outermost to third pair; dark area immediately anterior to white tip broader than the other dark bars;
bill yellow; culmen blackish and tinged with reddish
on mandible; iris brown; eye-ring red; feet and toes
brownish yellow.

Adult Female   Lores, cheeks, and throat black; rest
of head, chest, back, rump, and upperparts olive-
brown; white breast band separates breast from red
underparts; thighs gray; middle rectrices rufous-
brown with black terminal band; second and third
pairs of rectrices same, but with black inner vanes;
outer three rectrices mostly black, with white freckling or vermiculations on inner vanes, fine white
barring on outer vanes and terminal halves of inner
vanes, and immaculate white tips; wings as in adult
male but vermiculated with buff; thighs blackish
gray; soft-part colors similar to those of male; bill
yellow.

Immature Male   Variably female-like, but with
some bronzy sheen present on basal half of median
rectrices, distal portions rufescent, sometimes
tipped with blackish bar; outer three rectrices with
sometimes coarse black barring, their margins
mixed with rufous buff and marked with black;
white feathers of breast band finely barred with
black; red underparts variably mixed with rufous
(the latter color probably retained from juvenal
plumage).

Immature Female   Outer rectrices as in immature
male, but median ones entirely rufous, without
black tips; under tail-coverts rufous buff; other parts
generally female-like.

Juvenile   Almost entirely buffy brown, but with
black mask and species-typical colors on wings and
tail (Fjeldså and Krabbe 1990); presumably with
white or buffy markings on wing-coverts.

IDENTIFICATION

In the Hand   The "mask" of the masked trogon is definitive only in females; the black face of males is shared with several other species and is somewhat obscured by the surrounding green of the head and lower neck. Females are similar to collared trogon females but have a black forehead, face, and throat and a white crescent at the rear of the eye; and the wing-coverts are more finely vermiculated than in the collared trogon. Also like the collared trogon, both sexes have drooping white breast bands, or "necklaces," between the green breast and the red abdomen and also have regularly barred black-and-white undertail patterns. However, the masked trogon shows fine barring of this type in both sexes; whereas in the collared trogon the barring is coarser in males, and in females the barring is replaced by indefinite freckling. Female collared trogons also have less definite facial "masks" than do masked trogons.

In the Field   The masked trogon is most likely to be confused with the collared and black-throated trogons; all have red underparts and a white breast band. The masked trogon has fine barring and vermiculations on the underside of the tail, and the collared has coarser barring in males and freckling in females. In the masked trogon, the underside of the tail is almost entirely slaty colored. Vocalizations are extremely similar to those of the collared and rufus trogons. See the vocalizations section.

GEOGRAPHIC VARIATION

Zimmer (1948) has made a complete review of the numerous races of the masked trogon, which seem to vary erratically in terms of bill size, color of the middle rectrices, barring and extent of white tips on the outer rectrices, and vermiculations on the upper wing-coverts and inner secondaries. All these races are associated with the subtropical zone but vary considerably in their altitudinal limits, the ranges generally becoming gradually lower from Colombia to Bolivia. The form *temperatus,* sometimes considered a separate species, has a smaller bill; females are more rufescent, and males have fine vermiculations on the undersides of their outer rectrices. In addition, the bars of the outer rectrices of *temperatus* are very narrow, indistinct, and confined to the outer web in males; in females the vermiculations are whitish rather than brownish, and the underparts may be less bright red. Tail barring on *assimilis* is also very poorly developed.

ECOLOGY

Habitats   Collectively the masked trogon occurs from 700–3,600 m (2,300–11,800 ft), in subtropical and temperate evergreen forest but occasionally extending to forest edges and secondary forest. In Colombia these birds occupy humid and wet forests, forest edges, and secondary forests, usually between 1,400 and 2,900 m (4,600–9,500 ft; extremes 700 and 3,600 m [2,300 and 11,800 ft]). The form *temperatus* occurs at highest altitudes, above 2,300 m (7,500 ft; Hilty and Brown, 1986). Fjeldså and Krabbe (1990) described the habitat as the interior of not too wet or too dense montane forest and cloud forest.

Foods and Foraging Ecology   Little specific information on foods is available, but the masked trogon apparently consumes a mixture of fruits and insects, as do related species. Haverschmidt (1968) reported the foods to consist of fruits, berries, and insects, including orthopterans (Locustidae). Of 22 stomachs examined by Remsen, Hyde, and Chapman (1993), 72.7 percent contained arthropods only, 13.6 percent contained mixed fruit and arthropods, and 13.6 percent contained fruit only.

BEHAVIOR

General, Social, and Sexual Behavior   Little specific information is available on behavior. De Schauensee and Phelps (1978) stated that masked trogons tend to stay well up in trees of rain forests, cloud forests, and open forests but are not found at tree-top level.

Vocalizations   The masked trogon's typical call is a soft and steady series of four to eight *kwaa* notes, which are of somewhat higher pitch than the collared trogon's. There is also a descending trill that may be repeated many times (Hilty and Brown 1986). Fjeldså and Krabbe (1990) described the male's call as a *zooorrh hr hr* and also described a series of many whistled *whyh* notes. The female was said to respond with a descending *trrrrr*. Vocalizations of this species were recorded in Peru by Ted Parker (Hardy, Reynard, and Coffey 1995). The recorded song sequences, consisting of three *keeow* notes of constant amplitude and uniform spacing, are virtually identical to those of the black-throated trogon and also essentially like the collared trogon's. See also the identification section.

BREEDING BIOLOGY

Chronology of Breeding   In Colombia breeding-condition masked trogons have been collected between April and August, and a nest was found in early March (Hilty and Brown 1986).

Nest Sites   The single masked trogon nest from Colombia was found in a cavity excavated in a partly rotted tree at about 3 m (10 ft) above ground (Fjeldså and Krabbe 1990; Hilty and Brown 1986).

Eggs and Incubation   The only masked trogon nest described so far contained two white eggs.

Brood Rearing   No information available.

CONSERVATION AND EVOLUTIONARY RELATIONSHIPS

The masked trogon occurs in the upland and montane forests of seven South American countries from northern Colombia to central Bolivia. It is allopatric or parapatric with the black-throated trogon, obviously its nearest lower-altitude relative. It is generally fairly common and does not appear to be in any immediate danger. Stotz et al. (1996) listed the species as fairly common, with a medium degree of sensitivity to disturbance.

BLACK-THROATED TROGON

Trogon rufus Gmelin 1788

Trogon rufus was described as *T. atricollis* by Gould and as *T. curucui* by Ridgway (1911) and others. Zimmer (1930, 295–296) provided arguments supporting the use of *rufus* as the specific epithet.

OTHER VERNACULAR NAMES

Colombian black-throated trogon (*cupreicauda*), graceful trogon (*tenellus*), Pacific black-throated trogon (*cupreicauda*), sulfury trogon (*sulphureus*), yellow-bellied trogon; sorocua amarillo, trogon cabeciverde, trogon graciosa (Spanish).

RANGE

Evergreen forests, second-growth, and plantations from eastern Honduras through Nicaragua, Costa Rica (Caribbean slope, Guanacaste, southern Pacific slope), Panama (mainly on Caribbean slope), Colombia (northwestern, and also Pacific coast to lower Cauca and middle Magdalena Valleys, eastern Andes), Ecuador (east and west), northeastern Peru, Venezuela (widespread), Guyana (widespread), Suriname (widespread in interior), French Guiana; Brazil (widespread; Amazonas, Mato Grosso, southwestern Bahia to Rio Grande do Sul), Bolivia, Paraguay, and northern Argentina (Misiones). See map 21.

SUBSPECIES

Trogon rufus tenellus Cabanis 1862: east-central Honduras (east of Sula River) to northwestern Colombia.

Trogon rufus cupreicauda (Chapman) 1914: western and central Colombia to western Ecuador. Includes *virginalis*.

Trogon rufus rufus Gmelin 1788: eastern Venezuela, Guyana, Suriname, French Guiana, and northern Brazil.

Trogon rufus sulphureus Spix 1824: eastern Colombia, eastern Ecuador, northeastern Peru, southern Venezuela, western Brazil.

Trogon rufus amazonicus Todd 1943: northeastern Brazil, southern Venezuela.

Trogon rufus chrysochloros Pelzeln 1856: southern Brazil, Paraguay, and northern Argentina.

MORPHOMETRICS

Measurements   Wing, males (of *tenellus*) 105–117 mm (4.1–4.6 in; average of 21, 110.4 mm [4.3 in]); females 105.5–115.5 mm (4.1–4.5 in; average of 19, 111.6 mm [4.4 in]). Tail, males (of *tenellus*) 124–144 mm (4.8–5.6 in; average of 21, 134.1 mm [5.2 in]); females 129.5–152 mm (5.1–5.9 in; average of 19, 138.4 mm [5.4 in]; Ridgway 1911). Egg of *chrysochloros* 29.8 × 22.3 mm (1.2 × 0.9 in; Schönwetter 1966). Egg of *tenellus* 28.1 × 22.8 mm (1.1 × 0.9 in; Wetmore 1968).

Weights   Thirteen birds of both sexes and various locations 40.6–58.0 g (1.4–2.0 oz; average 52.8 g [1.8 oz]; s.d. 4.78 g [0.2 oz]; Dunning 1993). Males from Suriname (unstated sample size) 49–54 g (1.7–1.9 oz), one female 56 g (2.0 oz; Haverschmidt 1968). Unspecified Costa Rica weight 57 g (2.0 oz; Stiles and Skutch 1989). Male from Rio Grande do Sul, Brazil, 57 g (2.0 oz); two females 54 and 58 g (1.9 and 2.0 oz; Belton 1984). One male of *sulphureus* 54.3 g (1.9 oz), four females of various subspecies 48–66.2 g (1.7–2.3 oz; average 55.7 g [1.9 oz]; specimens, MVZ). One male of *amazonicus* 48 g (1.7 oz), one of *sulphureus* 51 g (1.8 oz; specimens, FMNH). Average mass (both sexes), visually estimated from graph, 53 g (1.9 oz; Remsen, Hyde, and Chapman 1993). Estimated egg weight 7.6 g (0.3 oz; Schönwetter 1966).

DESCRIPTION (of *tenellus*, adapted from Ridgway 1911; see plate 26)

Adult Male   Crown, hindneck, and most upperparts bright iridescent green to golden green or bluish green; face, chin, and throat black; middle pair of rectrices iridescent bluish green, tipped with black; next two pairs of rectrices with similar outer vanes, but inner vanes blackish; three outer pairs of rectrices broadly tipped white, subterminal portion of inner vanes and most of outer vanes broadly

Map 21. Distribution of the black-throated trogon.

barred with white; most upper wing-coverts and outer vanes of secondaries finely vermiculated black and white; primaries black, edged with white; posterior underparts iridescent green; chest pure orange-yellow, anterior margin paler or intermixed with white, forming an indistinct breast band adjacent to breast; thighs sooty; bill light apple-green or chromium-green, sometimes becoming yellow on culmen and tip of maxilla; iris dark brown; eye-ring light blue or glaucous; feet and toes lead-gray.

Adult Female   Crown and most upperparts plain brown, slightly darker on crown and lighter on rump and upper tail-coverts; middle pair of rectrices cinnamon-rufous to chestnut, narrowly tipped with black and with a more buffy subterminal band; next two pairs of rectrices similar but with inner vanes blackish brown; three outer pairs of rectrices exten-

sively white terminally and on most of outer vanes, black basally, white portion broadly barred with black except for an immaculate white tip; median and greater wing-coverts paler than back, finely vermiculated with dusky; ring of white feathers surrounding eye broader behind eye, where margined by a black bar; most of facial region plain brown; rest of underparts as in adult male, but yellow usually less intense and sides of breast mixed with brownish gray; bill blackish, but lower basal portion of maxilla and mandible more yellow, entire maxilla sometimes yellowish; other soft-parts probably as in adult male.

Immature Male   Intermediate between adult male and female; upperparts with varying amounts of golden green, but head, throat, chest, and wing-coverts more female-like; tail as in adult female, but six middle rectrices green basally (middle pair lacking black tip), with distal edging of cinnamon or dull rufous on outer vanes; outer rectrices with narrower black bars than in adult; upper tail-coverts rufous distally. According to Skutch (1983), males may remain in an immature plumage stage for about a year and presumably do not breed until their second year. At this time the male's breast is still mostly brown, and the wing and tail are female-like. During that plumage stage, there is a visible white bar between the brown breast and yellow abdomen, a bar similar to that evident in the mountain trogon; but this bar is inconspicuous in adult males of this species.

Immature Female   Much like adult female, but outer rectrices narrower, more pointed, and more coarsely barred.

Juvenile   Upperparts warm cinnamon to russet; middle rectrices chestnut, tipped with black; wing-coverts and secondaries brown, vermiculated with dusky, most of these feathers with roundish buffy spots and dusky margins; rest of wing and outer rectrices as in adult female; most of face, chin, throat, and breast cinnamon-brown to russet; narrow white eye-rung and white crescents below and in front of each eye; rest of underparts light tawny, deeper anteriorly and paler posteriorly; gape pinkish; oral flanges well developed and white; bill dark gray, with lighter gray base and tip.

IDENTIFICATION

In the Hand   In essentially every respect except underpart and eye-ring color, the black-throated species resembles the masked and collared trogons. In the black-throated trogon males have contrasting pale blue eye-rings (versus inconspicuous reddish ones in the others) and bright yellow underparts similar to those of the violaceous trogon. The fine barring on the undertail of the black-throated trogon in both sexes closely resembles that of the masked trogon, but females lack the distinct "mask" of that species; and unlike the violaceous trogon, black-throated trogons have brown rather than gray head and neck coloration.

In the Field   The black-throated trogon is distinctive in that it is the only yellow-bellied trogon in its range with a green (males) or brown (females) head, a yellow bill (less apparent in females), and a bluish eye-ring. In both sexes the undertail pattern is finely barred with black and white; the violaceous trogon has a similar undertail pattern but also has a pale grayish bill and a yellow eye-ring. Black-throated trogons occupy wet to humid lowland forests (up to 1,000 m [3,300 ft] in Costa Rica, 900 m [3,000 ft] in Panama), and their usual territorial vocalization consists of from two to five plaintive descending *kyow* whistles that remain constant or gradually descend in pitch, the vocalization resembling that of the collared trogon. See also the vocalizations section.

GEOGRAPHIC VARIATION

Males of the northern race of the black-throated trogon (*tenellus*), at least in Panama, differ from those of western Colombia (*cupreicauda*) in having bluish or bluish-green rather than greenish or coppery middle rectrices. However, examples of *tenellus* from Nicaragua and Costa Rica vary from green to blue in this trait, perhaps thus making the blue-versus-green distinction of little taxonomic significance. In nominate *rufus* these feathers range from bronzy to bluish, while in *sulphureus, amazonicus,* and *cupreicauda* they tend to be distinctly coppery. In *chrysochloros* the tail color resembles that of *rufus;* but *chrysochloros* are larger, and the vermiculations of the wing area are finer. Coarse vermiculations are typical of *amazonicus* and *cupreicauda*. Tail lengths

also vary geographically to some degree, being shorter in *cupreicauda* and longer in *sulphureus* and *chrysochloros*. Generally, Central American populations have middle rectrices that are green, glossed with peacock-blue; whereas in South American forms these feathers are glossed with bronze, and the barring on the outer rectrices is narrower. In southernmost forms the underparts are orange-yellow. The white bars on the outer rectrices of *tenellus* are somewhat wider than those of typical *rufus*.

ECOLOGY

Habitats   Collectively the black-throated trogon occurs in tropical lowland evergreen forest (e.g., rain forest), secondary forest, and open woodlands from sea level to 1,400 m (4,600 ft) elevation. In Costa Rica this is one of the commonest and most widespread species, at least at lower altitudes. It occurs from near sea level to about 1,000 m (3,300 ft) but has been seen fairly commonly as high as 1,400 m (4,600 ft). It preferentially occupies primary forests of the subtropical zone but sometimes extends into stands of advanced secondary forest (Carriker 1910; Slud 1964). In Panama it occurs in primary forests and shaded secondary forests up to about 900 m (3,000 ft; Ridgely and Gwynne 1989) and in Colombia extends to 1,100 m (3,600 ft), in humid forest and secondary forest (Hilty and Brown 1986). In Venezuela it occupies similar humid to wet forest habitats, as well as forest edges and open woodlands, between 100 and 900 m (300–3,000 ft; de Schauensee and Phelps 1978).

Foods and Foraging Ecology   Black-throated trogons tend to feed in middle to lower canopy heights of the forest, sometimes joining other birds while following swarms of army ants. They also are attracted to fruiting figs and to berries of the mangabe (*Didimoponax*). A bird shot in Panama had in its stomach the remains of caterpillars, locustids, and a larval neuropteran, as well as a few seeds (Wetmore 1968). The species has been observed following army ants to feed on insects that are stirred up by the ants (Willis, Wechsler, and Kistler 1982). Of twenty-seven stomachs examined by Remsen, Hyde, and Chapman (1993), sixteen contained arthropods only, four contained mixed fruit and arthropods, six contained fruit only, and one contained unidentified materials.

BEHAVIOR

General, Social, and Sexual Behavior   The black-throated trogon is a rather solitary species, often seen perched alone on branches that are situated rather low in the forest. Its disturbance call, a cackling *churrr*, is accompanied by a slow lifting of the tail (to the level of the head or higher) followed by a rapid return to the resting position (Slud 1964). Karr (1971) described the species as having a dominance-overlap territorial system and as being associated with higher canopy heights of moist forest.

Vocalizations   Slud (1964) stated that the usual call of the black-throated trogon is a series of two to five deliberate, equally spaced, whistled notes that sound like *kyow, aow, owp, pow,* or *poo,* uttered in slightly descending pitch. The call is similar to that of some antbirds but is slower than corresponding calls of the collared and orange-bellied trogons, being uttered at the rate of about one call per second or even slower. The black-throated also utters a cackling call of decreasing intensity and a similar excited and repeated *chik* note when flying from branch to branch. Its disturbance cackle was mentioned in the general, social, and sexual behavior section (Slud 1964). Vocalizations of this species were recorded in Costa Rica by Gary Stiles (Hardy, Reynard, and Coffey 1995). The individual songs on the recording have two- or three-noted, high-pitched, and perhaps slightly down-slurred *keeow* components, each of constant amplitude and uttered at the uniform rate of slightly more than one per second. See also the identification section above for possible call variations.

BREEDING BIOLOGY

Chronology of Breeding   Black-throated trogons breed in Costa Rica from February until June, with a probable peak in April (Skutch 1983; Stiles and Skutch 1989). In Panama active nests have been found from February to June (Wetmore 1968; Willis and Eisenmann 1979), and in Colombia breeding-condition birds have been obtained from February to May (Hilty and Brown 1986). A nest with eggs in Brazil was recorded in May (Oriki and Willis 1983).

Nest Sites   Skutch (1983) located 14 black-throated trogon nests in Panama and Costa Rica. Most of these were in old forests or forest edges, but one was in a plantation 30 m (98 ft) from forest. The nests ranged in height from 0.8 m to 6 m (3–20 ft) but half were located between 1.2 and 2.4 m (4 and 8 ft) above ground. The average height of 13 was 2.4 m (8 ft). The birds seem to avoid massive trunks and often choose sites well below the top of a low stump. The cavity is a shallow niche with a roughly pear-shaped entrance, which is higher than it is wide (11.4–16.5 × 6–7 cm; 4.5–6.5 × 2.3–2.8 in). The excavation below the entrance averaged about 5 cm (2.0 in) in Skutch's observations, and the size of the nesting cavity averaged about 9 cm (3.5 in) wide and 16.5–20 cm (6.5–7.9 in) high, measured to the top of the entrance. This vertical height provides space for the adult's tail, which is tilted up and forward above the sitting bird's back.

A nest in Brazil was found in the cavity of a stump 15 cm (5.9 in) in diameter and 4 m (13 ft) high, the opening of the nest at 3.5 m (11 ft) above ground (Oriki and Willis 1983). Carriker (1910) found a nest in the hollow trunk of a small palm that had broken off about 1.5 m (5 ft) above ground. He also observed a nest excavated in an old termitarium.

Eggs and Incubation   Nearly all nests of the black-throated trogon have been found to contain two eggs; Skutch (1983) knew of no larger clutches. The egg-laying interval is normally 2 days. Incubation is shared by both sexes; the male may begin his shift as early as 7:00 A.M. or as late as about noon. The female may return to relieve him any time between 3:00 P.M. and 5:30 P.M., according to Skutch's observations. In one nest under his observation the incubation period was at least 18 days.

Brood Rearing   Skutch (1983) found the black-throated trogon young to be hatched naked, with tightly closed eyes. Very young chicks were brooded almost constantly, by both parents. Feeding was also done by both parents, the food items including insects (beetles, probable grasshoppers); initially the insects were small but later became quite large. No fruit or berries were seemingly fed. At 6 days the pinfeathers began to sprout, and the sheaths of these feathers opened at 11 days, producing the juvenal plumage (the rectrices and some crown feathers did not break free of their sheaths until 2 days later). Fledging occurred at 14 or 15 days. Among twelve nests studied by Skutch, only three young survived long enough to fledge. In seven of the twelve, the eggs disappeared before hatching.

CONSERVATION AND EVOLUTIONARY RELATIONSHIPS

The black-throated trogon is an extremely wide-ranging species that occurs in four Central American and 11 South American countries from Honduras to extreme northeastern Argentina. The center of its range is Brazil, where it is widely distributed and fairly common. Stotz et al. (1996) listed the species as uncommon and patchily distributed, with a medium degree of sensitivity to disturbance.

SURUCUA TROGON

Trogon surrucura Vieillot 1817

OTHER VERNACULAR NAMES

Brazilian trogon, orange-breasted trogon (*aurantius*); sorucua comun (Spanish).

RANGE

Montane forests from southern Brazil (Bahia and Mato Grosso south) to Paraguay and northern Argentina (Misiones, Corrientes, and eastern Formosa and Chaco). See map 22.

SUBSPECIES

Trogon surrucura aurantius Spix 1824: eastern Brazil, from Bahia to Rio de Janeiro and eastern Minas Gerais.

Trogon surrucura surrucura Vieillot 1817: southern Brazil (Rio de Janeiro and Minas Gerais to Rio Grande do Sul and southern Mato Grosso), Paraguay, and northern Argentina.

MORPHOMETRICS

Measurements   Wing, two males 127 and 135 mm (5.0 and 5.3 in); tail, 140 and 145 mm (5.5 and 5.7 in; specimens, FMNH). Wing, one male 144.7 mm (5.7 in); one female 144.7 mm (5.7 in). Tail, one male 147.3 mm (5.8 in); one female 147.3 mm (5.8 in; Ogilvie-Grant 1892). Egg of *surrucura* 29.8 × 23.1 mm (1.2 × 0.9 in; Schönwetter 1966).

Weights   Two males from Rio Grande do Sul, Brazil, 72 and 78 g (2.5 and 2.7 oz); one female 70 g (2.5 oz; Belton 1984). Estimated egg weight 8.8 g (0.3 oz).

DESCRIPTION (see plate 27)

Adult Male   Head, neck, throat, and breast blue (sometimes appearing green or violet); back, rump, and upper tail-coverts bright iridescent green, becoming bluish on rump; no white breast band separating green chest from red underparts; alular feathers and primaries blackish brown, primaries mostly with white edgings; upper wing-coverts, inner secondaries, and tertials vermiculated or freckled black and white; abdomen and under tail-coverts bright orange or crimson-red (varying among subspecies); middle pair of rectrices bluish green, tipped with black; next pair of rectrices bluish green on outer vane, black inwardly; all six inner rectrices tipped with black; outermost pair of rectrices black from base to middle of feather, white distally; next pair of rectrices black for basal two-thirds; third pair (from outside) black except for white on tip of outer vanes and at feather tip; thighs gray; bill yellowish white to olive; eye-ring yellow to orange; iris reddish brown to sooty; feet and toes lead-gray to dark brown.

Adult Female   Similar to male, but predominantly gray above; head and breast dark sooty gray, with small white spots in front of and behind eye, sometimes forming a complete white eye-ring; lower belly and under tail-coverts light crimson; upper wing-coverts brownish black, vermiculated or narrowly barred black and white; six central rectrices brownish gray tipped with black; three outer pairs of rectrices similar to those in male, but amount of white patterning on outer rectrices reduced relative to black, outer vanes margined with white and inner vanes much more narrowly so; bill gray to slate-gray; other soft-part colors similar to those of male.

Immature   Immature males (probably those nearly adult) reportedly have wide, shiny black tips on their middle rectrices, tips that are lacking in females (Sick 1993). The upper wing-coverts and secondaries are more coarsely speckled or vermiculated with white or buffy than in adults; and the outer rectrices are also narrower and more rounded or pointed in immatures (and juveniles) of both sexes. In young males the outermost pair of rectrices is also extensively patterned with black on the inner vanes, whereas the second and third rectrices are marked with black on both vanes. Some immatures have complete white eye-rings, and immatures may have variable amounts of gray on their underparts, presumably carried over from the juvenal plumage.

Juvenile   No information available.

IDENTIFICATION

In the Hand   There are two races of surucua trogons, and they have different underpart colors. Both sexes of the southern (nominate) race have a red abdomen, whereas the northern race has an orange abdomen. In both forms the three outer rectrices are mostly white and contrast with the longer middle rectrices, which form a black tail-tip when seen from below; and the blackish bases of the outermost pair of rectrices are also visible behind the under tail-coverts. The head and breast color of males is iridescent blue, and both sexes have eye-rings that vary from yellow (in the southern race) to orange (in the northern race). The male surucua trogon is similar in appearance to the male blue-crowned trogon; but the former is larger and lacks a white breast band, and its outermost pair of rectrices is black from the base to the middle of the feather. The third pair (from outside) is almost entirely black, tipped with white. Female surucuas differ from the blue-crowned species by being darker, lacking a white breast band, having wider and more regular barring on the wing-coverts and secondaries, and having uniformly white outer vanes on the three outer pairs of rectrices.

In the Field   The surucua is the only trogon in its range having red (in southeastern Brazil) abdomen color and almost entirely white undertail patterning (the three outer rectrices white except basally, where they are black). In southern Brazil the quite similar (but yellow-bellied) white-tailed trogon also occurs in contact with the red-bellied form of the surucua trogon; but male white-tailed trogons have light blue rather than yellow eye-rings, and females show more black at the bases of their outer three rectrices, which are also barred with black for much of their length. Vocalizations include an ascending sequence of 14–20 or more *duis, koaw,* or *kwas* notes, with the final note in the series lower in pitch. See also the vocalizations section.

GEOGRAPHIC VARIATION

The northern race of the surucua trogon (*aurantius*) has a distinctive orange abdomen and yellow eye-

Map 22. Distribution of the surucua trogon. The dashed line approximately separates the zones of the two subspecies.

ring, whereas the southern race (*surrucura*) has a red abdomen and an orange eye-ring.

ECOLOGY

Habitats   Collectively the surucua trogon is associated with xeric subtropical forest, woodland, and savanna, from under 500 to 2,000 m (1,600–6,600 ft). It is a forest-dwelling species that also occurs in the more xeric cerradao. (The cerrado consists of savanna-like areas of Brazil dominated by sparsely distributed arboreal cacti, low trees, and bushes having twisted branches, thick bark and leathery leaves, often supporting termite nests. The cerradao is a more forest-like variant.) Holt (1928) observed the species in second-growth forest, at an altitude of about 900 m (3,000 ft).

Foods and Foraging Ecology   Of fourteen surucua trogon stomachs examined by Remsen, Hyde, and

Chapman (1993), nine contained arthropods only, three contained mixed fruit and arthropods, and two contained fruit only. Little additional information is available, but Belton (1984) observed a pair feeding on the bark surface of a cactus, perhaps eating lichens or aphids. One bird also plucked a green blackberry while in flight but later dropped it.

BEHAVIOR

General, Social, and Sexual Behavior   Belton (1984) saw a female surucua trogon sunning with spread tail and near-side wing, fluffed feathers, and gaping mouth. He noted that the species can swivel its neck to a remarkable degree.

Vocalizations   Belton (1984) described the surucua trogon's call as a series of fairly high-pitched *quiow* notes, all identical except for a final simple *coe*. The birds also utter a rattling call sounding like the noise produced by a woodpecker pecking on resonant wood. During this call the tail moves from vertical to nearly horizontal in a rhythmic, pulsating manner. Vocalizations of this species were recorded in Brazil by William Belton (Hardy, Reynard, and Coffey 1995). The song sequences on the recording consist of 17–20 *qoaw* notes uttered at the rate of about three per second; the notes tended to remain at the same loudness or increase slightly in amplitude, except for the one or two final notes, which drop in loudness and perhaps slightly also in pitch. See also the identification section.

BREEDING BIOLOGY

Chronology of Breeding   Breeding activities or breeding-condition surucua trogons have been recorded in southern Brazil from October to January. Both eggs and nestlings have been seen in De-

cember (Belton 1984). Two Argentine nests were found in late October and mid-December, and a Paraguayan nest was found in November.

Nest Sites   Surucua nests mentioned by Belton (1984) were observed in rotted trees and in a tall cactus. Locations ranged from 5 to 12 m (16–39 ft) above ground. Both sexes have been seen working at nest sites, each working alternately. It would seem likely that termitaria might also be used commonly.

Eggs and Incubation   Clutches of three surucua trogon eggs have been found, and broods of three nestlings have been seen in Brazil (Belton 1984). One Argentine nest excavated by the birds and containing three eggs was about 2 m (7 ft) up in a dead tree stub (Western Foundation of Vertebrate Zoology), while another was found in a tree 5.5 m (18 ft) high. The latter cavity was 17 cm (6.7 in) deep and 7 cm (2.8 in) wide and contained four eggs (de la Pena 1987). Two-egg clutches have also been reported. A nest in Paraguay was found during November in an excavated ants' nest (Chubb 1910). No information on incubation periods is available.

Brood Rearing   No detailed information is available.

CONSERVATION AND EVOLUTIONARY RELATIONSHIPS

The surucua trogon ranges through three countries (southern Brazil, Paraguay, northeastern Argentina) and is not considered especially rare. It is widely sympatric with and certainly a very close relative of the blue-crowned trogon, tending to replace it toward the south and east in Brazil. Stotz et al. (1996) listed the species as common, with a medium degree of sensitivity to disturbance.

BLUE-CROWNED TROGON

Trogon curucui L. 1766

Trogon curucui has been described as *T. variegatus* by Gould and most later authors. Linnaeus's name (*curucui*) is of doubtful validity (Todd 1943) and has also often been judged as pertaining to *collaris* (Ridgway 1911, 764).

OTHER VERNACULAR NAMES

Behn's trogon (*behni*), Bolivian trogon (*peruvianus*), Peruvian trogon (*peruvianus*), purple-breasted trogon; sorucua pecho purpura (Spanish).

RANGE

Tropical forests, from southeastern Colombia (east of eastern Andes, from Meta south to Amazonas) to eastern Ecuador, eastern Peru, and all but westernmost Bolivia; also south and east through much of Brazil (Amazonas, central Brazil, eastern Brazil from Maranhão south to Rio de Janeiro), western Paraguay, and northwestern Argentina (northern Salta, Jujuy, and western Formosa and Chaco). See map 23.

SUBSPECIES

Trogon curucui peruvianus Swainson 1838 (1837): south-central Colombia, Ecuador, Peru, and western Brazil (*bolivianus* Ogilvie-Grant 1892 is a synonym; the race *peruvianus* is not yet known from Bolivia). Sometimes accorded species distinction (as the Peruvian trogon), for example by Chapman (1917) and Todd (1943).

Trogon curucui curucui L. 1766: eastern Brazil, Paraguay, and Bolivia.

Trogon curucui behni Gould 1875: eastern Bolivia, southern Brazil, Paraguay, and northern Argentina

MORPHOMETRICS

Measurements Wing, males (of *peruvianus*) 131–138 mm (5.1–5.4 in; Zimmer 1949). Wing, males (of *peruvianus* ["*bolivianus*"]) 122–128 mm (4.8–5.0 in; average of 8, 124.9 mm [4.9 in]); females 121–129.5 mm (4.7–5.1 in; average of 6, 125.8 mm [4.9 in];

Bond and de Schauensee 1943). Wing, males (of *curucui*) 121–127 mm (4.7–5.0 in; average of 8, 123.6 mm [4.8 in]); males (of *behni*) 131–136 mm (5.1–5.3 in; average of 6, 132.6 mm [5.2 in]); males (of *peruvianus*) 121–129 mm (4.7–5.0 in; of 5, average 124.6 mm [4.9 in]; Hellmayr 1929). Tail, males (of *curucui*), 127–135 mm (5.0–5.3 in; average of 8, 129.9 mm [5.1 in]), males (of *behni*) 140–151 mm (5.5–5.9 in; average of 6, 145.3 mm [5.7 in]), males (of *peruvianus*) 124–132 mm (4.8–5.1 in; average of 4, 130 mm [5.1 in]; Hellmayr 1929). Wing, males (of *curucui*) 115–125 mm (4.5–4.9 in; average of 11, 120.6 mm [4.7 in]); females 122–127 mm (4.8–5.0 in; average of 4, 123.5 mm [4.8 in]). Tail, males (from Brazil) 123–138 mm (4.8–5.4 in; average of 11, 129.8 mm [5.1 in]); females 120–143 mm (4.7–5.6 in; average of 4, 133.2 mm [5.2 in]; Pinto 1950). Egg of *behni* 27.5 × 20.7 mm (1.1 × 0.8 in; Schönwetter 1966).

Weights Six birds of both sexes from Panama average 52.2 g (1.8 oz; Hartman 1961). One male of *peruvianus* 59.3 g (2.1 oz; specimen, MVZ). Eight males of *behni* 45–65 g (1.6–2.3 oz; average 54.4 g [1.9 oz]), five females 38–64 g (1.3–2.2 oz; average 48.8 g [1.7 oz]); three males of *peruvianus* 58–63 g (2.0–2.2 oz; average 59.9 g [2.1 oz]), one female 54 g (1.9 oz; specimens, FMNH). Average mass (both sexes), visually estimated from graph, 55 g (1.9 oz; Remsen, Hyde, and Chapman 1993). Estimated egg weight 6.2 g (0.2 oz; Schönwetter 1966).

DESCRIPTION (see plate 28)

Adult Male Head and breast iridescent violet to blue-green; breast band white, narrowly separating breast from carmine-rose underparts; upperparts otherwise mostly iridescent coppery green, becoming blue-green on upper tail-coverts and middle rectrices; upper wing-coverts narrowly vermiculated with black and white; three outer pairs of rectrices regularly banded with narrow black-and-white bars, and with wider white tips; bill yellow to pearly slate; eye-ring yellow to orange; iris brown; feet and toes brownish to blackish, powdered with white.

Map 23. Distribution of the blue-crowned trogon.

Adult Female   Upperparts, head, and breast dark slate-gray, except back lighter gray and lower breast grayish white; abdomen and under tail-coverts carmine-rose, former sometimes separated from gray breast by a very narrow white breast band; upper wing-coverts and inner (median) remiges black with narrow white vermiculations; three median pairs of rectrices slate gray; outer three pairs of rectrices barred with black and white on outer vanes; maxilla brown; mandible pearly; eye-ring inconspicuous and visually replaced by white feathering around (mainly in front of and behind) eye; feet and toes brown.

Immature   Not obviously different from those of the surucua trogon

Juvenile   No information available.

IDENTIFICATION

In the Hand   Adult males of the blue-crowned trogon are most similar to those of the collared trogon (both have a red abdomen, a white breast band, and a black-and-white barred tail) but differ in having a bluish-green (not true green, except in *behni*) head and breast, a conspicuous yellow (not an inconspicuous reddish to whitish) eye-ring, and finer vermiculations on the wing-coverts. Females of the blue-crowned trogon have gray (not brown, as in the collared trogon) heads and upperparts.

In the Field   The only other species in the blue-crowned trogon's range likely to cause confusion is the collared trogon, whose differences are indicated in the previous section. The blue-crowned trogon is a lowland species, whose vocalizations are still poorly known but certainly closely resemble those of the white-tailed trogon. In the blue-crowned trogon the calling sequence tends to accelerate, then suddenly stop. See also the vocalizations section.

GEOGRAPHIC VARIATION

As noted in the measurements section, there are some minor differences in wing and tail lengths among the races of the blue-crowned trogon. The nominate form has a bronzy-green back, as compared to *behni*, which has a more coppery green back with a golden gloss. This latter form also is slightly smaller and has a more greenish head and breast sheen. These differences in iridescence are probably of little taxonomic significance.

ECOLOGY

Habitats   Collectively the blue-crowned trogon ranges from less than 500 m (1,600 ft) to approximately 2,000 m (6,600 ft). The species' primary habitat is tropical and subtropical zone forests and woodlands, but the birds extend into subtropical savanna. In Colombia the birds inhabit humid forests, tall secondary growth, and forest edges from the lowlands (including varzea forests) to about 500 m (1,600 ft).

Foods and Foraging Ecology   Of 36 blue-crowned trogon stomachs examined by Remsen, Hyde, and

Chapman (1993), 19 contained arthropods only, 16 contained mixed fruit and arthropods, and 1 contained unidentified materials.

BEHAVIOR

General, Social, and Sexual Behavior   No information available.

Vocalizations   Vocalizations of the blue-crowned trogon are still little known, but some songs were recorded in Brazil by Jacques Viellard (Hardy, Reynard, and Coffey 1995). The song examples are nearly identical in tone, pitch, and speed to those of the surucua trogon but are more extended in duration. One sequence of 17 notes lasted 6 seconds, and another sequence of at least 130 notes lasted nearly 30 seconds. This latter sequence varied somewhat in loudness through the series. See also the identification section.

BREEDING BIOLOGY

No information available.

CONSERVATION AND EVOLUTIONARY RELATIONSHIPS

The blue-crowned trogon ranges through seven South American countries, from southern Colombia to northern Argentina and Paraguay. It is seemingly rather uncommon over much of its range, and surprisingly little is known of its biology. Stotz et al. (1996) listed the species as fairly common, with a medium degree of sensitivity to disturbance.

VIOLACEOUS TROGON

Trogon violaceus Gmelin 1788

Trogon violaceus has been designated as *Trogon caligatus* by Gould. The confusing taxonomic history of this species has been discussed at length by Zimmer (1948).

OTHER VERNACULAR NAMES

Amazonian gartered trogon (*crissalis*), Bahia trogon (*crissalis*), booted trogon (*caligatus*), Deville's trogon (*ramonianus*), gartered trogon (*caligatus*), lesser yellow-bellied trogon, little trogon, Ramon de la Sagra's trogon, southern trogon (*violaceus*), western gartered trogon (*concinnus*); sorocua violetta, trogon violaceo (Spanish).

RANGE

Forests, forest edges, and plantations from eastern Mexico (southern Veracruz and Yucatán region) south through Belize, Guatemala (both slopes), El Salvador (widespread), eastern Honduras, Nicaragua (both slopes), Costa Rica (both slopes), Panama (widespread except perhaps Azuero Peninsula), and Colombia (widespread); from Venezuela (widespread in west, locally to Bolívar and Amazonas), east and south to Guyana (widespread), Suriname (widespread), French Guiana, Trinidad (common) and much of Brazil (Amazonas, northern Mato Grosso, eastern Para) southwest to eastern Ecuador, eastern Peru, and northern Bolivia. See map 24.

SUBSPECIES

Trogon violaceus braccatus (Cabanis and Heine) 1856: central Mexico to Costa Rica. Previously this form was often referred to as *sallaei* Bonaparte, and this epithet was considered as taxonomically correct by Todd (1943).

Trogon violaceus concinnus Lawrence 1862: Panama to Colombia, northwestern Peru, and western Ecuador.

Trogon violaceus caligatus Gould 1838: northern Colombia to western Venezuela. Considered by Gould to be specifically distinct (the gartered trogon). Includes *columbianus* Chapman 1914.

Trogon violaceus violaceus Gmelin 1788: Venezuela, Guyana, Suriname, French Guiana, northern Brazil, and Trinidad.

Trogon violaceus ramonianus Deville and Des Murs 1849: upper Amazonia of western Brazil and eastern Peru. Considered specifically distinct by Gould, as well as by Cory (1919).

Trogon violaceus crissalis (Cabanis and Heine) 1862–1863 (1863): eastern Brazil, northern Peru, eastern Ecuador, Colombia, and southern Venezuela

MORPHOMETRICS

Measurements Wing, males (of *braccatus*) 110–134 mm (4.3–5.2 in; average of 53, 121.1 mm [4.7 in]); females 113–130.5 mm (4.4–5.1 in; average of 37, 121.5 mm [4.7 in]). Tail, males (of *braccatus*) 112–135 mm (4.3–5.3 in; average of 53, 121.1 mm [4.7 in]); females 114.5–135.5 mm (4.5–5.3 in; average of 37, 126.1 mm [4.9 in]; Ridgway 1911). The wings and tails of *concinnus* average about 5 mm (0.2 in) shorter (Wetmore 1968). Wing, fifteen males from Nicaragua to Trinidad 104.5–120 mm (4.1–4.7 in); tail, 113.5–127.5 mm (4.4–5.0 in; Chapman 1914). Egg of *braccatus* 29.1 × 22.6 mm (1.1 × 0.9 in), of *violaceus* 30.3 × 23.4 mm (1.2 × 0.9 in; Schönwetter 1966).

Weights Five males from Trinidad 49–55 g (1.7–1.9 oz; average 52.1 g [1.8 oz]); six females 48–53 g (1.7–1.9 oz; average 51.0 g [1.8 oz]; ffrench 1991). Males from Suriname (unstated sample size) 47–54 g (1.6–1.9 oz), females 48–50 g (1.7–1.8 oz; Haverschmidt 1968). Unspecified Costa Rica weight 56 g (2.0 oz; Stiles and Skutch 1989). Three males from Belize 55.7–58.8 g (1.9–2.1 oz; average 57.2 g [2.0 oz]; Russell 1964). Sex and sample size unspecified 53.75–59 g (1.9–2.1 oz; average 56 g [2.0 oz]; Smithe 1966). One male of *concinnus* 59 g (2.1 oz); four females of various subspecies 49.7–63 g (1.7–2.2 oz; average 56.2 g [2.0 oz]; specimens, MVZ). One male of *concinnus* 50 g (1.8 oz), one female 49 g (1.7 oz; Weidenfield, Schulenberg, and Robbins 1985). Two males of *braccatus* 60 and 66 g (2.1 and 2.3 oz), four females 55.3–63.7 g (1.9–2.2 oz; average 59.5 g [2.1 oz]), three males of *ramonianus* 44–46 g (1.5–1.6 oz; average 44.7 g [1.6 oz]; specimens, FMNH). Estimated egg weight of *violaceus* 8.7 g (0.3 oz), of *braccatus* 7.9 g (0.3 oz; Schönwetter 1966).

DESCRIPTION (of *braccatus*, adapted from Ridgway 1911; see plate 29)

Adult Male Head and neck black, sometimes becoming iridescent blue or violet-blue on lower hindneck; back, scapulars, upper rump, and anterior lesser wing-coverts bright iridescent green or golden green, becoming pure iridescent green, bluish green, or nearly greenish blue on lower rump, upper tail-coverts, and middle pair of rectrices, the latter abruptly tipped with black; the next two pairs of rectrices with outer vanes similar to those of middle pair but inner vanes wholly uniform black; three outer pairs of rectrices broadly tipped with white, remaining portion black, for the most part barred with white; outermost rectrix with outer vane barred for whole length, inner vane also barred for greater part, but white bars becoming gradually narrower and shorter proximally; third rectrix (from outside) with outer vane barred for more than half its length, but inner vane with not more than one or two complete bars, white bars continuous on subterminal portion of two outer rectrices and there nearly equal in width to black interspaces; wing-coverts and secondaries minutely vermiculated with black and white; primaries dull black, longer ones edged with white for basal half or more; chest iridescent bluish green, blue, or violet-blue (rarely black), glossed, more or less, with iridescent bluish green or blue; rest of underparts rich pure orange-yellow or yellowish orange, paler, sometimes slightly intermixed with whitish, along anterior margin; outer portion of sides and flanks gray, tinged or washed with orange-yellow; thighs black; bill pale olive-gray, gray, dull glaucous, or ashy blue; eye-ring orange-yellow to bright yellow; iris dark brown; feet and toes neutral gray or lead colored. Adults undergo a complete molt from July to September (in El Salvador), this molt followed in spring by a limited molt on the head and foreparts (Dickey and van Rossem 1938).

Adult Female Upperparts except for wings plain slate, becoming blackish slate on crown; middle rec-

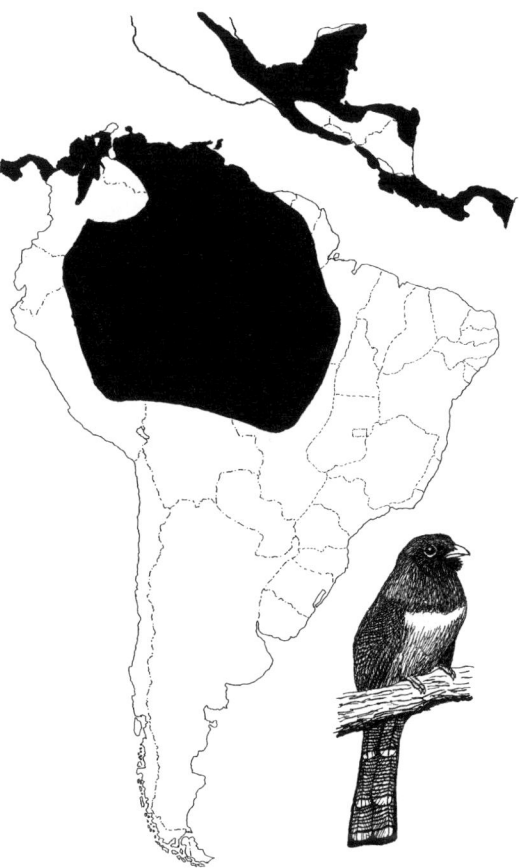

Map 24. Distribution of the violaceous trogon.

trices abruptly and rather broadly tipped with black; a broad orbital edging of white feathers, interrupted on upper and lower eyelids; wings black or slate-black, wing-coverts and secondaries narrowly barred with white transverse lines, longer primaries edged with white; three outer rectrices with inner vanes black tipped with white, the latter extending, wedge-like, for a variable distance next to shaft, outer vanes broadly barred with white; side of head and neck, chin, throat, and chest plain slate or blackish slate; outer portion of sides and flanks lighter slate or slate-gray, tinged with orange-yellow; tibial and tarsal feathers black; rest of underparts orange-yellow, paler along anterior margin; bill light grayish to mostly blackish on maxilla; mandible light grayish to greenish blue; eye-ring inconspicuous; iris dark brown; feet and toes grayish, as in male.

Immature Male   Similar to adult male but outer rectrices narrower, less regularly barred; wing-coverts more coarsely vermiculated; and chest more or less mixed with dull slate color. In younger males, the slate color of upperparts and chest is intermixed with few iridescent green feathers; the upper wing-coverts are browner than in adult; and the undertail pattern similar to that of adult female. Soft-part colors probably less bright than in adults.

Immature Female   Upperparts blackish slate, with white crescent markings in front of and behind each eye; outer rectrices relatively narrow, more pointed, and more coarsely barred with black and white than those of adults; inner rectrices tipped with white; bill dark above.

Juvenile   Closely resembling adult female; exposed parts of outer two rectrices white, and the rest blackish; head brown, with crescent of whitish feathers behind each eye, and a smaller one in front; bill bluish gray, with a lighter tip; eye brown, with a surrounding ring of bare yellow skin.

IDENTIFICATION

In the Hand   The violaceous trogon is one of the smallest of New World trogons. Both sexes have yellow abdomens, which color is separated from the bluish-violet (males) or blackish (females) breast and head by a drooping white "necklace." In adult males the yellow eye-ring is conspicuous. The name "booted trogon" refers the fact that in this species the tarsal surface is not clearly subdivided into separate scutes along its anterior edge. The origin of the term "gartered trogon" is less obvious but perhaps refers to the decorative black-and-white banding pattern on the rectrices.

In the Field   Because of its broad range, the violaceous trogon overlaps with most other yellow-bellied trogons. To the north, in Mexico and Guatemala, it may be distinguished from both sexes of citreoline and black-headed trogons by its strongly barred undertail pattern and from males of the same by its head color (violet rather than the black present in males of these two species). In Costa Rica and Panama the violaceous trogon differs from the male black-throated trogon in its violet-tinted (not green) head and breast and from the female in its gray (not brown) head and breast. Its yellow eye-ring helps separate the male of this species from males of the white-tailed trogon farther south, which have a blue eye-ring and an unbarred tail pattern. Vocalizations of the violaceous trogon include a long series of down-slurred *kow*, *kew*, or *kwer* notes on the same pitch, which may accelerate toward the end. Alarm or disturbance notes include a rattling, resonant *krrr* and a more nasal *nyrrrp*, which is uttered while the bird raises and lowers the tail. See also the vocalizations section.

GEOGRAPHIC VARIATION

Birds of the northern race of the violaceous trogon (*braccatus*) average larger than *concinnus* from Panama (Wetmore 1968). Zimmer (1948) reported that Costa Rican birds have measurements closer to those of *braccatus* than to those of *concinnus*, and he noted that the former show little tendency toward the coarser wing-covert vermiculation pattern typical of *concinnus*. Male birds from Amazonian Brazil and southward vary somewhat in the colors of the crown, breast, and tail of the males, but there seems to be no clear trend in these variations, which are largely or entirely based on minor shifts in iridescence hues. The form *caligatus* is notable for its blackish head, chin, and throat, all of which have only a slight purplish sheen in males.

ECOLOGY

Habitats   Collectively the violaceous trogon is one of the most widespread and ecologically tolerant of all trogons, and it occurs in virtually all forests, woodlands, forest edges, clearings, and partially wooded areas, from sea level to 1,800 m (5,900 ft). In Mexico it ranges from sea level to 1,800 m (5,900 ft), in humid semi-arid forests, forest edges, plantations, and mangroves (Howell and Webb 1995). In Guatemala it has essentially the same vertical ranges and occurs in humid forests, brushy woodlands, and plantations (Land 1970). In Honduras it ranges to 1,400 m (4,600 ft), in lowland and montane rain forest, but prefers secondary forest and open forest (Monroe 1968). In El Salvador it reaches 1,250 m (4,100 ft), from the lower tropical zone to the oaks of the upper tropical zone (Dickey and van Rossem 1938). In

Vespiary nest of violaceous
trogon. Photograph by
A. F. Skutch.

Costa Rica it extends from lowlands to 1,200 m
(3,900 ft), in evergreen gallery forest, humid forest
edges, tall secondary forest stands, and clearings
with scattered tall trees (Slud 1964; Stiles and Skutch
1989). In Panama it occurs from the lowlands to
900 m (3,000 ft) regularly and occasionally to 1,280 m
(4,200 ft), in gallery forests, forest edges, clearing
edges, or rather open situations (Ridgely and Gwynne
1989; Wetmore 1968). In Colombia it ranges to at
least 1,000 m (3,300 ft), in forest edges, secondary
forest, and clearings with scattered trees (Hilty and
Brown 1986). In Venezuela it ranges to 1,200 m
(3,900 ft), in forests, clearings, secondary forest,
gallery forests, and open woodlands (de Schauensee
and Phelps 1978). On Trinidad Island it occupies

forests and cocoa plantations (Herklots 1961), and in
Suriname it occurs in savanna forests as well as interior forests (Haverschmidt 1968).

Foods and Foraging Ecology   Stomachs of violaceous
trogon specimens taken in El Salvador contained
mistletoe berries (four stomachs), fruit pulp and
small caterpillars (one), fruit pulp and small seeds
(one), fruit pulp (one), and berries and caterpillars
(one). See Dickey and van Rossem 1938. Haverschmidt (1968) reported that foods include fruits,
berries, arachnoids, and insects, including larval lepidopterans and locustids. Wetmore (1968) similarly
listed orthopterans, large caterpillars, earwigs, an
ant, and berries among the materials he found in

stomachs. Slud (1964) saw violaceous trogons perching near an active wasp nest and frequently flying out to hawk these insects. Skutch (1972) saw adults eating wasps, the fruiting spikes of *Cecropia* trees, and various unidentified fruits and insects. Of nineteen stomachs examined by Remsen, Hyde, and Chapman (1993), eleven contained arthropods only, five contained mixed fruit and arthropods, and three contained fruit only.

BEHAVIOR

General, Social, and Sexual Behavior   Skutch (1972) reported that violaceous trogons are most vocal during the dry season, at which time several males may assemble and vocalize, apparently in competition for mates. This behavior mainly occurs during the first quarter of the year, but may extend sporadically until late September. The mate may also sing softly while its partner is at work excavating a nest site.

Vocalizations   Skutch (1972) described the call of the violaceous trogon as a long series of soft, clear *cow* notes, all very similar in tone and loudness and uttered rapidly at a uniform cadence or slightly accelerated toward the end. Slud (1964) described the voice of this species as higher pitched and lighter than the voices of the slaty-tailed, lattice-tailed, and Baird's trogons and consisting of repeated *hoop, chow, poo,* or *chyow* notes that accelerate and increase slightly in intensity. The violaceous trogon also utters a "cooing cackle" and a liquid *churr* that is accompanied by a slow raising of the tail and possibly a nodding of the head. Skutch (1972) similarly mentioned that a trogon responding to the presence of an owl uttered sharp rattling notes, each time slowly raising his tail above the back. According to Davis (1972), the song of this species consists of a series of whistled notes uttered at the rate of about three per second and resembles the call of the ferruginous pygmy-owl (*Glaucidium brasilianum*) but is much higher in pitch. Vocalizations of this species were recorded in Guatemala by Ben Coffey Jr. (Hardy, Reynard, and Coffey 1995). Five recorded sequences consisted of eight to thirteen *kow* notes uttered rapidly, these notes tending to become both louder and faster, with the sequences lasting about 3 or 4 seconds each. Similarities to the songs of surucua and blue-crowned trogons are very apparent. See also the identification section above.

BREEDING BIOLOGY

Chronology of Breeding   In Mexico, violaceous trogon nests have been found in March, April, and May, and chicks have been seen in June (Binford 1989; Skinner 1901; Sutton and Burleigh 1940). In El Salvador breeding occurs from early May to early July (Dickey and van Rossem 1938). Breeding in Costa Rica extends from February (nest excavation) to June (brooding young; Skutch 1972), and in Colombia breeding-condition birds have been taken in March. On Trinidad Island breeding has been reported not only from April to July (Herklots 1961) but also in November, February, and March (ffrench 1991).

Nest Sites   Ten violaceous nest sites seen by Skutch (1972) were all in turban-shaped vespiaries in trees. These vespiaries were typically attached at their tops to the horizontal tree branch located between 4.6 and 32 m (15–105 ft) above ground (usually between 9 and 16 m [30 and 52 ft]). The sites were usually at forest edges or at some distance from the forest, in a pasture, plantation, streamside tree, or cane brake. As usual, both sexes worked on the excavation but neither bird would work in the absence of its partner. A similar vespiary site was seen in Panama by Wetmore (1968), who also reported the use of a termitarium as a nest. Skutch (1999) observed a pair using an ants' nest and watched another pair excavating the roots of an arboreal fern. Herklots (1961) reported that on Trinidad Island holes in dead trees are used. A nest found by Sutton and Burleigh (1940) was in burrow made in a clump of orchid roots on the branch of a large tree, about 9 m (30 ft) above ground. The two nests found by Skinner (1901) in Mexico were in decaying stumps with entrances 2.5 and 4.7 m (8 and 15 ft) high; the cavities were about 15–20 cm (5.9–7.9 in) in diameter, and the bottom of the cavity was located only slightly below the base of the entrance.

Eggs and Incubation   All reports indicate that the normal clutch size of the violaceous trogon is two or three eggs, more commonly two. Skutch (1972) observed the usual trogon pattern of incubation, namely, the male incubating during most of the day and the female at night. The incubation period has not been established.

Brood Rearing   Skutch (1972) reported that young violaceous trogons are fed insects and small fruits. He estimated the fledging period as at least 17 days.

CONSERVATION AND
EVOLUTIONARY RELATIONSHIPS

The violaceous trogon is a wide-ranging species comparable to the collared trogon in that it extends through eight Central American and 10 South American countries, from central Mexico to Bolivia and central Brazil. Since it is not forest dependent, it is tolerant of cleared areas, edges, and other habitats that are becoming increasingly frequent in Latin America. Determining its nearest living relative is difficult, but the black-throated trogon is a possibility. Stotz et al. (1996) listed the species as fairly common, with a medium degree of sensitivity to disturbance.

Asian Trogons (Tribe Harpactinini)

The tribe Harpactinini consists of 11 small to mostly fairly large species (total length 24–38 cm [9.4–14.9 in], adult mass 35–125 g [1.2–4.4 oz]) having notched but nonserrated maxillae, oval nostrils that lack opercula and are largely hidden by anteriorly oriented bristles, partially feathered tarsi, and relatively small feet. The rectrices are long and broad, the more interior ones having distinctly truncated tips in adults. The outer three pairs of rectrices are graduated in length and extensively white to (rarely) pinkish or yellowish. They rarely have extensive black spotting or barring but have black freckling in one species. Red to brownish-red colors predominate in the dorsal plumages of adults of most species (but the blue-tailed trogon is entirely and uniquely iridescent green to bluish dorsally), especially in males. In both sexes, the upper wing-coverts and inner secondaries, including the tertials, are always patterned with dark and light vermiculations or similar but somewhat coarser wavy barring. Coarser barring in these areas is also typical of young birds. Carotenoid reds, pinks, and yellows, or sometimes cinnamon tones, predominate on the underparts of adult males; whereas females of most species tend to be cinnamon toned on the underparts. The bill colors are primarily blue to purplish, but in one species the bill is orange-red. Prominent blue eye-rings are present in adults of nearly all species, and the iris is brown in all. All species of the subgenus *Harpactes* are believed to be almost entirely insectivorous, but some limited consumption of leaves, fruit, or berries has also been reported, these items perhaps providing a valuable source of carotenoids pigments for red feathers.

The Asian trogons are widely distributed in southern and southeastern Asia as well as throughout the Greater Sunda Islands, occurring both in lowland forests and in montane forests (mountainous regions in this area are shown in figure 18), but since the 1960s their forest habitats have been rapidly disappearing. As of the 1960s, the Malay Peninsula was about 68 percent forested, Sumatra about 59 percent, and Java about 22 percent (Heske 1973). The figures for various parts of Borneo include Kalimantan 77 percent, Sarawak 74 percent, Sabah 79 percent, and Brunei 75 percent. Figure 19 shows areas forested as of the 1960s. The droughts of the mid-1990s have had devastating effects on the humid forests, as have massive logging activities throughout Southeast Asia (see figure 20 for currently forested regions).

Current deforestation rates of tropical forest in Asia are the highest in the world, averaging 30–40 percent higher than those in South America, and 20–30 percent higher than those in Africa. Over the period 1960–1970 the deforestation rate in Asia averaged about 6.5 percent, versus about 8 percent over

the period 1980–1990 (World Resources Institute data). As of 1990, Asia and the Pacific Islands had 3.110 billion hectares (7.685 billion acres) of tropical forests remaining, the largest percentage of which (44 percent) was in insular Southeast Asia. Annual recent deforestation rates in the general region have averaged 3.9 million hectares (9.6 million acres), with the largest percentage (64 percent) occurring in tropical rain forests (FAO 1993). Although Borneo was perhaps still over 50 percent forested as of the early 1990s, the current logging rate in Indonesia collectively is 1.2 million hectares (3.0 million acres) annually, an area roughly the size of Switzerland. On the Malay Peninsula nearly all nonprotected forested areas have already been logged, and logging continues at a rate of 700,000 hectares (1.7 million acres) annually. Likewise, on Java at least 90 percent of the primary forests have long since been logged,

Figure 18. Mountainous and elevated regions (above 2,000 m [6,600 ft]) of Southeast Asia and the Greater Sunda Islands (inked), after various sources.

Figure 19. Recently (ca. 1960s) forested regions (inked areas mostly evergreen tropical forests and dipterocarp forests; stippled areas swamp forests) of the Malay Peninsula and the Greater Sunda Islands, adapted from Heske 1973.

Figure 20. Currently (1990s) forested regions of the Malay Peninsula and the Greater Sunda Islands (montane rain forests inked, lowland and swamp forests stippled), after various sources.

and on some islands of the Philippines virtually all the forested areas have disappeared (Wheatley 1996a). As of the 1970s, the percentages of land mass still covered by primary forest in the Greater Sunda Islands region were: Malay Peninsula 42 percent, Sumatra 57 percent, Java 6 percent; and Kalimantan 70 percent. Estimates for 1990 indicated that 60.5 percent of Indonesia, 53.5 percent of Malaysia, and 26.3 percent of the Philippines were still forested (FAO 1993), although much of this forested area is probably secondary forest. The most recent (1990s) available estimates (Brooks et al. 1999) of total forested areas still remaining in insular Southeast Asia are: Sumatra, 50 percent; Java, 10 percent; Lesser Sundas, 18 percent; Borneo, 72 percent; and the Philippines, 25 percent. Only 10 percent of Indonesia's land area is now protected, whereas in peninsular Malaysia nearly half of the remaining forest area is under some degree of protection (Johnsgard 1999).

Of 31 critical, endangered, or vulnerable species in Malaysia, 19 (61 percent) are forest dependent and 9 more include forests in their range of habitats. In Indonesia the corresponding figures are 104 species, with 76 (73 percent) classified as forest dependent and another 18 that include forests in their range of habitats. Indonesia has more threatened species than almost any country in the world, but the situation is several times worse in the Philippines, which has an area equal to 16 percent of Indonesia's total area. This island nation is still being subjected to massive logging, and some of the smaller islands have been essentially denuded of their forests. A total of 86 Philippine Islands species are classified as critical, endangered, or vulnerable, and of these, 70 (81 percent) are forest dependent and another 8 include forests in their range of habitats (Collar, Crosby, and Stattersfield 1994).

ASIAN TROGONS (Genus Harpactes Swainson 1833)

Harpactes, the single genus in the tribe Harpactinini, has the same traits as the tribe. The genus is divided into two subgenera, *Hapalarpactes* and *Harpactes*; however, the former has often been distinguished as a separate genus (*Hapalarpactes*) because its single species, the blue-tailed trogon, has unique iridescent greenish adult plumage and no black terminal band on the three interior pairs of greenish-blue rectrices (a characteristic that is also typical of the African genus *Apaloderma*). In the ten other species of the genus *Harpactes,* these three pairs of rectrices are instead chestnut and, in adults, have a black terminal band. Interestingly, the submicroscopic structure of the feather barbules that is responsible for the iridescence of the blue-tailed trogon's plumage is also more similar to that of *Apaloderma* species than to that of *Trogon* species (Durrer and Villiger 1966). For these reasons *Hapalarpactes* is here recognized as a distinct subgenus.

On the basis of gene-sequencing studies, Espinosa de los Monteros (1998) considered the tribe's closest relatives to be the New World trogons (tribe Trogonini), of which they represent a sister taxon. He was able to study only three species (*H. diardii, H. ardens,* and *H. oreskios*) and thus learned little about the intratribal relationships.

Subgenus *Hapalarpactes* Cabanis and Heine 1863

The monotypic subgenus *Hapalarpactes* differs from the subgenus *Harpactes* in that both sexes of the former are largely iridescent green, and their three central pairs of rectrices are bluish green and lack a terminal black bar. The throat is also uniquely yellow in adults of both sexes.

BLUE-TAILED TROGON

Harpactes reinwardti (Temminck) 1822

OTHER VERNACULAR NAMES

Blue-billed trogon, Macklot's trogon (*mackloti*), Reinwardt's trogon (nominate *reinwardti*).

RANGE

Montane evergreen forests (between 1,000–2,500 m [3,300–8,200 ft]) of Sumatra (Barisan Range) and western Java, east to Gunung Papandayan (south of Bandung) and the twin peaks Gunung Gede and Pangrango. See map 25.

SUBSPECIES

Harpactes reinwardti mackloti (S. Müller) 1835: Sumatra (Barisan Range).
Harpactes reinwardti reinwardti (Temminck) 1822: Western Java (to ca. long. 108° E).

MORPHOMETRICS

Measurements   Wing, one male (of *reinwardti*) ca. 145 mm (5.7 in); one female ca. 140 mm (5.5 in). Tail, one male ca. 190 mm (7.4 in); female ca. 180 mm (7.0 in; specimens, FMNH). Wing, one male (of *mackloti*) ca. 129 mm (5.0 in); female ca. 129 mm (5.0 in). Tail, one male ca. 165 mm (6.4 in); one female ca. 152 mm (5.9 in; Ogilvie-Grant 1892). Egg of *reinwardti* 32.0 × 24.7 mm (1.24 × 1.0 in; Schönwetter 1966).

Weights   No information on adult weights available. Estimated egg weight 10.0 g (0.4 oz; Schönwetter 1966).

DESCRIPTION (see plate 30)

Adult Male   Crown and nape olive-brown with a greenish wash; scapulars and posterior upperparts glossy greenish blue; middle three pairs of rectrices greenish blue to iridescent blue, without black tips; three outer pairs of rectrices similar, but with white edges and broad white tips; remiges and wing-coverts mostly black, primaries edged anteriorly with white, and upper wing-coverts closely barred or vermiculated with green and yellow; throat, abdomen, and other underparts yellow, breast and abdomen separated by a broad olive breast band;

Map 25. Distribution of the blue-tailed trogon.

bill orange-red; eye-ring bright blue; iris brown; feet and toes orange to yellow.

Adult Female   Very similar to male, but dark green upper wing-coverts barred with brown rather than yellow; throat, breast, and remaining underparts less brilliant. Soft-part colors similar to those of adult males.

Immature   Young males with rufous brown tint to the olive-colored breast; upper wing-coverts and secondaries coarsely barred yellowish, with darker edging, rather than closely barred or vermiculated with yellow and dusky. Iridescent greenish feathers appear first on rump, gradually spreading forward. Soft-part colors probably duller than those of adults.

Juvenile   Generally brownish on head, back, and upper tail-coverts, with variable amounts of blue-green feathers present; belly cinnamon-brown, mixed with white; black on outer vanes of outer tail feathers extending nearly as far up on inner vanes as in adults, but outer two rectrices barred with white; bill brown in quite young birds; other soft-part colors duller than in adults.

IDENTIFICATION

In the Hand   Both sexes of the distinctively plumaged blue-tailed trogon are iridescent blue to green on the upperparts and have bright yellow underparts. The bill is bright orange-red in adults. Immatures are generally much more brownish, but older ones, at least, show some green feathers on the back.

In the Field   The blue-tailed trogon is the only Asian trogon having distinctly iridescent coloration on the tail and upperparts and is the only one with a yellow throat and abdomen separated by an olive-green breast band. The species is limited to montane rain forests at 1,000–2,500 m (3,300–8,200 ft) elevation and utters loud, hoarse *chierr* notes.

GEOGRAPHIC VARIATION

The Sumatran race (*mackloti*) of the blue-tailed trogon is somewhat smaller than the Javan form, and the former's lower back color is dark chestnut to maroon rather than bluish green in males. Sumatran birds also have a smaller yellow throat patch than do males of the nominate race.

ECOLOGY

Habitats   The little-known blue-tailed trogon is confined to montane rain forests above 1,000 m (3,300 ft) and is uncommon even in these preferred habitats.

Foods and Foraging Ecology   No specific information is available on foraging, but the blue-tailed trogon is reportedly insectivorous.

BEHAVIOR

General, Social, and Sexual Behavior   Like trogons generally, blue-tailed trogons tend to perch upright on horizontal forest branches, looking for insects or calling occasionally.

Vocalizations   Little is known beyond what is mentioned in the identification section.

BREEDING BIOLOGY

No information available.

CONSERVATION AND EVOLUTIONARY RELATIONSHIPS

The blue-tailed trogon seems to be a relatively isolated form with no apparent near relatives. It has a highly restricted range in Sumatra and western Java, and its situation should be closely watched, especially in Java.

Subgenus *Harpactes* Swainson 1833

The subgenus *Harpactes* differs from *Hapalarpactes* in that the three central pairs of rectrices in both sexes of the former are chestnut and have a terminal black bar. The throat is never yellow. There are ten species in the group, which includes *Pyrotrogon* Bonaparte 1854.

MALABAR TROGON

Harpactes fasciatus (Pennant) 1769

OTHER VERNACULAR NAMES

Central Indian trogon (*legerli*), Ceylon trogon (*fasciatus*), fasciated trogon.

RANGE

Lowland and hill forests of western peninsular India (from southern Gujarat and northern Maharashtra south through the Western Ghats of Goa, Karnataka [Mysore], Tamil Nadu, and Kerala), eastern peninsular India (from Chota Nagpur plateau south and west through Madhya Pradesh and Orissa to Anhdra Pradesh), and Sri Lanka (Ceylon). See map 26.

SUBSPECIES

Harpactes fasciatus malabaricus (Gould) 1834: western and southern India.

Harpactes fasciatus legerli Koelz 1939: east-central India. This subspecies was not recognized by Peters (1945), who included it within *malabaricus*.

Harpactes fasciatus fasciatus (Pennant) 1769: Sri Lanka (includes *Harpactes fasciatus parvus* Deraniyagala 1956).

MORPHOMETRICS

Measurements Wing, fourteen males (of *legerli*) 130–142 mm (5.1–5.5 in); eight females 125–135 mm (4.9–5.3 in). Tail, fourteen males 155–184 mm (6.0–7.2 in); eight females 143–176 mm (5.6–6.9 in). Wing, sixteen males (of *malabaricus*) 122–129 mm (4.8–5.0 in); ten females 122–131 mm (4.8–5.1 in). Tail, sixteen males 156–180 mm (6.1–7.0 in); ten females 162–167 mm (6.3–6.5 in). Wing, both sexes (of *fasciatus*, sample size unstated) 118–126 mm (4.6–5.0 in; Ali and Ripley 1983). Egg of *malabaricus* 26.7 × 23.4 mm (1.0 × 0.9 in), of *fasciatus* 25.6 × 24.4 mm (1.0 × 1.0 in; Schönwetter 1966).

Weights One male of *malabaricus* 62 g (2.2 oz; Ali and Ripley 1983). Estimated egg weight of *malabaricus* 8.2 g (0.3 oz), of *fasciatus* 8.5 g (0.3 oz; Schönwetter 1966).

DESCRIPTION (see plate 31)

Adult Male Head, neck, and breast blackish brown to grayish black; mantle, scapulars, back, rump, upper-tail-coverts, and middle rectrices yellowish brown, sometimes suffused with chestnut; wing-coverts and inner remiges blackish with fine white vermiculations; drooping white breast band, or "necklace," separates the blackish chest from the crimson-pink underparts, which include the flanks and under tail-coverts; outer rectrices white, outermost pair with barely visible black bases as seen from below; bill blue, with black culmen and tip; iris dark brown; eye-ring and adjoining bare facial skin bright cobalt-blue; feet and toes paler blue.

Adult Female Duller throughout than male; head, neck, and chest dull orange-brown; crimson-pink underparts of male replaced in female by orange-brown; upper tail-coverts and rump bright rufous orange; upper wing-coverts vermiculated as in male, but with yellowish buff replacing the white; soft-part colors similar to those of male, but less bright.

Immature In young males, underparts pinkish salmon or buffy brown, variegated with some red (adult) feathers; primaries, secondaries, primary coverts, and some lesser coverts retained from juvenal plumage; rectrices narrower than in adults; middle pair of rectrices with rounded, not truncated, tips and lacking black terminal band (Ali 1949).

Juvenile No information available.

IDENTIFICATION

In the Hand Males of the Malabar trogon are the only ones of this genus found in India that have a head that is blackish (in males) or dull orange-brown (in females). Unlike the red-headed trogon, female Malabar trogons lack any red tones on the underparts, which instead are olive-brown.

In the Field Only one other trogon, the red-headed trogon, occurs within the Malabar trogon's range.

The Malabar trogon has a blackish (males) or dull brown (females) head and chest color, whereas the red-headed has a red (males) to cinnamon-rufous (females) head and chest color. Both species are limited to evergreen or moist deciduous forests at lower elevations (reaching 1,200 m [3,900 ft] in India, 1,800 m [5,900 ft] in Sri Lanka). The typical male vocalization is a series of rather musical *cue* or *mew* notes, often uttered in a series of from three (usually) to five elements. This vocalization has also been described as a whistled and repeated *h'yoch*, with the "y" pronounced as in 'you' and each syllable sharply cut off. See also the vocalizations section.

GEOGRAPHIC VARIATION

The Sri Lankan race of the Malabar trogon is distinctly smaller than *malabaricus,* which in turn is slightly smaller than *legerli.* The Sri Lankan from is also slightly paler, mainly on the head and breast of the male. The brightest-colored race is *legerli,* males being bright yellowish brown and paler above than *malabaricus,* which has a darker head and neck than *fasciatus.*

ECOLOGY

Habitats   In India, semi-evergreen and moist deciduous forests from the low plains to 1,200 m (3,900 ft) in central India are favored habitats of the Malabar trogon. In Sri Lanka the birds similarly prefer evergreen and moist deciduous forests, as well as mixed bamboo jungles, from the plains to more than 1,800 m (5,900 ft; Ali and Ripley 1983). Ali (1949) stated that the birds prefer secondary jungle, with an abundance of seedlings, rattan brakes, and bamboo.

Foods and Foraging Ecology   Foods of the Malabar trogon include caterpillars, beetles, grasshoppers, cicadas, and other insects, as well as some leaves and berries. The birds are said to be semicrepuscular, sometimes hunting until well after sunset. The birds may catch insects in flight by snatching them from leaves while hovering or clinging briefly or by momentarily swooping down to the ground (Ali and Ripley 1983). Henry (1971) also observed them eating moths and a stick insect.

Map 26. Distribution of the Malabar trogon.

BEHAVIOR

General, Social, and Sexual Behavior   Much like the other trogons in its general demeanor, the Malabar trogon has the typical trogon trait of bending forward and flicking its tail upward when alarmed and simultaneously uttering a mewing twitter. The birds are usually seen alone or in well-separated pairs, perching silently on tree branches for long periods. They may flit from one site to another, flicking their tail open and shut as they go, briefly flashing the outer white feather tips. (Ali and Ripley 1983).

Vocalizations   Little about the vocalizations of the Malabar trogon is known beyond what has been mentioned in the identification section. Besides the three- to five-note *mew* call, presumably the territorial announcement, there is also a low, rolling creak, *krr-r-r,* that is used as an alarm note.

BREEDING BIOLOGY

Chronology of Breeding   In India the nesting season of the Malabar trogon extends from February to May, and in Sri Lanka from March to June but mainly March and April (Ali and Ripley 1983). There is a clutch of four eggs in the Western Foundation of Vertebrate Zoology collection that was obtained in Travancore (now part of Kerala), India, on March 1.

Nest Sites   The usual nest site of the Malabar trogon is an unlined natural cavity, in a broken tree stump or snag in dense forest, usually located less than 6 m (20 ft) above ground (Ali and Ripley 1983). Ali (1949) described once finding a nest that was a dove-like platform of twigs, on a cane stem about 2.4 m (8 ft) above ground. This nest he described sounds remarkably like an actual dove nest, and no trogon has ever been known to construct a nest of twigs.

Eggs and Incubation   The clutch of the Malabar trogon ranges from two to four eggs, which are incubated by both parents. The incubation period is still unknown (Ali and Ripley 1983).

Brood Rearing   No information available.

CONSERVATION AND EVOLUTIONARY RELATIONSHIPS

The moderately wide range of the Malabar trogon and its seeming preference for second-growth forests should cushion it somewhat from conservation difficulties.

RED-NAPED TROGON

Harpactes kasumba (Raffles) 1822

OTHER VERNACULAR NAMES

Kasumba trogon.

RANGE

Lowland forests (to ca. 600 m [2,000 ft]) of Malay Peninsula (north to Pattani in Thailand, south formerly to Singapore, more recently to Johore), Sumatra, and Borneo. See map 27.

SUBSPECIES

Harpactes kasumba kasumba (Raffles) 1822: Malay Peninsula north to ca. lat. 8°30' N in southern Thailand, also Sumatra.
Harpactes kasumba impavidus (Chasen and Kloss) 1931: Borneo. Includes *usa* Harrisson and Hartley 1934.

MORPHOMETRICS

Measurements   Wing, sex and sample size unspecified 140–152 mm (5.5–5.9 in); tail 165–178 mm (6.4–6.9 in; Chasen 1939). Wing, one male 145 mm (5.7 in); one female ca. 145 mm (5.7 in). Tail, one

male ca. 180 mm (7.0 in); one female ca. 183 mm (7.1 in; Ogilvie-Grant 1892). Wing, males (of *kasumba*) 142–148 mm (5.5–5.8 in; average 144.6 mm [5.6 in]); females 142–149 mm (5.6–5.8 in; average 145.6 mm [5.7 in]). Wing, ten males (of *kasumba*) 145–149 mm (5.7–5.8 in); ten females 144–152 mm (5.6–5.9 in). Tail, ten males 164–178 mm (6.4–6.9 in); ten females 153–180 mm (6.0–7.0 in; Wells 1998). Wing, males (of *impavidus*) 131–141 mm (5.1–5.5 in; average 133.5 mm [5.2 in]); females 133–142 mm (5.2–5.5 in; average 138 mm [5.4 in]; Harrisson and Hartley 1934). No information on eggs is available.

Weights   One male from Borneo 72.1 g (2.5 oz; Thompson 1966). An adult male of *kasumba* 115 g (4.0 oz), three adult females 95.5–105 g (3.3–3.7 oz; Wells 1998).

DESCRIPTION (see plate 32)

Adult Male   Head, throat, and upper breast velvety black to very dark gray, red nape-stripe extending from malar area and ear-coverts around hindneck; scapulars, mantle, back, and remaining upperparts below stripe entirely golden brown except for (sometimes) reddish tones on rump; upper wing-

Map 27. Distribution of the red-naped trogon.

coverts and inner secondaries vermiculated black and white; primaries dark brown, with white edgings; middle three pairs of rectrices black-tipped but otherwise golden brown; outer three pairs of rectrices white, patterned on inner vanes with black; white breast band separates blackish upper breast from rosy red abdomen, under tail-coverts, and flanks; bill mostly cobalt-blue, black at tip and on culmen; fleshy margin of mouth also cobalt-blue; iris dark brown to reddish chestnut; eye-ring violet-blue (lighter above, more lavender below); lower eyelid dirty pinkish; feet and toes blue-gray to violet-blue, sometimes pinkish.

Adult Female   Similar in patterning to male, but black of head, neck, and breast replaced by medium gray; red nape-stripe and underparts replaced by golden brown or (on posterior underparts) brighter, yellowish brown; black-and-white wing-covert and secondary vermiculations replaced by similar dark brown and yellow patterning; no white breast band; bill black on distal half and culmen, otherwise dull blue; iris brown; bare eye-ring dull blue; feet and toes blue-gray.

Immature   Older immature males intermediate in plumage between adult males and females; immature females different from adults in having wing-coverts and secondaries more widely barred and margined with buff; middle rectrices pointed and lacking black terminal bars.

Juvenile   Female-like, with buff barring on wing-coverts about same width as intervening black bars; middle rectrices pointed and lacking black terminal bars.

IDENTIFICATION

In the Hand   The red-naped trogon is one of the most distinctively colored Asian trogons, having a red partial collar around the back of the black head and neck, and a drooping white breast band, or "necklace," separating the black chest from the red underparts. The species is most similar to *H. diardii neglectus*, but the former has a scarlet rather than pink nape, a black rather than maroon crown, and a white rather than pink "necklace" and lacks the black stippling on the white-tipped outer rectrices.

Females are much less distinctive but show a clear color break separating the grayish brown chest from the more cinnamon-toned abdomen and, unlike the other variably smaller cinnamon-breasted species, have a dark grayish-olive breast, and the face is as dark as the rest of the head.

In the Field   As noted above, the bright red (not pinkish) partial collar around the nape of the otherwise black head and its white (not pink) breast band are distinctive for male red-naped trogons, but females have few diagnostic field marks. They closely resemble orange-breasted and cinnamon-rumped females in having cinnamon underparts and brown breasts, but the transition between these colors is more abrupt in the red-naped than in the other two species. The red-naped trogon is limited to lowland evergreen forests up to 600 m (2,000 ft) elevation. Its vocalizations are still poorly described, but a soft, chuckling *purr* or *churr* is said to be uttered at intervals as the tail is moved rapidly back and forth. See also the vocalizations section.

GEOGRAPHIC VARIATION

The Bornean race of the red-naped trogon (*impavidus*) differs from the nominate form only in being slightly smaller. There is no constant color difference among males, but females of *impavidus* tend to have a grayer throat.

ECOLOGY

Habitats   In Sarawak the red-naped trogon has been seen as high as about 625 m (2,100 ft), mostly in primary forests (Harrisson and Hartley 1934) and occasionally also in logged forests. On the Malay Peninsula it occurs in lowland jungle as high as 470–625 m (1,500–2,100 ft, according to Robinson 1928) or up to 560 m (1,800 ft, according to Medway and Wells 1976). It occupies the midstratum and shaded parts of the upper canopy of lowland evergreen forests but also occurs in peatswamp forest that has reached a closed-canopy stage (Wells 1998).

Foods and Foraging Ecology   Like other trogons the red-naped trogon still-hunts from perches and makes short sallies out to catch prey from leaves or other vegetation (Wells 1998).

BEHAVIOR

General, Social, and Sexual Behavior   Little information on the behavior of the red-naped trogon is available; it is a relatively solitary species.

Vocalizations   Little is known of the vocalizations of the red-naped trogon beyond what is mentioned in the identification section. MacKinnon and Phillipps (1993) diagrammed three vocalizations of this species, all of which consist of repeated slurred notes, the sequence having either constant or descending pitch ranges. Medway and Wells (1976) described the male's vocalization as a descending sequence of three or four harsh *kau* notes, and the female's as a harsh rattle. Wells (1998) also more recently stated that the advertising call of both sexes never has more than six *taup* or *kaup* notes, which are spaced and even toned. The male's voice is lower in pitch than the female's, and his notes are uttered more slowly.

BREEDING BIOLOGY

Chronology of Breeding   No information on breeding chronology is available for the red-naped trogon, but primary or secondary molt has been seen from late March to late November, with some comple-tions by mid-August, suggesting a spring and summer breeding period. One active clutch was monitored from July 30 to August 30 (Wells 1998).

Nest Sites   One red-naped trogon nest was found 1.2 m (4 ft) high in a partially rotted stump 1.6 m (5 ft) tall and 18 cm (7.1 in) in diameter (Wells 1998).

Eggs and Incubation   Red-naped trogon eggs are said to be plain white, with one incubated clutch having two eggs. Daytime visits over a week-long period indicated that only the male was incubating at that time. (Wells 1998).

Brood Rearing   No information available.

CONSERVATION AND EVOLUTIONARY RELATIONSHIPS

The red-naped trogon is a generally uncommon to occasional species, with rather strong plumage similarities to Diard's trogon, its probable nearest relative. The former's fairly large overall distribution should help in its conservation, as should its apparent ability to colonize mature second-growth forests. In peninsular Malaysia it is local and uncommon to sparse in the north, but in the south it is still more or less common (Wells 1998).

DIARD'S TROGON

Harpactes diardii (Temminck) 1832

OTHER VERNACULAR NAMES

Blasius's trogon (*sumatranus*), Temminck's trogon.

RANGE

Lowland forests (to ca. 950 m [3,100 ft]) on the Malay Peninsula, Sumatra, and Borneo. See map 28.

SUBSPECIES

Harpactes diardii neglectus Forbes and Robinson 1899: Malay Peninsula, south formerly to Singapore, more recently to Johore, also north through peninsular Thailand to about lat. 9° N. Not accepted by Peters (1945) or Wells (1998) but recognized by Riley (1938).

Harpactes diardii sumatranus Blasius 1896: Sumatra.

Harpactes diardii diardii (Temminck) 1832: Borneo, Bangka Island, and Lingga Archipelago.

MORPHOMETRICS

Measurements   Wing, sex and sample size unspecified 142–147 mm (5.5–5.7 in); tail, 160–170 mm (6.2–6.6 in; Robinson 1928). Wing, one male ca. 145 mm (5.7 in); one female ca. 142 mm (5.5 in). Tail, one male ca. 168 mm (6.6 in); one female ca. 170 mm (6.6 in; Ogilvie-Grant 1892). Wing, one male (of *sumatranus*) 134 mm (5.2 in); one female 135 mm (5.3 in). Tail, one male 163 mm (6.4 in). Wing, one male (of *diardii*) 131.5 mm (5.1 in); one female 132 mm (5.1 in). Tail, one male 158 mm (6.2 in; specimens FMNH). Wing, nineteen males (from Malaysia) 140–153 mm (5.5–6.0 in); eleven females 140–151 mm (5.5–5.9 in). Tail, males 158–175 mm (6.2–6.8 in); females 155–167 mm (6.0–6.5 in; Wells 1998). Egg 28.7 × 26.5 mm (1.1 × 1.0 in; Chasen 1939).

Weights   Five birds from Malaysia and Borneo average 101.0 g (3.5 oz; Thompson 1966; Wong 1986). Six males from Malaysia 90.7–110.8 g (3.2–4.3 oz), five females 87.5–106.3 g (3.1–3.7 oz; Wells 1998). Estimated egg mass 10.4 g (0.4 oz).

DESCRIPTION (see plate 33)

Adult Male   Head, chin, throat, neck, and upper breast black; nuchal collar pink (located as in red-naped, but less distinct); dark claret-colored patch

Map 28. Distribution of Diard's trogon.

on rear of crown; below black upper breast a whitish to pink breast band (continuous posteriorly with pink nuchal collar); upperparts from mantle backwards golden brown; primaries dark brown with white edging; upper wing-coverts and inner secondaries vermiculated black and white; middle rectrices chestnut brown with black tips; next two pairs of rectrices black; three outer pairs of rectrices white, heavily freckled with black; flanks, abdomen, and under tail-coverts pinkish crimson; bill (including fleshy margins of mouth) dark cobalt-blue; culmen and cutting edges black; iris brown to hazel-red; eye-ring mauve-blue, heliotrope, or violet, with dirty white patch on lower eyelid and pale sky-blue on cheeks; feet and toes brownish lilac to lavender or grayish lavender.

Adult Female   Similar to male in general patterning, but duller; head, neck, breast, scapulars, mantle, back, rump, and upper tail-coverts rusty brown to olive-brown, lighter and more olive toned on back, which color separates darker brown breast from pinkish (not crimson, as in males) underparts as an indistinct breast band corresponding to that of the male; vermiculated areas of wing-coverts and inner secondaries dark brown and light olive-brown; rectrices as in male, outer three pairs similarly freckled with black; soft-part colors similar to those of male, but bill area duller blue.

Diard's trogon, adult female. Photograph by author.

Immature   Similar to adult female, but spotted with white on wing-coverts and paler pink below. In immature males rectrices more pointed than in adults; white areas of outer rectrices less freckled with black; black of chest mixed with sandy rufous; upper wing-coverts more female-like (dark brown and olive-brown). Immature females similar to adult females except for white spotting and heavier barring on upper wing-coverts and more-pointed rectrices.

Juvenile   Both sexes female-like, but with paler pink underparts, an entirely brown breast, and wing-coverts and secondaries coarsely and more evenly barred with buff.

IDENTIFICATION

In the Hand   Diard's trogon very closely resembles the Malabar trogon of India but has a pale pink nuchal collar and a narrow pink (not white) necklace separating the black throat and chest from the pinkish-crimson underparts. The three outer rectrices are white with distinct black vermiculations or stippling in both sexes, this characteristic separating it from other trogons of the region. Females are dark brown on the head, chest, and upperparts but are distinctly pinkish below, with no trace of the pale necklace present in males.

In the Field   Like several other species of the region, male Diard's trogons have black heads and crimson underparts, the two colors separated by a narrow pink breast band that also extends back around the head and hindneck. Females lack the pink breast band, but are distinctly pinkish below their brown chest. This is a species of moist forests at low to moderate elevations (to 900 m [3,000 ft]). See also the vocalizations section.

GEOGRAPHIC VARIATION

Riley (1938) found that Diard's trogons from the Malaysian mainland are less strongly marked with red on the crown than are those from Sumatra and Bangka Island, and he thus recognized *neglectus* as distinct.

ECOLOGY

Habitats   Diard's trogon is largely limited to heavy lowland jungle, only infrequently ascending to elevations as high as about 2,800 m (9,300 ft) on the Malay Peninsula (Robinson 1928). It favors the middle to lower strata of semi-evergreen and evergreen forests up to about 600 m (2,000 ft) in Thailand but to 900 m (3,000 ft) in the southern part of the Malay Peninsula. Up to two pairs were found in one 15-hectare (37-acre) plot (Wells 1998).

Foods and Foraging Ecology   Little detailed information on the foods of Diard's trogon is available, but caterpillars, stick insects, locustids, and fruit have all been reported (Smythes 1968). Wells (1998) stated that, like other trogons, Diard's trogon sallies out from perches to take prey from vegetation and that one instance of fruit eating (large figs) has been reported.

BEHAVIOR

General, Social, and Sexual Behavior   Diard's trogon is a solitary species, and there are no reported social interactions. One adult female was retrapped 63 moths after initial capture, providing the only life span data yet available (Wells 1998).

Vocalizations   Little about Diard's trogon vocalizations is known beyond what is mentioned in the identification section. MacKinnon and Phillipps (1993) illustrated two vocalization sequences of this species, both consisting of up-slurred or down-slurred sequences of notes that diminish in pitch and duration. Medway and Wells (1976) stated that the male's usual call is a cadence of ten to twelve *kau* notes, the second higher than the first and the last ones either falling off or all evenly spaced. Wells (1998) added that this advertising call is uttered by

both sexes and is faster in cadence (but with all the notes distinct) than the similar calls of other Malaysian species.

BREEDING BIOLOGY

Chronology of Breeding   Eggs of Diard's trogon have been found in May and June, and nestlings have been reported in March. Extrapolations suggest that laying occurs from February to at least mid-May. Wing molt has been observed among adults from late March to mid-October, with the apparent peak coming between June and August (Wells 1998).

Nest Sites   Diard's trogon nests found in Perak have been located 1.25–2.5 m (4–8 ft) above ground, in old, rotted tree stumps. In one instance the opening was only about 11 cm (4.3 in) in diameter, and this diameter and the distance to the base of the cavity from the bottom of the entrance hole were about the same (Chasen 1939). Wells (1998) stated that three nests were all in rotted stumps, the cavities 1.2–3 m (4–10 ft) above ground. One cavity was no more than 12 cm (4.7 in) in maximum dimension.

Eggs and Incubation   Diard's trogon eggs are said to be white, but the clutch size seems to be unreported. A brood of two fledglings has been seen. On two occasions the male was attending the nest during afternoon hours (Wells 1998).

Brood Rearing   No information available.

CONSERVATION AND EVOLUTIONARY RELATIONSHIPS

Like its probable nearest relative (the red-naped trogon), Diard's trogon is a lowland-forest-dependent species that is no doubt declining with deforestation. But like the red-naped, it has a broad range, which should make it somewhat less of a conservation problem than, for example, the blue-tailed trogon. Wells (1998) stated that Diard's trogon is fairly common to (in the north) uncommon but needs closed-canopy forests and accepts well-regenerated secondary growth. He suggested that it may form a superspecies with *H. ardens*.

PHILIPPINE TROGON

Harpactes ardens (Temminck) 1826

OTHER VERNACULAR NAMES

Rosy-breasted trogon.

RANGE

Philippine Islands forests, including Luzon, Polillo, Catanduanes, Marinduque, Samar, Leyte, Bohol, Dinagat, Mindanao, and Basilan. See map 29.

SUBSPECIES

Harpactes ardens ardens (Temminck) 1826: Mindanao, Basilan, and Dinagat Islands.
Harpactes ardens minor Manuel 1958: Polillo Island.
Harpactes ardens linae Rand and Rabor 1959: Bohol, Leyte, and Samar Islands.
Harpactes ardens herberti Parkes 1970: northeastern Luzon and Marinduque Islands.
Harpactes ardens luzoniensis Rand and Rabor: 1952: southern and central Luzon Island.

MORPHOMETRICS

Measurements   Wing, males (of *linae*) 139–145 mm (5.4–5.7 in; average of 6, 143.5 mm [5.6 in]); females 138–148 mm (5.4–5.8 in; average of 5, 142.2 mm [5.5 in]; Rand and Rabor 1960). Tail, both sexes (of *ardens*) 170–181 mm (6.6–7.1 in; average of 3, 175.3 mm [6.8 in]; various sources). Wing, adults (of *minor*) 128–138 mm (5.0–5.4 in; average of 11, 133.1 mm [5.2 in]); adults (of *herberti*) 135–141 mm (5.3–5.5 in; average of 8, 137.5 mm [5.4 in]); of *luzoniensis* 134–143 mm (5.2–5.6 in; average of 19, 137.7 mm [5.4 in]; Parkes 1970). Egg (of *ardens*) 29.9 × 24.2 mm (1.2 × 0.9 in; Schönwetter 1966).

Weights   Eleven birds of both sexes 82.6–114 g (2.9–4.0 oz; average 96.3 g [3.4 oz]; Dunning 1993). Three males of *linae* 82.6–99 g (2.9–3.5 oz; average 93.1 g [3.3 oz]), five females 88.3–98.6 g (3.1–3.5 oz; average 92.3 g [3.2 oz]; Rand and Rabor 1960). Four males of *ardens* 88.6–107.32 g (3.1–3.8 oz; average 96.0 g [3.4 oz]), seven females 81.6–99.8 g (2.9–3.5 oz; average 93.9 g [3.3 oz]); nine males of *linae* 88.6–100 g

(3.1–3.5 oz; average 93.1 g [3.3 oz]), ten females 88.3–102.5 g (3.1–3.6 oz; average 94.2 g [3.3 oz]); 27 males of *luzoniensis* 72.1–94.2 g (2.5–3.3 oz; average 81.0 g [2.8 oz]),.27 females 71.3–100 g (2.5–3.5 oz; average 87.7 g [3.1 oz]; specimens, FMNH). Estimated egg weight 9.5 g (0.3 oz; Schönwetter 1966).

DESCRIPTION (mainly adapted from Rabor 1977; see plate 34)

Adult Male   Head black, gradually shading to dark reddish purple on crown, nape, and ear-coverts; chin and throat black; upperparts orange-brown, lighter on rump and tail-coverts; primaries black with white feather edgings; wing-coverts and inner secondaries with fine black-and-white vermiculations; chest pink; remaining underparts and flanks scarlet, sometimes separated from pink breast by narrow whitish band; three outer pairs of rectrices white with black bases and dusky inner edges, next two pairs black, innermost pair light brown with black tips; bill yellow with a greenish base; iris dark brown; eye-ring blue to purple; feet and toes olive.

Adult Female   Upperparts similar to those of male, but head, nape, and ear-coverts olive; shoulders and back olive-brown; rump and tail-coverts orange-brown; chin and throat black; underparts mostly light rusty buff, becoming paler on abdomen and darker on under tail-coverts; wings like those of male, but vermiculations light brown instead of white; tail like that of male, but with less white on outer rectrices; bill yellowish, with brownish tip; other soft-part colors similar to those of male.

Immature   Immature males resemble adult female but have much wider brown barring on secondaries and lack black terminal bars on middle rectrices; rectrices probably also narrower and more pointed than in adults; young females closely resemble adult females, except for rectrix and wing-covert differences just mentioned as being typical of young males.

Juvenile   Probably similar to immature.

Map 29. Distribution of the Philippine trogon, showing islands for which specimens are known (inked). Dashed lines indicate known limits of currently accepted subspecies.

IDENTIFICATION

In the Hand   The Philippine trogon is notable for its blackish-purple head, which contrasts with a pink throat and chest, a narrow, dark breast band, and a scarlet lower breast and abdomen. Females have a black head and are bright cinnamon-yellow from the lower throat to the under tail-coverts. Both sexes have white outer rectrices.

In the Field   The Philippine trogon is the only trogon occurring on the Philippine Islands and thus should be easily identified. It is almost as large the largest pigeons on the islands, but there are no pigeons with scarlet present below, although pink may occur. This trogon's usual vocalization is a mew-like hoot rather than the cooing sounds typical of doves and pigeons.

GEOGRAPHIC VARIATION

In Philippine trogons, geographic variation is noticeable in plumage coloration (especially upperpart colors) and to some extent in bill length and wing length, with *luzoniensis* having a distinctly shorter bill than *ardens* and with *linae* closer to the latter. The race *minor* also has a short bill and is the only race having a shorter wing length than the others. The race *herberti* has a slightly longer bill than *luzoniensis*, but wing lengths do not differ (Parkes 1970).

ECOLOGY

Habitats   Philippine trogons are associated with deep forests, especially the dipterocarp and middle montane types. The birds may occasionally leave these forests to forage in dense areas of secondary forest (Rabor 1977). They have been collected at altitudes of 100–600 m (300–2,000 ft).

Foods and Foraging Ecology   The foods of the Philippine trogon are said to consist predominantly of insects (Rabor 1977).

BEHAVIOR

General, Social, and Sexual Behavior   Like other trogons, Philippine trogons sit quietly on branches of the understory and are usually not seen until they take flight. In spite of being one of the most beautiful Philippine bird species, very little more is known of it than was the case when Gould monographed the family more than a century ago.

Vocalizations   Little about Philippine trogon vocalizations is known beyond what is mentioned in the identification section.

BREEDING BIOLOGY

Chronology of Breeding   No detailed information is available, but breeding birds have been collected in April.

Nest Sites   No information available.

Eggs and Incubation   No information available.

Brood Rearing   No information available.

CONSERVATION AND EVOLUTIONARY RELATIONSHIPS

Only 2 percent of the primary forests of the Philippines still existed by the mid-1990s, and the Philippine trogon, a forest-dependent species, has certainly suffered greatly as a result. Among the few locations where the species could still be reliably found in the mid-1990s are Quezon National Park and Angat (Luzon), Rajah Sikatuna National Park (Bohol), and Mangagoy (Mindanao; Wheatley 1996a). The species' relationships are uncertain, but Wells (1998) has suggested that this form and *H. diardii* may form a superspecies. One might imagine that the Philippine trogon's nearest relatives could be expected to occur in Borneo.

WHITEHEAD'S TROGON

Harpactes whiteheadi Sharpe 1888

OTHER VERNACULAR NAMES

None in general use.

RANGE

Montane forests (above 1,000 m, 3,300 ft) of northern Borneo, from Mount Kinabalu (Sabah) south along the spinal range to Gunung Mulu, Usun Apau Plateau, and Mount Dulit (Sarawak). See map 30.

SUBSPECIES

None recognized.

MORPHOMETRICS

Measurements Wing, male (type specimen) ca. 132 mm (5.1 in); tail ca. 165 mm (6.4 in). Wing, female (type) ca. 140 mm (5.5 in); tail ca. 178 mm (6.9 in; Sharpe 1888). Wing, one male ca. 132 mm (5.1 in); one female ca. 137 mm (5.3 in). Tail, one male ca. 165 mm (6.4 in); one female ca. 163 mm (6.4 in; Ogilvie-Grant 1892). No information on egg measurements is available.

Weights No information on adult or egg weights is available.

DESCRIPTION (mainly after Sharpe 1888; see plate 35)

Adult Male Crown brilliant scarlet; lores black; ear-coverts scarlet, like the head; throat blackish; chest pearly gray, gray color extending in crescent-like pattern up sides of throat, lateral feathers tipped with scarlet; upperparts generally bright cinnamon; scapulars like back; wing-coverts and alular feathers black with fine white vermiculations; primaries black with distinct white feather margins; secondaries black, patterned on outer vanes with white vermiculations, like wing-coverts; upper tail-coverts like back; two middle rectrices deep cinnamon with broad black tip, next pair black with chestnut shaft, remainder black with black shafts and some white near tips of outer vanes, outermost feathers white for nearly the terminal half and a considerable additional distance along outer vanes; remainder of underparts brilliant scarlet, deeper in hue just below gray chest; abdomen slightly paler and more rose colored; thighs blackish with cinnamon feather tips; under tail-coverts like back; under wing-coverts black; bill and bare cheek areas blue; eye-ring blue; iris reddish brown; feet and toes dull brownish pink to pale gray.

Adult Female Differs from male in having cinnamon head, as well as cinnamon underparts from chest downwards (areas that are scarlet in adult male); upper wing-coverts and secondaries narrowly barred or vermiculated with ochreous brown instead of white; soft-part colors apparently similar to those of male.

Immature Undescribed, but probably similar to those of related species, with narrower, more pointed rectrices than in adults, buffy edging or spots on inner flight feathers and wing-coverts, and coarser barring on these coverts.

Juvenile No information available.

IDENTIFICATION

In the Hand Whitehead's trogon, a species of the dark, primary cloud forests of northern Borneo is notable for its two-toned head (a black chin and throat, contrasting with a scarlet [males] or cinnamon [females] forehead, crown, nape, and hindneck). Female scarlet-rumped trogons are fairly similar, but they have red rather than cinnamon rumps and flanks.

In the Field Of the five trogon species occurring in Borneo, Whitehead's trogon is the only one having a gray chest in both sexes. The adult male is unique in having a bright red head but a black chin and throat; in females the red is replaced by olive brown, which color also replaces the red present on the male's abdomen. The species is limited to dark and damp montane forests above 1,000 m (3,300 ft). Virtually unstudied in the field, its usual call is said to consist of growling notes, but it probably has a repeated advertisement call of some type.

GEOGRAPHIC VARIATION

None described.

ECOLOGY

Habitats   The little known Whitehead's trogon is limited to montane forests above 1,000 m (3,300 ft) and is usually encountered at about 1,200–1,300 m (3,900–4,300 ft), in wet cloud forest. It occurs at higher altitudes than any of the other five trogons of Sabah, with the orange-breasted typically extending about 300 m (1,000 ft) lower than Whitehead's (Gore 1968).

Foods and Foraging Ecology   Whitehead's trogons are known to eat grasshoppers, ants, and large green leaf insects, often capturing prey in flight (Smythes 1968). Oddly, stones have been found in stomach contents, suggesting that perhaps some seeds might be eaten too.

BEHAVIOR

General, Social, and Sexual Behavior   No specific information available.

Vocalizations   Nothing is known beyond what is mentioned in the identification section.

BREEDING BIOLOGY

No information available.

CONSERVATION AND EVOLUTIONARY RELATIONSHIPS

Whitehead's trogon has the smallest distribution of any of the Asian trogons, and indeed it is one of the

Map 30. Presumed distribution of Whitehead's trogon (dashed line); dots indicate specific locality records. Inset map shows regions above 1,000 m (3,300 ft) in northern Borneo (inked), where occurrence is most probable.

smallest of all trogons. It is protected in Kinabalu National Park; but the majority of Borneo's cloud forests are under concerted attack by loggers, and the status of this little-known species could already be precarious. Its evolutionary affinities are uncertain; Wells (1998) suggested that *H. erythrocephalus* may be its nearest relative; however, *erythrocephalus* now occurs no closer than Sumatra, and Diard's trogon and the red-naped trogon would seem to be more likely candidates on the basis of their current distributions.

CINNAMON-RUMPED TROGON

Harpactes orrhophaeus (Cabanis and Heine 1863)

OTHER VERNACULAR NAMES

Malacca trogon.

RANGE

Tall primary lowland forests of the Malay Peninsula (to ca. 200 m [700 ft]) and midlevel montane forests to 1,500 m [4,900 ft]) of Sumatra and northern and central Borneo. See map 31.

SUBSPECIES

Harpactes orrhophaeus orrhophaeus (Cabanis and Heine) 1863: Malay Peninsula (north to Nakhon Si Thammarat in peninsular Thailand, south to Malacca and Johore), Sumatra.
Harpactes orrhophaeus vidua Ogilvie-Grant 1892: northern Borneo (1,000–1,500 m; 3,300–4,900 ft).

MORPHOMETRICS

Measurements   Wing, unspecified sex and sample size 107–112 mm (4.2–4.4 in); tail, 132–142 mm (5.1–5.5 in; Chasen 1939). Wing, one male ca. 109 mm (4.3 in); one female ca. 109 mm (4.3 in). Tail, one male ca. 129 mm (5.0 in); one female ca. 127 mm (5.0 in; Ogilvie-Grant 1892). Wing, one male (of *orrhophaeus*) 100 mm (3.9 in); tail, 122 mm (4.8 in, Riley 1938). Wing, fifteen males (of *orrhophaeus*) 104–111 mm (4.1–4.3 in); fifteen females 106–114 mm (4.1–4.4 in). Tail, fifteen males 121–135 mm (4.7–5.3 in); fifteen females 121–136 mm (4.7–5.3 in; Wells 1998). No egg measurements are available.

Weights   Twelve birds of unspecified sex from Malaysia average 52.7 g (1.8 oz; Wong 1986). Ten males from Malaysia 46.8–60.6 g (1.6–2.1 oz), seventeen females 45.7–60.0 g (1.6–2.1 oz; Wells 1998).

DESCRIPTION (see plate 36)

Adult Male   Head, neck, and throat all black, color abruptly changing to medium brown on scapulars, mantle, and back and continuing down to rump and upper tail-coverts as cinnamon-brown; primaries dark brown, edged with white; inner secondaries and upper wing-coverts vermiculated black and white; middle pair of rectrices medium cinnamon-brown with black tips, the next two pairs black, and three outer pairs predominantly white; underparts

Map 31. Distribution of the cinnamon-rumped trogon.

and flanks crimson-red from edge of black throat downward (no white breast band), the crimson paler posteriorly becoming pinkish on under tail-coverts; bill (including fleshy edge of mouth) bright cobalt-blue with black culmen and tip; iris brown; eye-ring bright blue; feet and toes gray.

Adult Female   Head and upper throat mostly dark brown except for rusty brown area surrounding eye; scapulars and back dark brown, becoming lighter on lower back, rump, and upper tail-coverts; primaries dark brown; upper wing-coverts and inner secondaries vermiculated with dark brown and yellow; middle pair of rectrices medium brown like back, next two pairs black, outer three pairs predominantly white; flanks and underparts below dark brown throat progressively lighter and more yellowish, becoming pale yellow on under tail-coverts; soft-part colors similar to those of male, eye-ring and fleshy edge of mouth paler blue.

Immature   Undescribed in detail, but presumably nearly identical to immatures of closely related species: with buffy edgings or spotting on wing feathers, rectrices narrow, more pointed than in adults, middle pair without black terminal bands. According to Wells (1998), face of immatures slightly brighter than in juvenile scarlet-rumped, and barred upper wing-coverts with less black and more cinnamon evident.

Juvenile   No information available.

IDENTIFICATION

In the Hand   In only two Asian trogons do the males have entirely black heads and scarlet underparts: the cinnamon-rumped trogon and the scarlet-rumped trogon, the latter usually easily separated by its bright red rather than cinnamon-brown rump area. Males of the former also are less bright below, and the markings on the wing-coverts and secondaries are narrower and more wavy. Females of the two species are quite similar, but the head of the cinnamon-rumped is two toned, with a paler rusty brown facial area surrounded by a darker brown forehead, crown, and hindneck. Furthermore, the chin and throat feathers of the cinnamon-rumped are mostly black; the rump, upper tail-coverts, and

underparts are all rusty buff; and the rust bars on the secondaries are more widely spaced. Linear measurements between these two species may overlap, but the bill of the cinnamon-rumped is distinctly more robust than is typical of the generally smaller scarlet-rumped.

In the Field   The differences listed above provide the best field marks for separating the cinnamon-rumped and scarlet-rumped trogons. Both are lowland-forest species of Malaysia, Sumatra, and Borneo, but at least in the Sunda Islands the scarlet-rumped is found only in low-altitude forests (below 600 m [2,000 ft]), whereas the cinnamon-rumped reaches higher altitudes (up to 1,000–1,500 m [3,300–4,900 ft]). Both utter descending-scale vocalizations, the cinnamon-rumped trogon's call typically consisting of three or four *taup* notes, and the scarlet-rumped trogon's of about twelve *yau* notes. See also the vocalizations section.

GEOGRAPHIC VARIATION

The race *vidua* differs from the nominate form of the cinnamon-rumped trogon in that the female of the former has an olive-brown rather than blackish-brown crown; in addition, the back color is also olive-brown rather than rufous-brown, the rump and upper tail-coverts are olive-buff rather than dull rust, the wing-coverts and secondaries are more widely barred, and the chin and throat are rust-red rather than mostly black (Ogilvie-Grant 1892).

ECOLOGY

Habitats   The cinnamon-rumped trogon is largely limited to lowland forests on the Malay Peninsula, at elevations under 200 m (700 ft; Medway and Wells 1976). There it is perhaps the commonest trogon in plains-level evergreen forests, but it is rare or absent from more seasonal forest types such as deciduous or semi-evergreen forests. A density of four pairs was estimated in one 15-hectare (37-acre) area (Wells 1998). In Borneo, however, the species has been found mainly in montane forests 1,000–1,400 m (3,300–4,600 ft) in elevation but sometimes also occurs in lowlands well away from mountains (Smythes 1968).

Foods and Foraging Ecology   No specific information on the foods of the cinnamon-rumped trogon is available. Like other trogons it still-hunts from branches in rather dense understory vegetation and is usually found within 2–3 m (7–10 ft) of the ground (Wells 1998).

BEHAVIOR

General, Social, and Sexual Behavior   The cinnamon-rumped trogon is a solitary species that seems to show no evidence of movement. Two of fourteen birds banded in one area were still present on their territories 109 and 115 months later, and another bird was retrapped after an interval of 156 months at the site where it had initially been banded (Wells 1998).

Vocalizations   Little about the vocalizations of the cinnamon-rumped trogon is known beyond what is mentioned in the identification section. An explosive *purrr* call, perhaps an alarm note, has been reported (MacKinnon and Phillipps 1993). Wells (1998) stated that this species' calls are much like those of the red-naped trogon, but weaker, and that the calls include a slightly declining series of three to four spaced and downwardly inflected *ta-aup* notes.

BREEDING BIOLOGY

Chronology of Breeding   Cinnamon-rumped trogon eggs have been found in early April and mid-June, and nestlings have been seen in mid-April (Wells 1998). Adults in primary molt have been reported from early June to mid-October, suggesting a spring breeding period (Wells 1998).

Nest Sites   Three cinnamon-rumped trogon nest sites have been found, all of which were cavities 1–1.5 m (3–5 ft) above ground in rotted forest stumps. One of the cavities was 10 cm (3.9 in) deep (Wells 1998).

Eggs and Incubation   Cinnamon-rumped trogon eggs are known to be plain white, and the clutch is evidently two (information based on one clutch and one brood; Wells 1998).

Brood Rearing   Little information on the brood rearing of cinnamon-rumped trogons is available. Both sexes are known to incubate and tend the nestlings (Wells 1998).

CONSERVATION AND EVOLUTIONARY RELATIONSHIPS

The cinnamon-rumped trogon is an uncommon to rare bird of lowland (Malay Peninsula) and mid-altitude (Sundas) forests, forests that are being rapidly destroyed in Southeast Asia; and the species warrants close attention from conservationists. It was considered "very rare" by Delacour (1947) in Sumatra and Borneo and "rare" by MacKinnon and Phillipps (1993). Wells (1998) stated that it is in the "front rank" of species dependent on low-altitude, closed-canopy forests and threatened by habitat loss. It is seemingly a very close relative of the smaller and more brightly colored scarlet-rumped trogon, which is much more common. However, Wells (1998) did not make this assumption. The undertail pattern is similar to that of both the scarlet-rumped and orange-breasted trogons.

SCARLET-RUMPED TROGON

Harpactes duvaucelii (Temminck) 1824

OTHER VERNACULAR NAMES

Duvaucel's trogon, red-rumped trogon.

RANGE

Lowland forests of the Malay Peninsula (to ca. 1,000 m [3,300 ft]) north to southwestern Thailand (lat. 14° N); Sumatra; Borneo; also Riau Archipelago and Banjak, Bangka, Belitung, and Natuna Islands. See map 32.

SUBSPECIES

None recognized.

MORPHOMETRICS

Measurements   Wing, one unsexed adult ca. 106.7 mm (4.2 in); tail ca. 142 mm (5.5 in; Robinson 1928). Wing, one male ca. 104 mm (4.1 in); one female ca. 107 mm (4.2 in). Tail, one male ca. 129 mm (5.0 in); one female ca. 127 mm (5.0 in; Ogilvie-Grant 1892). Wing, eight males from Malay Peninsula 99–103 mm (3.9–4.0 in; average 101 mm [3.9 in]); four males from Sumatra and Bangka Island 103.5–109 mm (4.0–4.3 in; average 106.5 mm [4.2 in]); eight males from Borneo 94.5–102.5 mm (3.7–4.0 in; average 99.9 mm [3.9 in]; Riley 1938). Wing, twenty males 101–109 mm (3.9–4.3 in); thirteen females 101–111 mm (3.9–4.3 in). Tail, twenty males 113–136 mm (4.4–5.3 in); thirteen females 113–132 mm (4.4–5.1 in; Wells 1998). Egg 23.7 × 19.9 mm (0.9 × 0.8 in; Schönwetter 1966).

Weights   Fourteen adult males 36.4–43 g (1.3–1.5 oz), eight adult females 33.8–43.1 g (1.2–1.5 oz; Wells 1998). Five birds (sexes mixed or unspecified) from Malaysia and Borneo average 35.2 g (1.2 oz; Thompson 1966; Wong 1986). One individual declined from ca. 41.5 g to ca. 38 g (1.5 oz to 1.3 oz, masses interpolated from published graph) during the molting period (Fogden 1972). Estimated egg weight 5.2 g (0.2 oz; Schönwetter 1966).

DESCRIPTION (see plate 37)

Adult Male   Head, neck, and throat black, changing abruptly to cinnamon-brown on scapulars, mantle, and back; primaries dark brown with white edging;

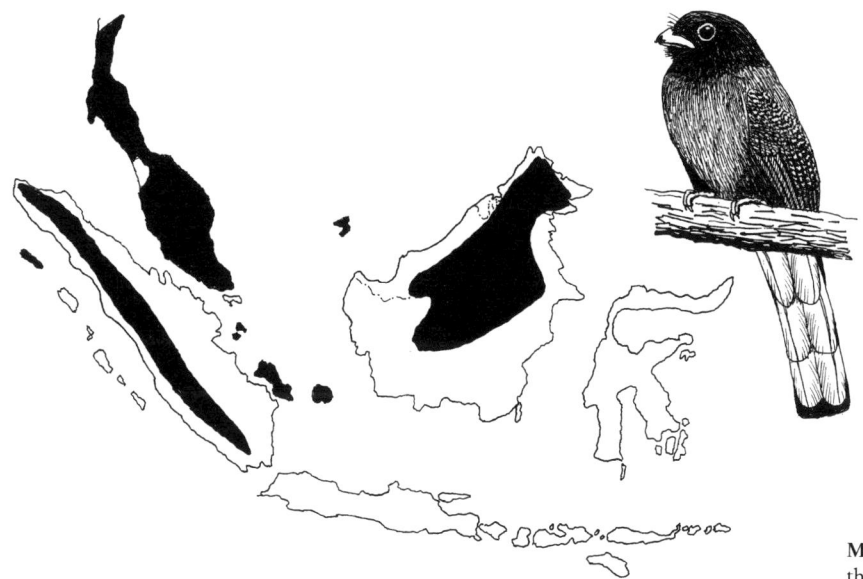

Map 32. Distribution of the scarlet-rumped trogon.

upper wing-coverts and inner secondaries vermiculated with black and white (tertials vermiculated on both vanes, more-distal secondaries on outer vanes only); rump and upper tail-coverts scarlet; middle pair of rectrices cinnamon-brown like back, with black tips; next two pairs of rectrices black, outer three pairs predominantly white; underparts and flanks, from throat back, scarlet, without a light band separating this area from black throat; thighs grayish; bill (including fleshy edge of mouth) bluish cobalt, with black culmen, tip, and cutting edges; iris brown; eye-ring narrow and silvery to grayish blue, upper lid forming separate broad, brow-like blue arch above eye; feet and toes variously described as pinkish, bluish, or (more unlikely) grayish black.

Adult Female   Head, hindneck, and sides of neck medium brown, becoming dark yellowish brown on throat, which color extends in paler tones to lower breast in a broad band but is replaced on abdomen, flanks, and under tail-coverts by pale pink; scapulars and back medium brown like head, becoming pinkish on rump and upper tail-coverts; upper wing-coverts and inner secondaries vermiculated with dark brown and yellowish brown; middle pair of rectrices brown like back with black tip narrow or absent, next two pairs black, and outer three pairs predominantly white; soft-part colors apparently similar to those of male, but bare eye skin clear pale blue and bill duller blue.

Immature   Similar to adult female but lacking pink on rump; abdomen and under-tail-coverts rufous buff rather than pinkish, and head dark brown, tinged with reddish; iris brown; upper eyelid blue; lower eyelid pink; bill purplish cobalt, maxilla with black culmen and cobalt-blue base; feet and toes dark pink (but also described as blue).

Juvenile   Like adult female but with coarser buff barring on upper wing-coverts, white to buff under tail-coverts, and no black tips on middle rectrices; rectrices pointed, middle pair without a black terminal band, and second pair sometimes with subterminal oval rufous spot. Immature males may pass through stage with wing-covert barring of adult width but with cinnamon rather than white bars (Wells 1998).

IDENTIFICATION

In the Hand   Both sexes of the scarlet-rumped trogon, the smallest Asian trogon, are unique in having rump areas that are red, which color also extends to the flanks of females and the entire underparts of males. The species' smaller size should separate it from Diard's trogons but perhaps not from the less brightly colored cinnamon-rumped. The latter species has a bill that is 9.2–10.4 mm (0.4 in) deep at the nostrils, as compared with 7.1–7.8 mm (0.3 in) in the scarlet-rumped (Wells 1998).

In the Field   Although the red rump color is diagnostic for both sexes of the scarlet-rumped trogon, this feature may not be easily visible in the field. Males have all black heads and a scarlet abdominal and breast color extending up to the throat, where it meets the all black head and neck. Females are reddish on the flanks, but their breast color is cinnamon-yellow, which color terminates at the brownish foreneck. Juveniles are rather uniformly brownish and much like juvenile cinnamon-rumped trogons but have a less contrasting face patch, and their wing-covert barrings are dominated by cinnamon rather than black (Wells 1998). The species is limited to lowland forests from sea level to 400 m (1,300 ft, in Thailand) or sometimes to 1,070 m (3,500 ft, southern Malay Peninsula), and its usual vocalization consists of about 12 descending *yau* notes that are uttered in accelerating cadence. See also the vocalizations section.

GEOGRAPHIC VARIATION

None described.

ECOLOGY

Habitats   The scarlet-rumped trogon is associated with lowland evergreen forests and, to some extent, secondary forest but does not extend into mangroves, cultivated areas, or open country. It shares these habitats with the chestnut-rumped trogon (Chasen 1939). The scarlet-rumped has been recorded as high as 1,100 m (3,600 ft; Robinson 1928). Wells (1998) described the habitats as including semi-evergreen and evergreen lowland forests, as well as mature second-growth forest.

Foods and Foraging Ecology   Foods of the scarlet-rumped trogon reportedly include moths, beetles, "bugs," stick insects, and locustid orthopterans, including larval forms (Robinson 1927). Caterpillars are also evidently eaten (Smythes 1968).

BEHAVIOR

General, Social, and Sexual Behavior   Scarlet-rumped trogons reportedly prefer to perch on the lower leafy branches of trees in deep jungle, especially in shady ravines, sometimes among creeping vines (Robinson 1927). Wells (1998) stated that the birds occupy middle and, less frequently, lower stratum forest levels, from which the birds sortie to catch prey from vegetational surfaces or sometimes from the air. The birds are usually solitary, but at times pairs occur together. They have also been seen following foraging flocks of babblers (Timaliidae).

Vocalizations   Little about the vocalizations of the scarlet-rumped trogon is known beyond what is mentioned in the identification section. MacKinnon and Phillipps (1993) diagrammed a long vocalization sequence consisting of numerous short nonslurred notes that descend in pitch and become both briefer and more frequent. Medway and Wells (1976) described the male as having a descending cadence of about 12 *yau* notes, accelerating with a diminuendo, and the female as uttering a quiet rattle. Wells (1998) added that the individual notes of this series may accelerate to the point that they seem to merge and that the call is uttered much more rapidly than that of other trogons. Besides a rattling whirr, a squirrel-like scolding call has also been noted.

BREEDING BIOLOGY

Chronology of Breeding   Scarlet-rumped trogon eggs have been found during May in Perak, Malaysia (Medway and Wells 1976). There are also records of hatched chicks in mid-March and May (Wells 1998). Adults in wing molt have been reported mainly between May and October, with an apparent peak from June to August (Wells 1998).

Nest Sites   A scarlet-rumped trogon nest found in Perak was located in the hollow of an old stump in evergreen forest (Chasen 1939). Wells (1998) seemingly questioned this record.

Eggs and Incubation   One scarlet-rumped trogon nest with two eggs has been (unreliably) reported; no other information is available.

Brood Rearing   Little information is available, but two scarlet-rumped trogon parents were observed tending their young and providing the majority of their chick's food as late as 17 weeks following hatching (Fogden 1972).

CONSERVATION AND EVOLUTIONARY RELATIONSHIPS

Generally considered relatively common, the rather widespread scarlet-rumped trogon is adapted both to primary and to logged forests and seems to be secure. Wells (1998) judged that the species is less threatened than are the cinnamon-rumped and red-naped trogons. It is curious that its apparent nearest relative, the cinnamon-rumped trogon, should be so much rarer and apparently so much more at risk.

ORANGE-BREASTED TROGON

Harpactes oreskios Temminck 1823

OTHER VERNACULAR NAMES

Malayan orange-breasted trogon (*uniformis*), mountain trogon, Stella's orange-breasted trogon (*stellae*), yellow-breasted trogon.

RANGE

Lowland and hill forests of southern China (southwestern Yunnan), Thailand, Laos, Cambodia, Vietnam, Myanmar, the Malay Peninsula, Sumatra, Nias Island, Java, and northern and central Borneo. See map 33.

SUBSPECIES

Harpactes oreskios stellae Deignan 1941: southern Burma, Cambodia, northern and eastern Thailand, Laos, southern Vietnam, China (Yunnan).
Harpactes oreskios uniformis (Robinson) 1917: southern Thailand, Malay Peninsula (75–1,250 m; 200–4,100 ft), and Sumatra (300–1,500 m; 1,000–4,900 ft).
Harpactes oreskios nias de Schauensee and Ripley 1939: Nias Island.
Harpactes oreskios oreskios Temminck 1823: Java (to ca. 1,200 m [3,900 ft]).
Harpactes oreskios dulitensis Ogilvie-Grant 1892: Borneo (300–1,500 m; 1,000–4,900 ft).

MORPHOMETRICS

Measurements   Wing, both sexes (of *uniformis*) 120–130 mm (4.7–5.1 in); tail, 164–185 mm (6.4–7.2 in; Delacour and Jabouille 1931). Tail, seven adults (of *uniformis*) 142–156 mm (5.5–6.1 in; Deignan 1941). Tail, twenty-two adults (of *stellae*) 158–179 mm (6.2–7.0 in; Deignan 1941). Wing, sixteen males (of *uniformis*) 120–127 mm (4.7–5.0 in); ten females 119–130 mm (4.6–5.1 in). Tail, sixteen males 150–167 mm (5.9–6.5 in); ten females 145–160 mm (5.7–6.2 in; Wells 1998). Averages of four males (from peninsular Thailand), wing 120.7 mm (4.7 in), tail 147.9 mm (5.8 in); averages of males (of *uniformis* from eastern Thailand), wing 118.5 mm (4.6 in), tail

163.8 mm (6.4 in); averages of three males (of *nias*), wing 119.7 mm (4.7 in), tail 140.7 mm (5.5 in; Riley 1938). Wing, males (of *stellae*) 122–126 mm (4.8–4.9 in; average of 4, 127.7 mm [5.0 in]); females 121–127 mm (4.7–5.0 in; average of 5, 124.7 mm [4.9 in]; Gyldenstolpe 1916). Egg of *uniformis* 26.5 × 21.3 mm (1.0 × 0.8 in), of *oreskios* 27.7 × 21.4 mm (1.1 × 0.8 in), of *dulitensis* 29 × 21 mm (1.1 × 0.8, Schönwetter 1966).

Weights   One adult male 57.3 g (2.0 oz; Wells 1988). Estimated egg weight of *oreskios* and *dulitensis* 6.8 g (0.2 oz), of *uniformis* 6.5 g (0.2 oz; Schönwetter 1966).

DESCRIPTION (see plate 38)

Adult Male   Head, chin, throat, neck, and upper breast grayish yellow to olive-green, grayer on crown and nape, more yellowish on throat and breast; scapulars, mantle, back, and rump rich chestnut; primaries and outer secondaries mostly black with some white marginal markings; upper wing-coverts and inner secondaries finely vermiculated with black and white; middle pair of rectrices brown, tipped with black, next two pairs black, outer three pairs mostly white with some black markings; lower breast orange, becoming bright yellow on flanks and remaining underparts; bill (including fleshy edge of mouth) cobalt-blue with black culmen and tip; eye-ring bluish; iris brown, gray, or grayish purple; feet and toes variously described as blue (usually), bluish green, or pale lead, soles pinkish.

Adult Female   Head, neck, and upper breast dull brownish olive to grayish brown; back dull olivaceous brown to grayish brown, shading to rufous on rump and upper tail-coverts; underparts and flanks lemon-yellow, becoming more orange on lower breast; secondaries and wing-coverts with pale or rufous buff barring, bars as broad as intervening dark bars between; soft-part colors similar to those of male, but blue tones duller.

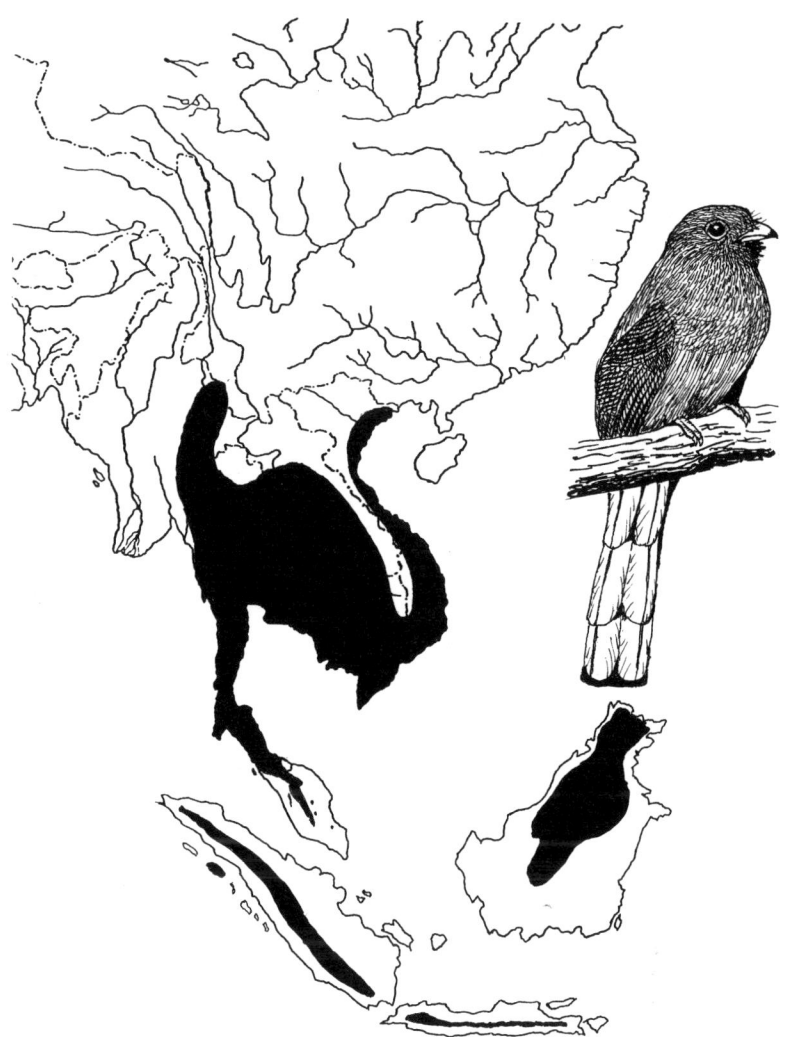

Map 33. Distribution of
the orange-breasted trogon.

Immature Male   Differs from adult in having outer
secondaries pale buff, irregularly barred and
blotched with black, and inner ones black, toothed
and margined with buff outwardly; rectrices
pointed, middle ones lacking terminal black band.
Immature female differs from adult in having outer
secondaries buff, blotched and barred with black,
and inner ones toothed and notched with very pale
buff; rectrices as in immature male.

Juvenile   Upper surface with mingled rusty buff
and olive-green; underparts behind breast yellowish
white to rusty buff; secondaries and upper wing-
coverts golden buff, with narrow blackish barring;
outer secondaries blackish, with pale buff edging
on outer vanes. One young fledgling entirely light
rufous above with pale cinnamon barring on wing-
coverts and inner secondaries wider than intervening black bars, other secondaries edged and tipped
with cinnamon-yellow; chin to breast grayish olive,
remaining underparts white with yellowish wash on
flanks (Wells 1998).

IDENTIFICATION

In the Hand   The orange-breasted trogon is the only
Asian trogon in which the males have a grayish-
green head and breast color and an orange abdomen.

Females are gray and cinnamon-yellow in these respective areas, with no trace of reddish or orange below.

In the Field   The orange-breasted trogon is the most widespread species of Asian trogons and one of the few (the blue-tailed being the other) that lacks red colors in the plumage of both sexes. The uniformly colored head and breast areas (greenish gray in males and dark gray to brownish in immatures and females) are also distinctive. The birds occur in lowland to midmontane forests, reaching 1,200 m (3,900 ft) in Java and 1,500 m (4,900 ft) in Sumatra and Borneo. The male's typical territorial vocalization is a five-note sequence, *kek tau-tau-tau-tau*. See also the vocalizations section.

GEOGRAPHIC VARIATION

There are minor wing-length and tail-length differences among several of the recognized races of the orange-breasted trogon, as indicated in the measurements section. Birds from the southern Malay Peninsula have closer barring (vermiculations) on their wing-coverts and secondaries than do those from farther north, while those from northern and eastern Thailand (*stellae*) are less dusky, are lighter above, and have longer tails than those from peninsular Thailand (*uniformis*). Additionally, birds from Nias Island (*nias*) closely resemble those from peninsular Thailand (Riley 1938).

ECOLOGY

Habitats   On the Malay Peninsula the orange-breasted trogon is associated with humid evergreen jungle, ranging from 625 to 1,250 m (2,100–4,100 ft) in southern parts of the peninsula but also occurring as low as only 60–90 m (200–3,000 ft) in more northerly areas (Robinson 1928). The species occurs in evergreen and semi-evergreen lowland forests, in addition to mature and disturbed swamp forests and lower montane forests (Wells 1998). In Myanmar the species favors moist evergreen forests, but birds have also been seen in thin tree jungle, bamboo forest, and isolated clumps of trees not far from heavy forest (Smythes 1953). On Borneo the species is mostly found along the spinal range between about 300 and 900 m (1,000–3,000 ft), but in the Kelabit uplands it

is fairly common above 900 m (3,000 ft; Smythes 1968).

Foods and Foraging Ecology   No detailed information on the foods of the orange-breasted trogon is available, but beetles and "bugs" have been noted (Smythes 1953), as have crickets, locustids, grasshoppers, lizards, ants, fruits, and mixtures of vegetable matter (Smythes 1968). The species forages at middle stratum and lower canopy levels but may roost in understory vegetation. It is also been observed as part of a mixed-species foraging flock (Wells 1998).

BEHAVIOR

General, Social, and Sexual Behavior   Little information on the behavior of the orange-breasted trogon is available, but the birds are said to favor perching on shorter trees, especially those well overgrown with epiphytic or parasitic vegetation. They have also been described as favoring middle- and upper-canopy trees. They are usually seen alone or in pairs; larger assemblages may be family groups.

Vocalizations   Little about the vocalizations of the orange-breasted trogon is known beyond what is mentioned in the identification section. MacKinnon and Phillipps (1993) diagrammed two vocalization sequence types of this species, one consisting of repeated notes of fairly constant pitch and duration and the other of a similar series of notes that gradually decline in pitch. Medway and Wells (1976) stated that the male has a four-noted *tu-tau-tau-tau* cadence. Wells (1998) stated that three notes (or occasionally four) are the usual number; the notes are said to be even pitched and may have a one- to three-note introductory phrase. The male may have a higher-pitched voice and more rapid utterance than typical of females.

BREEDING BIOLOGY

Chronology of Breeding   In Tenasserim the breeding season of the orange-breasted trogon extends from at least mid-February to the end of April. A half-grown chick was found in mid-February in Tenasserim (Smythes 1953). On the Malay Peninsula eggs have been reported from late January and early April, and a recently fledged bird was observed in mid-February (Wells 1998). There is also an egg record for Thailand in mid-March. Wing molt by adults has been reported nearly throughout the year, with a possible clustering between March and October (Wells 1998).

Nest Sites   In Tenasserim orange-breasted trogon nests were usually in hollow stumps less than a meter above ground, but some were in dead bamboos (Robinson 1928; Smythes 1953). Two nests on the Malay Peninsula were 1.5 and 2 m (5 and 7 ft) above ground, in rotted stumps or tree trunks (Wells 1998). In the Kelabit uplands of Borneo nesting occurs early in the year, with eggs reported in February and March (Smythes 1968). A Thailand nest site was in an open hole of a decayed tree and was unlined (Gyldenstolpe 1916).

Eggs and Incubation   The usual clutch of the orange-breasted trogon is two or three eggs, but four are said occur very rarely (Robinson 1928). Wells (1998) stated that two eggs constitute the full clutch.

Brood Rearing   No information available.

CONSERVATION AND EVOLUTIONARY RELATIONSHIPS

The relatively large distribution of the orange-breasted trogon, its tolerance of open forests and woodlands, and its generally common occurrence seem to place it out of immediate danger. It seems to be a somewhat "faded" version of the scarlet-rumped trogon, its apparent nearest relative, but is adapted to drier, cooler, and more upland forests. Wells (1998) believed it not immediately threatened by loss of habitat on the Malay Peninsula.

RED-HEADED TROGON

Harpactes erythrocephalus (Gould) 1834

OTHER VERNACULAR NAMES

Annamese red-headed trogon (*annamensis*),
Burmese red-headed trogon (*erythrocephalus*),
Hodgson's trogon (*hodgsoni*), Mishmi red-headed
trogon (*helenae*), Nepal red-headed trogon
(*hodgsoni*), East Pakistan red-headed trogon
(*erythrocephalus*).

RANGE

Montane forests of northeastern India (Kumaun
east to Arunachal Pradesh and then south to Assam, Tripura, Mizoram, Manipur, and Nagaland),
Bangladesh, Nepal, Bhutan, southern China
(north to Szechwan, Kweichow, Hunan, Kwangtung, Kiangsi, and Fukien, also Hainan Island),
Myanmar, Thailand, northern Laos, eastern Cambodia, Vietnam, Malay Peninsula, and Sumatra.
See map 34.

SUBSPECIES

Harpactes erythrocephalus hodgsoni (Gould) 1838:
northeastern India (Assam), eastern Nepal (where
uncommon and declining), and Bhutan.

Harpactes erythrocephalus helenae Mayr 1941: southern China (western Yunnan) and northern Burma
(above 2,200 m [7,200 ft]).

Harpactes erythrocephalus yamakanensis Rickett 1899:
southeastern China (Szechwan, Fukien, and
Kwangtung); includes *Harpactes erythrocephalus
rosa* (Stresemann) 1929

Harpactes erythrocephalus hainanus Ogilvie-Grant
1900: Hainan Island (China).

Harpactes erythrocephalus intermedius (Kinnear) 1925:
northern Laos, northern Vietnam, and southern
China (Yunnan).

Harpactes erythrocephalus erythrocephalus (Gould)
1834: eastern Himalayas of Burma and northwestern Thailand (south to Tak).

Map 34. Distribution of
the red-headed trogon.

Harpactes erythrocephalus annamensis (Robinson and Kloss) 1919: eastern Thailand, southern Laos, and Vietnam.

Harpactes erythrocephalus klossi (Robinson) 1815: southern Thailand and western Cambodia.

Harpactes erythrocephalus chaseni Riley 1934: Malay Peninsula (300–900 m [1,000–3,000 ft]; from Negeri Sembilan, Malaysia, north to at least Nakhon Si Thammarat, Thailand).

Harpactes erythrocephalus flagrans (S. Müller) 1835: Sumatra (above 700 m [2,300 ft]).

MORPHOMETRICS

Measurements Wing, six males (of *erythrocephalus*) 167–177 mm (6.5–6.9 in); ten females 146–168 mm (5.7–6.6 in). Tail, six males (of *erythrocephalus*) 218–233 mm (8.5–9.1 in); five females 210–228 mm (8.2–8.9 in; Ali and Ripley 1983). Wing, females (of *annamensis*) 141–148 mm (5.5–5.8 in; average of four, 145.2 mm [5.7 in]); tail 174–178 mm (6.8–6.9 in; average of 5, 175.2 mm [6.8 in]; Robinson and Kloss 1919). Wing, two birds of unspecified sex (of *klossi*) 134 mm (5.2 in) and 135 mm (5.3 in); tail 177 mm (6.9 in) and 177 mm (6.9 in; Robinson 1915). Wing, males (of *hodgsoni*) 146–155 mm (5.7–6.0 in; average of 4, 159.5 mm [6.2 in]); females 144–151 mm (5.6–5.9 in; average of 4, 148.2 mm [5.8 in]; Ali 1962). Wing, twenty-nine males (of *annamensis*) 134–155 mm (5.2–6.0 in); eighteen females 135–153 mm (5.3–6.0 in); thirty-two males (of *intermedius*) 137–167 mm (5.3–6.5 in); twenty-five females 140–149 mm (5.5–5.8 in); five males (of *klossi*) 132–140 mm (5.2–5.5 in); four females 131–137 mm (5.1–5.3 in; Delacour and Jabouille 1931). Wing, seven males (of *chaseni*) 136–139 mm (5.3–5.4 in); three females 132–142 mm (5.2–5.6 in). Tail, males 163–174 mm (6.4–6.8 in); females 153–166 mm (6.0–6.5 in; Wells 1998). Egg of *erythrocephalus* 28.6 × 24.0 mm (1.1 × 0.9 in; Schönwetter 1966).

Weights Four unsexed birds from Malaysia 75–84 g (2.6–2.9 oz; average 80.0 g [2.8 oz]; McClure 1964). Five birds of both sexes from Malaysia 75–86.6 g (2.6–3.0 oz), the heaviest a female (Wells 1998). A sample of males (number unstated) from India 85–110 g (3.0–3.9 oz); one female 76 g (2.7 oz; Ali and Ripley 1983). Estimated egg weight 9.2 g (0.3 oz; Schönwetter 1966).

DESCRIPTION (see plate 39)

Adult Male Head, neck, throat, and breast deep carmine, below which an indistinct white or brown-and-white crescent usually separates breast from slightly lighter crimson to rosy pink abdomen, flanks, and under tail-coverts; back and scapulars rusty brown to golden brown, becoming ferruginous on upper tail-coverts and rump; middle pair of rectrices chestnut tipped with black, next two pairs black, three outer pairs mostly white with some black patterning toward feather bases; upper wing-coverts and inner secondaries black, vermiculated with white; primaries dark brown with white feather edgings; bill purplish blue with black tip and culmen, becoming cobalt-blue at base; eye-ring and eyelid cobalt-blue with a purplish tinge to pinkish violet; iris hazel-brown (also described as light red or crimson); feet and toes fleshy blue, lavender, or perhaps (the toes?) fleshy pink.

Adult Female Similar in patterning to male, but carmine tones of head, neck, throat, and breast replaced by golden brown to orange-brown, bright crimson underparts usually replaced by slightly lighter hues, and black-and-white vermiculations of upper wing-coverts and inner secondaries replaced by dark brown and golden brown; breast with indistinct whitish crescent; soft-part colors similar to those of male, but blue tones duller.

Immature Males initially female-like, but with more-pointed rectrices, more-pinkish underparts, and wing-coverts and secondaries edged with white and coarsely toothed with rufous buff; immature females similar to young males until the latter develop carmine underpart and head coloration.

Juvenile Wing-coverts broadly edged with golden buff and lacking vermiculations; outer vanes of secondaries barred with buff and black; and red underparts much paler than in adults.

IDENTIFICATION

In the Hand Male red-headed trogons are the only Asian trogons with completely red heads, which color also covers most of the underparts except for (usually) a narrow white to brownish breast band. Females also exhibit the narrow breast band and red underparts, but they are brown where the males are red on the head and breast.

In the Field   The large and widespread red-headed trogon is the only Asian trogon having a white breast band in both sexes and also the only one having a completely red (in males) or completely brown (in females) head and breast, as well as bright crimson underparts in both sexes. These trogons are hill-forest birds, especially teak (*Tectona*) forests. The birds usually occur between 700 m (2,300 ft) and 2,000 m (6,700 ft) in Burma and Sumatra and are more generally found between 300 m (1,000 ft) and 1,500 m (4,900 ft) throughout most of their range. However, they have been seen above 2,000 m (6,700 ft) at the northern edge of their range. Their usual vocalization is a repeated, mellow *tiaup* or *kew*. See also the vocalizations section.

GEOGRAPHIC VARIATION

As noted in the measurements section, there are some definite size differences among the many races of the red-headed trogon. In the Indian subregion *helenae* is the largest, and nominate *erythrocephalus* is the smallest. Other races vary in color and in size; *chaseni* is smaller and darker above than nominate *erythrocephalus*, and males of the former have darker and duller red on the throat; whereas *flagrans* of Sumatra is smaller and brighter than the nominate form. In *annamensis* from the Malay Peninsula the upperparts are more ochreous brown rather than the rufous chestnut of the nominate race, and the vermiculations of the wing-coverts are much coarser. *Annamensis* differs slightly from *yamakanensis* in its head and breast color. The race *intermedius* is intermediate in size between *hainanus* and *yamakanensis*, and the male *intermedius* is more scarlet below than either of the latter. The race *klossi* is smaller than the nominate race to the north, and it has narrower vermiculated barring on the wing-coverts. The insular race *hainanus* is browner above in both sexes than is nominate *erythrocephalus* and has duller crimson hues on the head and breast; in addition, the white tips of the outer rectrices of males are shorter.

ECOLOGY

Habitats   The Nepalese race *hodgsoni* of the red-headed trogon occurs from the foothills up to 1,800 m (5,900 ft), in dense secondary evergreen jungle. The nominate race occurs from the foothills to similar altitudes, also in evergreen jungle (Ali and Ripley 1983). In Burma the nominate race reaches 1,900 m (6,200 ft), and the race *helenae* attains at least 2,200 m (7,200 ft) at the Yunnan-Burma border (Smythes 1953). In Nepal the species inhabits dense broad-leaved evergreen forests from about 150 m (500 ft) to nearly 2,000 m (7,200 ft), and breeding has been observed at 1,830 m (6,000 ft; Fleming, Fleming, and Bangdell 1975; Inskipp and Inskipp 1985). Damp, shady gullies seem to be favorite habitats (Robinson 1928). On the Malay Peninsula the species occurs in closed-canopy evergreen lowland, lower montane, and upper montane forests, from about 700 m (2,300 ft) to 1,680 m (5,500 ft; Wells 1998).

Foods and Foraging Ecology   Foods of the red-headed trogon are said to include insects and their larvae, such as grasshoppers, beetles, and stick insects (Phasmatidae), as well as leaves and berries (Ali and Ripley 1983). Leaves, seeds, and other vegetable matter are evidently eaten with some regularity (Smythes 1953); among the vegetable materials reportedly consumed are bamboo culms and fruit. Foraging occurs in the middle stratum and sometime at the lower edge of forests, where the birds perch on open perches and make short sallies out to catch prey from vegetation.

BEHAVIOR

General, Social, and Sexual Behavior   Red-headed trogons are solitary birds that have so far been little studied.

Vocalizations   Little about red-headed trogon vocalizations is known beyond what is mentioned in the identification section. A rather plaintive but rich series of *tyaw* or *cue* notes (six to ten) uttered at the rate of about two notes per second is said to be the male's typical (territorial?) vocalization. The series is repeated at intervals of a minute or so. Croaking or chattering calls have also been heard, apparently as alarm notes. Medway and Wells (1976) mentioned a mellow and repeated *tiaup* note, as well as a rattled *tewirr*. Wells (1998) stated that the usual call consists of four to five mellow but resonant and well-spaced notes, and the species also produces a light rattle.

BREEDING BIOLOGY

Chronology of Breeding   On the Indian subcontinent, including Nepal, Sikkim, and Bhutan, the breeding season of the red-headed trogon extends from April to July, with the peak falling in May and June (Ali and Ripley 1983). Baker (1927) stated that in Assam the species breeds from April to August, and in Nepal breeding occurs between mid-March and mid-July (Inskipp and Inskipp 1985). In Burma eggs have been found from March to May (Smythes 1953). On the Malay Peninsula juveniles have been seen in March and May, but there are no egg records (Wells 1998).

Nest Sites   The red-headed trogon nest site is the usual natural hollow in a rotted tree, or a vacant woodpecker hole, and is located from about 1.5–5 m (5–16 ft) above ground (Ali and Ripley 1983). A nest site of this type was described for Thailand (Gyldenstolpe 1916) as well as for other areas of Indochina (Delacour and Jabouille 1931). The birds are also said to sometimes excavate the cavity themselves (Robinson 1928).

Eggs and Incubation   Red-headed trogon clutch sizes range from two or (usually?) three to four buff-colored to light brown eggs. Gyldenstolpe (1916) mentioned a clutch of two, and there are three sets (of two, three and four eggs) in the collection of the Western Foundation of Vertebrate Zoology. All of these were obtained between March 11 and May 24 in India, Burma, and Thailand. Both sexes incubate, but the incubation period is unknown.

Brood Rearing   No information available.

CONSERVATION AND EVOLUTIONARY RELATIONSHIPS

The red-headed trogon is a relatively widespread species of the Indian subcontinent, and it ranges over a dozen countries and, at least in some areas, remains fairly common. On the Malay Peninsula it has only very local protection and like other closed-canopy, montane-forest species is threatened by logging (Wells 1998). On the basis of plumage similarities, its nearest relatives are probably the red-naped and Diard's trogons.

WARD'S TROGON

Harpactes wardi Kinnear 1927

OTHER VERNACULAR NAMES

Red-fronted trogon.

RANGE

Forests of central and eastern Bhutan (where uncommon), India (Arunuchal Pradesh, where recently observed), northern Myanmar (formerly common, but there are no recent records) east probably through northern Laos (status unknown) to northwestern Vietnam (northwestern Tonkin; there are no recent records), and southern China (northwestern Yunnan, where very rare). See map 35.

SUBSPECIES

None recognized.

MORPHOMETRICS

Measurements   Wing, six males 167–177 mm (6.5–6.9 in); five females 169–175 mm (6.6–6.8 in). Tail, six males 218–233 mm (8.5–9.1 in); five females 210–228 mm (8.2–8.9 in; Ali and Ripley 1983). Wing, nineteen adult males 160–176 mm (6.2–6.9 in); eleven adult females 160–171 mm (6.2–6.7 in; Delacour and Jabouille 1931). No information on eggs is available.

Weights   Two males 115 g (4.0 oz) and 120 g (4.2 oz), two females each 120 g (4.2 oz), average of all four 119 g (4.2 oz; Ali and Ripley 1983).

DESCRIPTION (see plate 40)

Adult Male   Crown and forehead bright vinous red, gradually becoming darker vinous brown on nape and hindneck; lores, malar area, ear-coverts, and chin still-darker reddish brown, which color grades into medium reddish brown forming broad breast band and then continues down scapulars and back as far as rump and upper tail-coverts; primaries dark brown, edged on outer vanes with white; concealed bases of secondaries white; inner secondaries, tertials, and upper wing-coverts finely vermiculated with black and white; three middle pairs of rectrices uniformly dark brown; outer three pairs of rectrices all predominantly pinkish white as seen from below and paler toward tips, outermost rectrix pinkish white except for black wedge-shaped stripe extending out from base to more than half the length of feather along inner vane, second rectrix with similar but wider and longer stripe, and third rectrix with a still wider and longer stripe; flanks, abdomen, and under tail-coverts carmine (anteriorly) to pink (posteriorly); most of maxilla and tip of mandible red, both with violet base; eye-ring blue; iris brown; feet and toes flesh colored.

Adult Female   Most of head olive-brown but primrose-yellow in front of eyes, becoming yellowish brown on forehead, which color extends back as a narrow superciliary stripe; darker brown on sides of face and ear-coverts; otherwise uniformly olive-brown on throat and upper and lower breast and down scapulars, mantle, back, and rump to upper tail-coverts; primaries as in male; secondaries with concealed white bases as in male; upper wing-coverts and inner secondaries, including tertials, finely vermiculated with dark brown and yellowish or buff; three middle pairs of rectrices medium to dark olive-brown, outer three pairs without black patterning as seen from below but distinctly tinted with primrose-yellow; flanks and remaining underparts behind brown breast band all orange-yellow to primrose-yellow; soft-part colors similar to those of male, but bill mostly horn-brown, becoming pinkish on sides of lower mandible.

Immature   Undescribed, but presumably similar to immatures of related species in having narrower, more pointed rectrices than adults, buffy edging on some remiges, and more coarsely vermiculated, barred, or spotted upper wing-coverts.

Juvenile   No information available.

Map 35. Distribution of Ward's trogon.

IDENTIFICATION

In the Hand   Ward's trogon is distinct in having a dark gray to brown head and breast, with a pink (in males) to yellowish (in females) superciliary stripe, which continues back from the forehead to above and behind each eye. Below the breast the underparts are either pink (males) or yellow (females), which colors extend as tints to the unpatterned outer rectrices.

In the Field   The traits mentioned above provide basic field marks for Ward's trogon, the largest of the Asian species and also the only Asian trogon with pale pink (males) or yellow (females) rather than white outer rectrices. The conspicuous pale superciliary stripe on both sexes is also a helpful field mark. The species' vocalizations include a soft *kew-kew-kew-tiree* and a whirring *whirr-ur* alarm note.

GEOGRAPHIC VARIATION

None described.

ECOLOGY

Habitats   Ward's trogon is a montane species, found in India and Bhutan between 1,500 m (4,900 ft) and

3,000 m (9,800 ft), frequenting the lower story, evergreen undergrowth, and bamboo layer of tall, subtropical forests dominated by oaks (*Quercus*), chinkapins (*Castanopsis*), and so on (Ali 1977; Ali and Ripley 1983). In Indochina the birds occur in humid forests of the high mountains, between 2,500 m (8,200 ft) and 3,500 m (11,500 ft; Delacour and Jabouille 1931).

Foods and Foraging Ecology   A variety of insects, including moths, stick insects (Phasmidae), grasshoppers, and "bugs" have been mentioned as foods of Ward's trogon, as well as berries, large seeds, fruit, and even acorns (Ali and Ripley 1983; Smythes 1953), although the last-named item seems unlikely.

BEHAVIOR

General, Social, and Sexual Behavior   Like other trogons, Ward's trogons are usually found either singly or in pairs, and they remaining silent most of the time (Ali and Ripley 1983).

Vocalizations   Little or nothing is known about the vocalizations of Ward's trogon beyond what is mentioned in the identification section.

BREEDING BIOLOGY

Chronology of Breeding   Breeding-condition birds have been collected in Bhutan in early April (Ali and Ripley 1983).

Nest Sites   No information available.

Eggs and Incubation   No information available.

Brood Rearing   No information available.

CONSERVATION AND EVOLUTIONARY RELATIONSHIPS

The most recently discovered of the Asian trogons, the little-known Ward's trogon is apparently quite rare everywhere except Bhutan. Several of its plumage traits are unique, and its relationships to the other Asian trogons seem uncertain at best. It was classified as vulnerable by the IUCN (Collar, Crosby, and Stattersfield 1994).

APPENDIX ONE

Derivations of Scientific and Vernacular Names of Trogons and Quetzals

In large part, the derivations of scientific and vernacular names in this appendix are after Jobling 1991.

Apaloderma: from the Greek *hapalos*, delicate, and *derma*, the skin.

aequatoriale: from Latin, associated with the equator.

narina: possibly from a Hottentot name for a kind of flower, but apparently from the name of a Hottentot girl from Knysna District, Cape Province, South Africa, ca. 1765–1782.

vittatum: from the Latin *vittatus*, banded, in reference to the tail. In this work, *vittatum* has been placed in the monotypic genus *Heterotrogon* (from Greek, a different trogon).

Euptilotis: from the Greek *eu-*, well, *ptilon*, feathered, and *-otis*, ears, referring to the ear-coverts of this genus.

neoxenus: from the Greek *neo-*, new, and *xenos*, stranger. Gould considered this species a "welcome" discovery, thus his name, welcome trogon.

Harpactes: from Greek, a robber (of fruit or other food).

ardens: from Latin, glowing or burning, in reference to the plumage color of this species.

diardii: after P.M. Diard (1795–1863), a French collector in the East Indies.

duvaucelii: after A. Duvaucel (1796–1824), a French collector in Sumatra.

erythrocephalus: from the Greek *erythros*, red, and *-kephalos*, headed. This species was originally named Hodgson's trogon by J. Gould, after Brian Hodgson (1800–1894), the official British resident in Katmandu for two decades.

fasciatus: from Latin, banded. The vernacular name of this species refers to the Malabar Coast of India.

kasumba: a Malay name for trogon, derived from the Sanskrit *kesumba*, a kind of tree that produces yellowish-red dyes. This trogon species was called Temminck's trogon by J. Gould, after C. J. Temminck (1778–1858), a Dutch ornithologist.

oreskios: from Greek, mountain bred.

orrhophaeus: from the Greek *orrhos*, the tail, and *phaios*, gray. The vernacular name Malacca trogon refers to the Strait of Malacca, east of Sumatra.

reinwardti: after C.G.C. Reinwardt (1773–1854), a Dutch ornithologist in Java. The Sumatran race is sometimes called Macklot's trogon, after the collector of the first specimens. The subgeneric designation *Hapalarpactes* is from the Greek *hapalos*, delicate, and *harpactes*, a robber.

wardi: after Captain F. Kingdom-Ward (1885–1958), English author and collector of biological specimens in Asia.

whiteheadi: after John Whitehead (1860–1899), English explorer of Borneo and the Philippines, who was the first explorer to reach the crest of Mount Kinabalu.

Pharomachrus: after the Greek *pharos*, a loose mantle, and *makros*, long, in reference to the long tail-coverts of the species in this genus. The vernacular "quetzal" is from the Nahuatl *quetzali*, properly pronounced with a strong accent on the second syllable, "ka-tzál" or "kat-zál." Apparently *quetzali* initially meant "tail feathers" but later implicitly also meant "precious," the long tail-covert feathers of these species having at times been valued more highly than gold. In some parts of Panama a quite similar native name is *guaco*, which is probably based on the species' loud double-noted call. Quetzalcoatl, the Aztec god of the winds and air, is sometimes depicted as a feathered serpent. Initially considered a quasi-historical Toltec hero-figure, Quetzalcoatl later was adopted by the Aztecs as a benevolent mythic god and a patron of the arts and music (see the introduction to this volume). In Central American folklore the blood-red feathers of the quetzal's underparts have been identified with the blood shed by native people at the hands of the Spanish conquerors.

antisianus: This species' describer, Alcide d'Orbigny (1803–1887), a French zoologist and explorer, referred to the Indians of the eastern Andes as "Anti-sien," and thus his 1839 name *antisiensis* probably means "of the Andes."

auriceps: from the Latin *aurum*, gold, and *-ceps*, headed.

mocinno: after Jose M. Mociño (1758–1819), Mexican naturalist.

pavoninus: from Latin, colored like a peacock.

Priotelus: from the Greek *prion*, a saw, and *telos*, finished, in reference to the concave shape of the tail feathers in the Cuban species.

roseigaster: from the Latin *roseus*, rosy, and *gaster*, the belly. The vernacular name Santo Domingo trogon refers to the capital of the Dominican Republic. The subgeneric name *Temnotrogon* is from the Greek *temno*, to cut, and *trogon*, presumably in reference to the sharp bill edge.

temnurus: from the Greek *temno*, to cut, and *-oura*, the tail, referring to the incised pattern of the tail feathers.

Trogon: from the Greek *trogon*, to gnaw or eat.

aurantiventris: from the Latin *aurantius*, orange colored, and *ventris*, the belly. The race *underwoodi* was named for British ornithologist C. F. Underwood, who collected the first specimens.

bairdii: after Spencer F. Baird (1823–1887), an American ornithologist and the secretary of the Smithsonian Institution when the species was described in 1868.

citreolus: diminutive of the Latin *citreus*, citrine or yellow.

clathratus: from the Latin *clathri*, a lattice, and *-atus*, possessing.

collaris: from Latin, collared. The subgenus *Trogonurus* is from the Greek *trogon*, to gnaw or eat, and *-oura*, the tail.

comptus: from Latin, adorned or neat. Sometimes called Zimmer's trogon, after its describer, American ornithologist John T. Zimmer (1889–1957).

curucui: a Tupi (Brazilian) Indian name for a small bird. The crested quetzal of Bolivia has been described as having a "couroucou" call. The subgenus *Curucujus* has the same origin. The race *behni* was named after a Professor Behn of Kiel, who provided John Gould with his first specimens.

elegans: from Latin, fine or choice. The form *ambiguus* was named the doubtful trogon by Gould because of its initially uncertain taxonomic status.

massena: after F. V. Massene, Prince d'Essling and Duc de Rivoli (1795–1863), a French ornithologist. The Colombian form is sometimes called Chapman's trogon, after F. M. Chapman (1864–1945), an American ornithologist and authority on Colombian birds.

melanocephalus: from the Greek *melos*, black, and *-kephalos*, headed.

melanurus: from the Greek *melanouros*, black-tailed.

mexicanus: associated with Mexico.

personatus: from Latin, masked.

rufus: from Latin, red or rufous.

surrucura: from a Paraguayan Indian name for a trogon.

violaceus: from Latin, violet colored. The vernacular name booted trogon as well as the Latin racial epithet *caligatus*, booted, refers to the species' tarsi, which lack a definite scute pattern. An older but now invalid species name, *braccatus*, is related and comes from Latin, wearing trousers.

viridis: from Latin, green. The species' older but now invalid name, *strigilatus*, is from Latin, possessing a scraper.

APPENDIX TWO

Keys to Genera and Species
of Trogons and Quetzals

KEY TO ADULT NEW WORLD TROGONS AND QUETZALS (Tribe Trogonini)

This key has been taken in part from Ridgway 1911; see figures 10 and 11 for male undertail patterns.

A. Ear-coverts elongated, either outcurved or hair-like and extending beyond nape.

    B. Eyelids bare, rectrices rounded at tips, Mexico: genus *Euptilotis* (eared trogons), eared trogon *E. neoxenus* (see figure 6C).

    BB. Eyelids feathered, rectrices truncated or concave at tips, West Indies: Genus *Priotelus* (West Indian trogons).

        C. Breast gray, secondaries narrowly barred with white, rectrices with truncated tips: Hispaniolan trogon *P. temnurus* (see figure 7B).

        CC. Breast white, secondaries with large white spots, rectrices with scalloped tips: Cuban trogon *P. roseigaster* (see figure 7A).

AA. Continental species lacking lengthened ear-coverts (see figure 10 for male undertail patterns).

    B. Bill edges notched, nostrils narrow, median wing-coverts elongated, curved, and glossy green: Genus *Pharomachrus* (quetzals).

        C. Abdomen brilliant red, outer rectrices never barred black and white (males).

            D. Forehead and nape somewhat crested, outer rectrices white.

                E. Crest highly developed: resplendent quetzal *P. mocinno* (see figure 6B).

                EE. Crest only moderately developed: crested quetzal *P. antisianus*.

            DD. Forehead not noticeably crested, outer rectrices all black or black tipped with white.

                E. Outer rectrices partly white: white-tipped quetzal *P. fulgidus*.

                EE. Outer rectrices all black.

                    F. Bill yellow: golden-headed quetzal *P. auriceps*.

                    FF. Bill red: pavonine quetzal *P. pavoninus*.

        CC. Anterior abdomen mostly brown or gray, outer rectrices barred (females).

            D. Crown and nape feathers lengthened: resplendent quetzal *P. mocinno*.

            DD. Crown and nape feathers of normal length.

                E. Head brown, with slight iridescence: crested quetzal *P. antisianus*.

                EE. Head at least partly green or bronze.

F. Abdomen and breast brown, bill red basally: pavonine quetzal *P. pavoninus*.

FF. Abdomen and breast reddish, bill not red basally.

    G. Outer rectrices with broad white tips: white-tipped quetzal *P. fulgidus*.

    GG. Outer rectrices either only narrowly tipped with white or black to tips: golden-headed quetzal *P. auriceps*.

BB. Bill edges serrated, nostrils rounded, wing-coverts of normal length and often barred or vermiculated: Genus *Trogon* (toothed trogons).

  C. Length ca. 28–35 cm (11–14 in), bill and feet robust, tarsus shorter than the longest front toe.

    D. Underparts yellow to orange, males with blackish upper wing-coverts.

      E. Crown and chest glossed with blue or violet (males), or wing-coverts narrowly barred with white (females).

        F. Posterior underparts orange to reddish orange: Baird's trogon *T. bairdii*.

        FF. Posterior underparts orange-yellow to yellow: white-tailed trogon *T. viridis*.

      EE. Crown and chest black to slate in both sexes.

        F. Outermost rectrices about one-third white (males) or with transverse anterior margins (females); adults with blue eye-ring, brown iris: black-headed trogon *T. melanocephalus* (see figure 7D).

        FF. Outermost rectrices at least half white (males) or with oblique lateral margins (females); adults with brownish eye-ring, yellow iris: citreoline trogon *T. citreolus*.

    DD. Underparts red, males with vermiculated upper wing-coverts.

      E. Outer rectrices black with white barring.

        F. Tail with narrow white bars, pale iris: lattice-tailed trogon *T. clathratus*.

        FF. Tail with broad white bars, brown iris: blue-crowned trogon *T. curucui*.

      EE. Outer rectrices slaty black or with only slight freckling.

        F. Upperparts emerald-green to bluish, underparts bright red (males).

          G. White band present on upper breast: black-tailed trogon *T. melanurus*.

          GG. No white band on upper breast.

            H. Facearea black, darker and more bluish on head, breast, and back; iris white to grayish blue: white-eyed trogon *T. comptus*.

            HH. Face green, lighter and more greenish above; iris dark brown: slaty-tailed trogon *T. massena* (see figure 7C).

        FF. Brown to sooty upperparts and breast (females).

          G. Sooty brown upperparts, with blackish upper wing-coverts, lower mandible yellow: white-eyed trogon *T. comptus*.

          GG. Upperparts less sooty, wing-coverts sometimes vermiculated with white.

            H. Wing at least 165 mm (6.4 in), lower mandible reddish: slaty-tailed trogon *T. massena*.

            HH. Wing no more than 161 mm (6.3 in), lower mandible orange-yellow: black-tailed trogon *T. melanurus*.

CC. Length ca. 23–28 cm. (9–11 in), with weak bill and feet, tarsus as long as the longest front toe.

    D. Upperparts brownish (females).

      E. With a conspicuous white ear-spot, lower breast whitish: elegant trogon *T. elegans*.

      EE. No white ear-spot, lower breast not white.

        F. Posterior underparts orange to pink.

          G. Outer rectrices mostly black: orange-bellied trogon *T. aurantiventris*.

          GG. Outer rectrices mostly white: surucua trogon *T. surrucura*.

        FF. Posterior underparts red.

          G. Lower breast brown: mountain trogon *T. mexicanus*.

          GG. Lower breast red, like rest of underparts.

        H.   Face and throat blackish, bill yellow: masked trogon *T. personatus.*

        HH. Head entirely dull brownish, maxilla blackish: collared trogon *T. collaris.*

    FF. Posterior underparts yellow

        G.   Head and breast dark gray: violaceous trogon *T. violaceus*

        GG. Head and breast brown: black-headed trogon *T. rufus*

DD. Upperparts bright emerald green to violet (males).

    E.   Outer rectrices black except for a white tip: mountain trogon *T. mexicanus.*

    EE. Exposed outer rectrices mostly or entirely white.

        F.   Outer rectrices entirely white: surucua trogon *T. surrucura.*

        FF. Outer rectrices barred or vermiculated with black and white.

            G.   Underparts yellow.

                H.   Outer rectrices black, broadly tipped with white: head and breast slate to black: black-headed trogon *T. melanocephalus.*

                HH. Outer rectrices mostly narrowly barred and tipped white.

                    I.   Bill mostly yellow: black-throated trogon *T. rufus.*

                    II. Bill blackish gray: violaceous trogon *T. violaceus* (see figure 8A).

            GG. Underparts red to orange.

                H.   Three outer rectrices narrowly barred with black and white.

                    I.   Underparts bright red: collared trogon *T. collaris* (Central American races).

                    II. Underparts orange-red: orange-bellied trogon *T. aurantiventris.*

                HH. Three outer rectrices broadly tipped with white.

                    I.   Outer rectrices predominantly white, barred or vermiculated black and white: elegant trogon *T. elegans* (see figure 8D).

                    II. Outer rectrices predominantly black, barred and tipped white.

                        J.   Outer rectrices and upper wing-coverts broadly barred with white: collared trogon *T. collaris* (South American races).

                        JJ. Outer rectrices and wing-coverts narrowly barred with white, darker toned: masked trogon *T. personatus.*

KEY TO ADULT OLD WORLD (AFRICAN AND ASIAN) TROGONS

This key has been taken in part from Delacour 1947; see figure 11 for male undertail patterns.

A. Bare skin areas above and below eyes: Genus *Apaloderma* (African trogons).

    B.   Outer rectrices barred black and white, small bare pinkish skin areas near eyes: bar-tailed trogon *A. vittatum.*

    BB. Outer rectrices essentially entirely white, larger areas of bare skin near eyes.

        C.   Bare skin between eye and maxilla base yellow and not subdivided, outer rectrices freckled basally: bare-cheeked trogon *A. aequatoriale.*

        CC. Bare skin between eye and mandible blue-green and subdivided by a line of feathers, outer rectrices unfreckled basally: Narina trogon *A. narina* (see figure 6A).

AA. No bare facial skin except for a (usually blue) eye-ring: Genus *Harpactes* (Asian trogons).

    B.   Philippine Islands, males with rose-pink breast, underparts bright red, females mostly cinnamon-yellow below, length ca. 40 cm (16 in): Philippine trogon *H. ardens.*

    BB. Distributed from Southeast Asia to the Greater Sundas, never with pink breast band.

        C.   Overall length no more than 25 cm (9.8 in), wing to 110 mm (4.3 in).

            D.   Rump and flanks scarlet to pink, female with olive-toned face, bill weaker: scarlet-rumped trogon *H. duvaucelii.*

            DD. Rump cinnamon, female with rusty-toned face, bill more robust: cinnamon-rumped trogon *H. orrhophaeus.*

        CC. Overall length 28–40 cm (11.0–15.7 in), wing usually well over 110 mm (4.3 in).

    D.   Belly yellow to cinnamon.

        E. Upperparts brown, wing to 130 mm (5.1 in), total length 28 cm (11.0 in): orange-breasted trogon *H. oreskios.*

      EE. Upperparts not brown, wing over 130 mm (5.1 in), total length 34–40 cm (13.4–15.7 in).

           F. Upperparts and tail greenish blue, outer rectrices blackish basally: blue-tailed trogon *H. reinwardti.*

          FF. Upperparts grayish, outer rectrices tinted yellow: Ward's trogon *H. wardi* (female).

  DD. Belly pinkish crimson or cinnamon.

        E. Abdomen cinnamon (females).

           F. Breast gray: Whitehead's trogon *H. whiteheadi.*

          FF. Breast brown.

              G. Narrow breast band present, India and Sri Lanka: Malabar trogon *H. fasciatus.*

             GG. No breast band, Southeast Asia: red-naped trogon *H. kasumba.*

      EE. Abdomen pinkish crimson.

          F. Breast red.

              G. Head red, Himalayas, China, and Southeast Asia: red-headed trogon *H. erythrocephalus* (see figure 8B).

             GG. Head black, India and Sri Lanka: Malabar trogon *H. fasciatus* (male).

          FF. Breast not red.

              G. Breast brown or pinkish.

                 H. Narrow white band separates breast from belly.

                    I. Head brown: red-headed trogon *H. erythrocephalus* (female, see figure 8B).

                    II. Head blackish: Malabar trogon *H. fasciatus* (immature male).

               HH. No white band on lower breast.

                    I. Underparts bright red, outer rectrices tinted pink, length 40 cm. (15.7 in): Ward's trogon *H. wardi* (male).

                    II. Underparts pale pink, outer rectrices speckled blackish, length 30 cm (11.8 in): Diard's trogon *H. diardii* (female).

             GG. Breast black or gray (males).

                 H. Breast gray: Whitehead's trogon *H. whiteheadi.*

               HH. Breast black, with a lower white border.

                    I. Outer three rectrices unmarked white, red nape-patch: red-naped trogon *H. kasumba* (see figure 8C).

                    II. Outer three rectrices speckled black and white, pink collar: Diard's trogon *H. diardii.*

GLOSSARY

Abdomen. The "belly" area of a bird, located behind the breast, below the sides, and in front of the under tail-coverts.

Adaptation. An evolved structural, behavioral, or physiological trait that increases an organism's individual fitness (its ability to survive and reproduce).

Advertising behavior. The social behaviors (signals, or displays) by which an animal announces (visually or acoustically or both in birds) its species and sex; these behaviors may also reveal its relative reproductive capacities, social status, and general vigor. See also *signals*.

Aftershaft. A secondary or rudimentary feather growing out from near the base of the main shaft; this feather may be rather down-like or similar to that of the main vane. Trogons have well-developed aftershafts on many of their contour feathers, which is considered a primitive trait.

Agonistic. Pertaining to the entire dominance/submission behavioral spectrum, comprising a response gradient ranging from attack to escape. See also *signals*.

Alcedinidae. A family of coraciiform birds that includes the kingfishers and is distantly related to trogons. See also *Coraciiformes*.

Alcedinoidea. A taxonomic category (proposed by Maurer and Raikow 1981) of coraciiform birds that includes the bee-eaters (family Meropidae) and kingfishers (family Alcedinidae). See also *Coraciiformes*.

Allopatric. Occurring in geographically isolated areas, at least during breeding. See also *parapatric* and *sympatric*.

Allospecies. Populations that are apparently distinct at the species-level but are allopatric in their distributions and thus impossible to assess as biological species.

Altitudinal migration. Seasonal migrations between elevations, usually (at least in tropical trogonids) related to seasonal changes in food supplies. North-south (latitudinal) migrations related to climate also occur in some temperate-zone trogons.

Altricial. Referring to the condition of being hatch in a relatively undeveloped stage, usually blind and without locomotory abilities. See also *nidicolous*.

Alular feathers. The small feathers, located at the anterior bend (wrist, alula) of a bird's wing, that are associated with the first digit and may serve as a secondary or supplemental airfoil, at least in some birds.

Antiphonal. Pertaining to alternation between two groups. Antiphonal vocalizations (calls or songs) are often uttered by the male and female of a pair or by males engaged in territorial encounters with competitors. See also *songs, calls,* and *vocalizations*.

Antrose. Oriented forwards, as opposed to retrose (oriented backwards).

Arachnoids. Members of the arthropod group Arachnoidea, including spider-like forms and their relatives.

Aril. A thin, fleshy covering of some seeds.

Auricular. Pertaining to the area behind and below the eyes and surround the ear openings. See *ear-coverts*.

Austral. Pertaining to the southern hemisphere.

Axillars (or axillaries). "Armpit" feathers, located between the under wing-coverts and the flanks.

Barbicels. The subdivisions of a feather barbule that interlock with those of nearby barbules.

Barbules. The subdivisions of a feather barb that support the barbicels.

Bee-eaters. See *Meropidae* and *Coraciiformes*.

Bill. The mandible and maxilla collectively; sometimes also called the beak. See also *culmen, mandible,* and *maxilla.*

Bromeliad. A member of the pineapple family (Bromeliaceae), usually epiphytic and arboreal; common in the forests of tropical America.

Brood. To cover and apply heat to hatched young (analogous to incubation of eggs); or a group of young tended by a single female or pair.

Caeca. Outpocketings of the digestive tract at the junction of the small and large intestines.

Calls. Avian vocalizations that are acoustically simple and often innately uttered and perceived, the production of which is not usually limited by age, sex, or season. Trogon calls, like those of most birds, are frequently uttered in situations of aggression or alarm, as well as in parent-offspring interactions, where simple acoustic signals that are not dependent upon learning are necessary. See also *songs* and *vocalizations.*

Carotenoids. Lipochrome pigment molecules, ranging from bright yellow through pink or orange to bright red, derived from plant precursors and often present in trogonid underpart feathers. In at least some trogons these include pigments such as zooerythrine in the carotenoid cantaxanthin group, which is responsible for the blood-red colors on the underparts of quetzals. These bright trogonid pigments are highly fugitive, prone to fading, even when kept away from bright light. See also *iridescent.*

Caterpillar. A nontechnical term for the worm-like larvae of lepidopterans (butterflies and moths); favored foods of trogons.

Cenozoic Era. The geologic era comprising the past 65 million years, the so-called Age of Mammals, which encompasses the known fossil and recent history of the trogons. See also *Eocene Epoch* and *Oligocene Epoch.*

Cerambycid. A member of the long-horned beetle family (Cerambycidae).

Cere. A protuberance or enlarged area at the base of the bill of a bird.

Cerrado. Semi-xeric grassland, savanna, or woodland of interior southeastern South America, rich in drought- and fire-adapted species. See also *savanna* and *woodland.*

Clade. A group of biological taxa that includes all descendants of one common ancestor.

Cloud forest. Temperate-zone, evergreen montane forest that occurs at an elevation where moisture-bearing wind currents form misty or cloudy zones that are extremely damp for much of the year. Located above tropical lowland evergreen forests, cloud forests are usually marked by abundant mosses, foliose lichens, ferns, and epiphytes. Cloud forests of Central America are also often rich in temperate-zone pines, oaks, and lauraceous trees. South American cloud forests are home to a great diversity of broad-leaved trees, as well as many tree ferns and palms. Elfin forests may occur at their cold and windy upper limits. See also *forest.*

Clutch. A complete set of eggs, normally laid by a single female and incubated simultaneously.

Coleopteran. A member of the beetle order of insects (Coleoptera).

Coliiformes. The order of birds that includes the African colies, or mousebirds (family Coliidae).

Columella. The middle ear ossicle (comparable to the mammalian stapes) of birds, reptiles, and amphibians.

Communal laying. The deposition of eggs in a single nest by more than one female, usually followed by shared incubation and caring for the young.

Conspecific. Belonging to the same species.

Cooperative nesting. The caring for eggs or young by "helper" individuals in addition to the parents.

Coraciiformes. The traditionally recognized avian taxonomic order that includes the kingfishers, bee-eaters, todies, and their relatives (coraciiform birds) and is characterized by a syndactyl toe arrangement. Some authorities include the trogonids in this order, either as suborder Trogones (e.g., Fry, Keith, and Urban 1988) or as a similar subgroup. See also *Alcedinoidea, Meropidae, Alcedinidae,* and *Todidae.*

Coverts. Feathers that lie above and below the flight feathers (wing-coverts), the tail feathers (tail-coverts), and the ear opening (ear-coverts, or auriculars).

Culmen. The dorsal edge of the maxilla. The corresponding lower edge of the bill is called the gonys, and the cutting edges are called the tomia.

Definitive plumage. The final or "adult" plumage pattern acquired by birds, which is marked by no further age-related plumage changes. See *plumage.*

Desmognathous. A descriptive term for the avian palate condition in which the vomers are small, and the maxillopalatine bones meet each other or the vomers at the midline. See also *schizognathous.*

Dichromatism. The presence of two distinct and genetically controlled plumage patterns or colors among adults of a species. Dichromatism is mostly limited among trogons to adult sexual differences, although in some species there may also be well-defined regional differences (so-called color phases) in carotenoid pigmentation. See also *dimorphism,* which is sometimes used synonymously.

Digits. The toes and fingers of vertebrates; birds such as trogons have four functional toes (two in front, two behind) and three wing digits (the first supporting the alular feathers, the others the primary feathers). See also *heterodactyl.*

Dimorphism. The presence of two absolutely or statistically distinct adult forms (morphs) of a species. Dimorphism is limited in trogonids to slight sexual differences in mass and linear measurements. Sex-related differences in plumage or in soft-part colors or patterns are more accurately described as sexual dichromatism, a subcategory of sexual dimorphism. See also *dichromatism.*

Display. See *advertising behavior, signals,* and *territoriality.*

Distal. Away from the center of the body. See also *proximal.*

Drupe. A fruit (such as a peach) consisting of a seed and a thick, fleshy covering. See also *aril.*

Ear-coverts. Feathers behind and below the eyes and surrounding the ear openings. In a few trogons (eared, West Indian) these are filamentous and variably elongated.

Elfin forest. The cold subalpine zone of stunted and wind-shaped trees that is located above montane evergreen forest and below timberline in tropical montane communities. Comparable to the krummholz zone of temperate forests.

Endemic. Native to and limited to a particular area or region.

Eocene Epoch. A major early subdivision of the Cenozoic Era, extending from about 60 to 37 million years ago. See also *Cenozoic Era.*

Epiphytic. Living on the surface of plants; epiphytic plants grow on other plants but are not necessarily parasitic upon them and are often found on tree limbs near the forest canopy, where light levels are higher than on the forest floor.

Evergreen forest. In the tropics, a broad category of broad-leaved forest types that receive sufficient moisture to remain leafy throughout the year. These forests include lowland evergreen forests, montane evergreen forests (e.g., cloud forests), evergreen swamp forests, mangrove forests, and other nondeciduous forest types. Whereas evergreen forests of temperate climates are dominated by only a few species of coniferous trees, those of the tropics typically comprise many species of broad-leaved trees (plus many epiphytes and woody vines), none of which is ecologically dominant. See also *forest.*

Evolution. Any gradual change. Biological or organic evolution results from changing gene frequencies, associated with biological adaptations, in successive generations and typically occurs through natural selection. See also *natural selection.*

Eye-ring. An area of bare skin (or, in many nontrogons, contrastingly colored feathers) surrounding the eye and sometimes extending to surrounding areas. Bare eye-rings are frequently brightly colored in trogons, and their relative size, color, or brightness may vary by age, sex, or season.

Flanks. The sides of a bird's plumage, from the base of the wings down to the abdomen. This term is sometimes also limited to the posterior part of the sides, especially if this area is distinctively colored.

Fledging success. The percentage of hatched young that are successfully fledged. See also *nesting success.*

Fledging. The initial acquisition of flight by a bird. The period from hatching to fledging is called the fledging period, which in trogons corresponds to the nestling period. See also *nestling.*

Fledgling. A newly fledged bird. See also *juvenile.*

Forest. A dense growth dominated by trees, typically having a closed canopy, in which the height of the tallest trees is greater than the distances between them. See also *cloud forest, elfin forest, evergreen forest, gallery forest, montane forest, primary forest, rain forest, riparian, secondary forest, savanna, terra firme forest, thorn forest, varzea forest,* and *woodlands*

Frugivorous. Having a diet composed largely of fruit, berries, and so on. See also *insectivorous.*

Fuscous. Brownish black.

Gallery forest. Narrow forest that follows rivers or other waterways out into nonforested habitats. See also *forest, riparian, savanna,* and *woodlands.*

Genus. A (literally) "general" Latin or Latinized name applied to a group of closely related living organisms. If a genus comprises a single species, it is called monotypic; otherwise it is polytypic. The genus name (capitalized and italicized) is the first component of a species' two-part scientific name. See also *species* and *Latin name.*

Gular area. The throat region.

Habitat. The place or environment in which a species survives; its natural "address" as opposed to its ecological profession, or niche. See also *niche.*

Hawking. The catching of insects or other aerial prey by birds in flight.

Hectare. Metric unit of area equal to 10,000 square meters (2.47 acres).

Hedgehog stage. The nestling phase common to trogons and coraciiform birds in which the otherwise still naked young have very long, spiny pinfeathers, the sheaths encasing the developing juvenal plumage during much of the feathers' growth.

Heterodactyl. Pertaining to a toe arrangement unique to trogons in which the first and second digit of the foot are directed posteriorly and the third and fourth are directed anteriorly. See also *syndactyl.*

Home range. The entire area used by an individual, pair, or family over a specified period. Trogons so far studied seem to defend their entire home range, making the term synonymous with territory. See also *territory.*

Immature. A transitional (and ill-defined) age class comprising fledged but sexually undeveloped birds; in trogons, at least, immatures carry a mixture of juvenal feathers and feathers acquired in the postjuvenal (first prebasic) molt. See also *fledgling* and *juvenile.*

Incubation. The parental application of body heat to eggs. See also *brood.*

Innate. Pertaining to genetically transmitted traits, especially behavioral ones.

Insectivorous. Having a diet composed mostly of insects and other arthropods. See also *frugivorous.*

Interspecific. Pertaining to interactions between species.

Intraspecific. Pertaining to interactions within species.

Iridescent. Descriptive of feather colors that depend on both reflection and refraction effects that split light into its component spectral colors and produce a shimmering or metallic effect. In trogons iridescent colors vary in hue from golden or emerald green to violet. Unlike feathers colored by carotenoid pigments, iridescent feathers do not fade under prolonged exposure to bright light, but they do depend on directed light sources. The term is derived from the name of the Greek goddess of the rainbow (Iris). See also *carotenoid.*

Isolating mechanisms. Genetically carried (intrinsic) and evolved traits (such as species-specific behavioral signals, anatomical differences, and ecological or temporal breeding restrictions) that serve to prevent the exchange of genes between individuals of different species and prevent the production of viable hybrids. Isolating mechanisms include various premating mechanisms (temporal, ecological, behavioral, or anatomical differences that serve to prevent attempted matings) as well as diverse postmating mechanisms (gametic mortality, embryonic mortality, hybrid inviability, hybrid sterility or subfertility, and hybrid adaptive inferiority). See also *reproductive isolation.*

Juvenal plumage. The feathers acquired during the nestling period, which are carried for a variable time after fledging. This plumage is later partly (at least in trogons) replaced during postjuvenal (or first prebasic) molt, but the juvenal tail feathers and most juvenal wing feathers may persist until the bird molts into its first adult (first basic) plumage. See also *fledging, juvenile,* and *molt.*

Juvenile. An age class comprising birds carrying predominantly feathers of the juvenal plumage; birds fledge during the juvenile period. See also *juvenal plumage.*

Kingfishers. See *Alcedinidae* and *Coraciiformes.*

Latin name. A species' "scientific name," usually a two-parted Latin or Latinized name given, from the time of Linnaeus onward, to a species when it is first officially described and by which it thereafter is technically known and properly identified. The name of the species' describer and the date of description are added in complete citations. See also *genus, species,* and *vernacular name.*

Lauraceous. Pertaining to members (including avocados) of the laurel plant family (Lauraceae). The fruits of this family are major food sources for quetzals and larger New World trogons.

Lore (adj., loreal). The area (feathered in trogons) between the eyes and the base of the bill.

Malar area. The jaw area behind and below the lower mandible. Some birds have a distinctively colored malar stripe, but trogons do not.

Mandible. The lower component of the bill. The term is sometimes used to refer to the entire bill. See also *maxilla.*

Mantid. A member of the praying mantis group of insects (Mantididae).

Mantle. The interscapular area of a bird, located behind the base of the neck between the scapulars.

Mating. A descriptive, nontechnical term often loosely applied to either initial pair bonding or copulation. "Pairing" is the more accurate term for the former activity, at least among bird species that pair bond, and "copulation" is the appropriate and self-evident term for the latter. See also *pair bond.*

Maxilla. The upper component of the bill. See also *culmen* and *mandible.*

Maxillopalatines. Paired bones of the palate laying lateral to the vomers.

Meropidae. The coraciiform family of Old World birds called bee-eaters, noted for their ability to hawk bees, wasps, and other flying insects. See also *Coraciiformes.*

Molt. The periodic sequential loss and replacement of feathers that produce a new plumage. A complete molt affects all feather tracts including the wings and tail and results in an entirely new plumage. An incomplete molt affects only certain feather tracts of the head and body, typically excluding the wing and tail feathers, which usually are molted only once annually. See also *plumage.*

Monogamous. Pertaining to a pair-bonding system in which a male and female remain together for part or all of a breeding season, or sometimes indefinitely. Monogamy is believed to be the only pair-bonding system present in trogonids.

Monophyletic. Having a single ancestral origin. See also *polyphyletic.*

Montane forest. Forest associated with mountains. See also *forest.*

Morph. Any of the phenotypic variants that result from sexual or nonsexual morphism (e.g., dimorphism, dichromatism). Sometimes also called "phase," which is semantically inappropriate for describing a permanent genetic condition. See also *dimorphism.*

Motmotidae. The family of New World coraciiform birds called motmots, nearly all of which have their two central rectrices elongated and racket tipped; these rectrices are often swung side-to-side in a pendulum-like manner. See also *Coraciiformes.*

Mousebirds. See *Coliiformes.*

Mutual. Descriptive of a behavior performed simultaneously by two or more individuals, often members of a pair. Mutual behaviors include mutual singing or calling, mutual preening (allopreening), and so on. See also *antiphonal.*

Nahautl. Pertaining to the people and language of various aboriginal Mexican and Central American cultures, including the Toltecs and Aztecs.

Natal plumage. The feather covering of a newly hatched bird, often downy. Trogons lack true natal down (they are hatched naked, or psilopaedic) and instead directly develop a juvenal plumage late in the nestling stage. See also *hedgehog stage.*

Natal. Pertaining to newly hatched (or, in mammals, newly born) young.

Natural selection. In a broad sense, the long-term changes in gene frequencies and associated traits in populations, resulting from differential survival and reproduction of the fittest individuals within such interbreeding populations. Also applied by Charles Darwin in a more restricted sense to refer only to selection based on those traits associated with differential survival in nature, with the term "sexual selection" being used to designate differential reproduction effects influencing only a single sex. See also *sexual selection.*

Neotropical region. The zoogeographic region of the Western Hemisphere that is located between the Mexican highlands and the southern tip of South America, including all associated offshore islands. The more northerly division of this hemisphere is called the Nearctic region.

Nesting success. The percentage of initiated nests that succeed in fledging one or more young. See also *fledging success.*

Nestling. A recently hatched bird still confined to the nest. See also *fledgling.*

Neuropteran. A member of the lacewing order of insects.

Niche. The behavioral, morphological, and physiological adaptations of a species to its habitat; also sometimes defined from an environmental standpoint as the range of ecological conditions under which a species potentially exists (fundamental niche), best survives (preferred niche), or actually survives (realized niche). See also *habitat.*

Nidicolous. Pertaining to birds hatched in a helpless condition, reared in the nest, and needing prolonged parental brooding and feeding. See also *altricial* and *nestling.*

Nonnuptial plumage. The nonbreeding plumage of a species, acquired (when present) by means of a postnuptial molt and partly or completely lost by means of a prenuptial molt. It is difficult to ascertain whether trogons have a distinct nonnuptial plumage; when present it appears to be virtually identical to the nuptial plumage. Often called the definitive basic plumage in

North America and sometimes also (but less appropriately for trogons) called the winter plumage. See also *molt* and *plumage*.

Nuchal collar. Distinct band in the nape region.

Nuptial plumage. The definitive breeding plumage of adult birds, typically acquired by means of a prenuptial molt of variable extent. Now generally termed definitive alternate plumage in North America. See also *molt, nonnuptial plumage,* and *plumage*.

Ochreous. Yellowish brown, ochre colored.

Odonata. The insect order of dragonflies and damselflies.

Oligocene Epoch. A major subdivision of the Cenozoic Era, extending from about 37 to 18 million years ago. See also *Cenozoic Era*.

Operculum. A shelf-like extension of the horny covering of the bill (rhamphotheca), which partially covers the nostril opening (nare).

Oral flanges. Swellings at the junction of the maxilla and mandible in nestling birds that make the chick's gape more conspicuous and are sometimes distinctively colored.

Orbital. Pertaining to the area of the eyes. Supraorbital, postorbital, and suborbital areas are located above, behind, and below the eyes, respectively. The term "superciliary" is sometimes used as a synonym for supraorbital, as in "superciliary stripes." See also *lore*.

Orthopteran. A member of the insect order Orthoptera, which includes mantids (Mantidae), walking sticks or stick insects (Phasmatidae), grasshoppers and locusts (e.g., Tettigonidae, Acrididae), crickets (Gryllidae), and others.

Pair bond. A (usually) monogamous and variably prolonged social bond between two individuals, typically for facilitating reproduction.

Pair-bonding and pair-forming behaviors. The "courtship" behaviors related to (1) mate choice and associated pair formation (pair-forming signals) and (2) pair-bond maintenance (pair-bonding signals). See also *mating*.

Parapatric. Pertaining to populations that come into limited or sometimes extensive contact with one another along some portion of their common border but do not actually overlap during the breeding season. Parapatric distributions are suggestive of competing, noninterbreeding populations, that is, of two separate species. See also *allopatric* and *sympatric*.

Pectoral. Pertaining to the lower breast (chest) area.

Phase. A time-related event, such as pair-forming, nesting, or brooding phase. See also *morph*.

Phasmatid. A member of the walking-stick or stick-insect group.

Phenotype. The visual properties of an individual produced both by its genetic makeup (genotype) and by environmental influences, such as diet, disease, age, accidents, and other influences not mediated by genetics.

Phylogeny. The evolutionary history or pathway of descent of an organism or group of organisms. See also *monophyletic* and *polyphyletic*.

Plumage. The feathers carried by a bird during a particular time period. Some birds acquire their plumage by means of a single (complete) molt, but others may also carry a compound plumage, representing a mixture of feather generations grown in during at least two separate (but incomplete) molts. See also *molt, definitive plumage, hedgehog phase, juvenal plumage, natal plumage, nuptial plumage,* and *nonnuptial plumage*.

Polyphyletic. Descriptive of assemblages of organisms believed to have been derived phyletically from two or more ancestral groups; thus an artificial (but perhaps convenient) grouping in an evolutionary sense. See also *monophyletic*.

Postnuptial molt. A molt that follows breeding and involves the loss of the nuptial plumage, resulting in a nonbreeding (nonnuptial) plumage. See *nonnuptial plumage*.

Prenuptial molt. A molt of the nonbreeding (nonnuptial plumage); precedes and produces the nuptial plumage.

Primaries. The outer ten pairs of flight feathers, attached to the hand bones of birds. Primaries are numbered from the innermost out, and in trogons molt occurs in that same (descendant) sequence. In trogons the sixth or seventh primary is the longest, and the tenth much the shortest.

Primary forest. "Virgin" forests that have not been exposed to prior fires, logging, or other major disturbance.

Primitive. Descriptive of generalized traits that appeared early in a group's phylogeny but were later replaced by more advanced (but not necessarily more complex) traits.

Proximal. Toward the center of the body. See also *distal*.

Quetzal. The vernacular English name of the large, mostly frugivorous trogons of the Neotropics (genus *Pharomachrus*) having ornate upper tail-coverts covering most or all of the rectrices. A variant older term is "quezal."

Radius. The more anterior of the two bones that make up the skeletal forewing of birds. See also *ulna*.

Rain forest. A forest type characterized by excessive amounts of rainfall throughout the year. The rainfall results in an "evergreen" floral condition, and, in the tropics, a multilayered canopy, high species diversity, rapid leaching of soil nutrients, and often complex patterns of plant succession and life histories. Includes lowland rain forests (terra firme and varzea forest types) and upland rain forests. See also *forest*.

Rectrices (sing., rectrix). The tail feathers, twelve (six pairs) being the consistent number in trogonids. Rectrices are numbered from the central pair out.

Remiges. The flight feathers (primaries, secondaries, tertials, and alular feathers) that provide the wing's airfoil surface. See also *alular feathers, primaries, secondaries,* and *tertials*.

Reproductive isolation. Isolation that prevents populations from interbreeding. Isolation can arise from either preventive environmental barriers (extrinsic isolating factors) or genetic restrictions (inherent isolating mechanisms). See also *isolating mechanisms*.

Riparian. Associated with rivers or streams. Gallery forests are a type of riparian community.

Savanna. A plant community type, most common in the tropics and subtropics, in which trees are relatively low, highly dispersed, and situated within a much larger matrix dominated by grasslands. A South American variant is the *cerrado*, in which the trees and other flora are highly xerophytic and thus are adapted to drought and fire. See also *cerrado, woodlands,* and *xeric*.

Scapulars. "Shoulder" feathers associated with the scapula bones and located between the wings and the middle of the back.

Schizognathous. Descriptive of a palate condition of birds in which the maxillopalatine bones do not meet each other or the vomers at the midline, which produces a longitudinal cleft in the bony palate. See also *desmognathous*.

Secondaries. The inner flight feathers whose bases are spatially associated with the ulna and radius (forewing) of birds. In trogonids there are typically eight to eleven secondaries, the larger figure including the three innermost and progressively shorter so-called tertials. Secondaries are numbered from the outermost (that adjacent to the innermost primary at the wrist junction) in. See also *tertials*.

Secondary forest. A forest that has regenerated after some kind of major disturbance (e.g., logging, fire, landslide); such regeneration is usually the result of natural plant-replacement sequences (secondary succession).

Sexual selection. A type of natural selection in which the evolution and maintenance of traits of one sex result from the social interactions that produce differential individual reproductive success. These include both interactions between members of the same sex (intrasexual, or agonistic, selection) and those between the sexes (intersexual, or epigamic, selection). See also *natural selection*.

Signals. Behaviors that have been evolutionarily modified (ritualized) to transmit information between members of a social group. Such signals (especially the visual ones) are often also called displays, but postures and vocalizations also constitute social signals. See also *agonistic* and *advertising behavior*.

Site fidelity (or site tenacity). The social attachment of an individual to a specific site (e.g., nest site, territory) in successive seasons or years.

Songs. Avian vocalizations that tend to be acoustically complex, prolonged, and often sex specific as well as unique to an individual. They frequently are uttered only during particular seasons and may have both intra- and intersexual signal value, such as territorial advertisement and sexual attraction. Like those of many cuckoos, songs of trogonids tend to be repetitive and nonmusical, with changes in loudness (amplitude) and duration more apparent than rapid or irregular changes in frequency (pitch) or cadence. See also *calls* and *vocalizations*.

Species. A group or groups whose members share the same isolating mechanism and thus are reproductively isolated from all other populations but are potentially capable of breeding freely among themselves. The nomenclatural category below genus and above subspecies. The species name is the second (specific) component of the two-part Latin or Latinized name and is never capitalized in that context. See also *Latin name* and *genus*.

Sphingid. A member of the insect family comprising sphinx moths or hawk moths (Sphingidae).

Subgenus. A subdivision of a genus containing one or more species that seem to belong to well-marked taxonomic groups within the genus. The subgeneric affiliation may be indicated by placing the name of the subgenus parenthetically between that of the genus and the species, as in *Apaloderma (Heterotrogon) vittatum*. See also *superspecies*.

Subspecies. A geographic race of a species. When a subspecies is recognized it becomes the third and final component of a species' scientific (Latinized) name; like the species name, the subspecies name is italicized but never capitalized. The race that shares its name with the species is the first-described, or nominate, subspecies.

Subterminal. Located just before the distal tip.

Superspecies. An assemblage or two or more essentially allopatric populations that are very closely related and might potentially represent biological subspecies; however, the test of reproductive isolation is lacking owing to their allopatric distributions. Superspecies affiliation may be indicated by placing the name of the species having taxonomic priority in parentheses between the generic and specific names of any additional members of the group, as in *Trogon (viridis) bairdii.* Species groups are assemblages of less closely related species that are often at least partly sympatric; these may constitute a subgenus. See also *subgenus.*

Sympatric. Pertaining to populations that overlap at least in part, especially during the breeding season. By definition, sympatrically breeding but noninterbreeding populations are never considered the same species. See also *allopatric, parapatric,* and *reproductive isolation.*

Syndactyl. A toe arrangement, developed to the highest degree in coraciiform birds but also evident in some trogonids, in which the third and fourth toes are variably fused for part or most of their length; this arrangement is apparently a digging adaptation.

Syrinx. The vocal organ of a bird, analogous in function to the larynx of mammals and located at the junction of the trachea and bronchi. Tension on the vibratory syringeal membranes (tympaniform membranes) is regulated by paired tracheal or syringeal muscles as well as by the nearby air sacs associated with the lungs, and vocal sound is produced by expelling air past these paired membranes. See also *vocalizations.*

Tail-coverts. The longer feathers (located behind the rump and abdomen) that partially or sometimes completely (as in quetzals) cover the true tail feathers, or rectrices. The term includes both the upper tail-coverts and under tail-coverts; the latter feather region is sometimes called the crissum.

Taxon (pl., taxa). A group of phyletically related organisms.

Taxonomy. A system of naming, describing, and organizing groups of organisms in ways that reflect perceived phyletic relationships.

Termitarium. A termite nest. See also *vespiary.*

Terra firme forest. Lowland tropical forests of South America that are not subject to seasonal flooding. See also *forest* and *varzea forest.*

Territoriality. The advertisement and agonistic behaviors associated with the establishment and defense of territory. In at least some trogons, the male initially establishes and advertises (vocally and visually) the breeding territory, but his mate later helps to advertise and defend it. See also *advertising behavior* and *signals.*

Territory. An area having resources defended or controlled by an animal against other members of its species (intraspecific territories) or, less often, against those of other species (interspecific territories). See also *advertising behavior* and *home range.*

Tertials. The three or so innermost secondaries that are of graded sizes and often are differently patterned than the more distal and more typical secondaries. However, no true tertials (feathers with bases associated with the distal head of the humerus rather than the proximal tip of the ulna) are present in trogons. The so-called tertials of trogonids are often strongly vermiculated and grade imperceptibly into the upper wing-coverts.

Thorn forest.

Thorn forest. A forest-like community of arid-adapted, usually thorny, trees and water-conserving plants such as stem- or leaf-succulents.

Todidae. A family of West Indian coraciiform birds called todies, which are small insectivorous species resembling motmots in their biology and breeding behavior. See also *Coraciiformes.*

Trait. A measurable phenotypic attribute (behavioral, structural, physiological), especially one that is at least in part genetically controlled.

Trogon. The collective vernacular English name for all the members of the family Trogonidae (except the quetzals). In its italicized form *Trogon* is also used as a generic name for the typical New World trogons.

Trogonid. A member of the family Trogonidae, which includes trogons and quetzals.

Truncated. Abruptly terminated or squared-off.

Ulna. The more posterior of the two bones that make up the skeletal forewing of birds. See also *radius.*

Varzea forest. Lowland forests of South America that are periodically flooded. See also *forest* and *terra firme forest.*

Vermiculations. Fine, wavy (worm-like) markings on feathers, especially on the wing-coverts and secondaries of most trogons.

Vernacular name. The common or English-language name of an organism (usually a species), as opposed to its scientific, or Latin, name.

Vespiary. A nest of wasps or hornets. See also *termitarium.*

Vocalizations. Utterances, including both songs and calls, generated (at least in birds) by the syrinx and sometimes modulated or resonated by nonsyringeal structures, such as the trachea or oral cavity. Nonvocal sounds include hissing, bill clapping, and feather-generated noises. See also *calls* and *songs*.

Vomers. Paired or fused bones of the median palate.

Wing-coverts. Variably sized and overlapping feathers covering the upper (upper wing-coverts) and lower (under wing-coverts) surfaces of the wing, located anterior to the primaries and secondaries. Major groups (front to back) are the marginal, lesser, median, and greater coverts of both primaries and secondaries. Among trogons, most wing-coverts of juveniles and immatures are usually edged and spotted with buffy or white markings, these being carried over from the juvenal plumage. In quetzals the median wing-coverts are unusually lengthened and outwardly curved in shape.

Woodlands. Relatively dry (xeric) but partially wooded habitats characterized by scattered, rather low trees (the height of the trees is typically less than the distance between the trees) in a matrix of grasslands or other nonforest vegetation. In savanna habitats the area of the grassland matrix is much more extensive than that of the widely scattered trees. See also *forest* and *savanna*.

Xeric. Desert-like or drought adapted.

Zygodactyl. The toe arrangement of woodpeckers and most other "yoke-toed" birds, in which the first and fourth toes are directed posteriorly and the second and third toes are anteriorly oriented. See also *heterodactyl*.

BIBLIOGRAPHY

Because no complete bibliography of the trogonids otherwise exists, these references include some sources that are not specifically cited in the text. Nearly all were consulted, and all are relevant to the Trogonidae literature.

Aldrich, J. W., and P. B. Bole Jr. 1937. The birds and mammals of the western slope of the Azuero Peninsula (Republic of Panama). *Scientific Publications of the Cleveland Museum of Natural History* 7:1–196.

Ali, S. 1949. *Indian hill birds*. Oxford University Press, Delhi, India.

———. 1962a. *The birds of Sikkim*. Oxford University Press, Madras, India.

———. 1962b. *The birds of Kerala*. Oxford University Press, Madras, India.

———. 1977. *Field guide to the birds of the eastern Himalayas*. Oxford University Press, Delhi, India.

Ali, S., and S. D. Ripley. 1968–1974. *Handbook of the birds of India and Pakistan*. 10 vols. Oxford University Press, Bombay, India.

———. 1983. *A pictorial guide to the birds of the Indian subcontinent*. Oxford University Press, Delhi, India.

Allen, A. A. 1942. An Arizona nest of the coppery-tailed trogon. *Auk* 61:640–642.

American Ornithologists' Union (AOU). 1998. *Check-List of North American birds*. 7th ed. American Ornithologists' Union, Washington, D.C.

Andrle, R. F. 1967. Birds of the Sierra de Tuxtla in Veracruz, Mexico. *Wilson Bulletin* 79:163–187.

Ash, J. S., and J. E. Miskell. 1998. *Birds of Somalia*. Pica Press, Mountfield (Sussex), U.K.

Avila, H., M. L. Hernandez O., and E. Velarde. 1996. The diet of resplendent quetzal (*Pharomachrus mocinno mocinno*: Trogonidae) in a Mexican cloud forest. *Biotropica* 28 (4B): 720–727.

Baepler, D. H. 1962. The avifauna of the Soloma region in Huehuetenango, Guatemala. *Condor* 64: 140–153.

Baicich, P. J., and C.J.O. Harrison. 1997. *A guide to the nests, eggs, and nestlings of North American birds*. 2d ed. Academic Press, San Diego, Calif.

Baker, E.C.S. 1927. *The fauna of British India, including Ceylon and Burma*. Vol. 4, *Birds*. Taylor and Francis, London.

Bangs, O. 1899. On some new or rare birds from the Sierra de Santa Marta, Colombia. *Proceedings of the Biological Society of Washington* 13:81–108.

———. 1909. Notes on some rare or not well-known Costa Rican birds. *Proceedings of the Biological Society of Washington* 22:29–38.

Bangs, O., and W. R. Zappey. 1905. Birds of the Isle of Pines. *American Naturalist* 39:179–215.

Bannerman, D. 1933. *The birds of tropical West Africa*. Vol. 3. Crown Agents for the Colonies, London.

Barbour, T. 1943. Cuban Ornithology. *Memoirs Nuttall Ornithological Club* 9:1–142.

Bates, C. L. 1927. Notes on some birds of Cameroon and the Lake Chad region: Their status and breeding times. *Ibis*, ser. 12, 3:1–64.

Belcher, C. F., and G. D. Smoker. 1934–1937. On the birds of the Colony of Trinidad and Tobago. Part 4. *Ibis*, ser. 13, 6:792–813.

Belton, W. 1984. Birds of the Rio Grande do Sul. Part 1, Rheidae through Funariidae. *Bulletin of the American Museum of Natural History* 178:371–631.

Bennett, P. M., and P. H. Harvey. 1987. Acting and resting metabolism in birds: Allometry, phylogeny, and ecology. *Journal of Zoology* (London) 213:327–363.

Bent, A. C. 1940. Life histories of North American cuckoos, goatsuckers, hummingbirds, and their allies. *Bulletin of the U.S. National Museum* 176:1–506.

Berlioz, J. 1955. Notes sur les couroucou du genre *Pharomachrus*. *Oiseau* 25:27–39.

Binford, L. C. 1989. A distributional survey of the Mexican state of Oaxaca. Ornithological Monographs, no. 43. American Ornithologists' Union, Washington, D.C.

Blake, E. R. 1950. Birds of the Acary Mountains, southern British Guiana. *Fieldiana: Zoology* 32:419–474.

———. 1956. A collection of Panamanian nest and eggs. *Condor* 58:386–388.

———. 1958. Birds of Volcan de Chiriquí, Panama. *Fieldiana: Zoology* 36:497–577.

———. 1962. Birds of the Sierra Macarena, eastern Colombia. *Fieldiana: Zoology* 44:69–112.

———. 1963. The birds of southern Surinam. *Ardea* 51: 53–72.

Bond, J. 1928. The distribution and habits of the birds of the Republic of Haiti. *Proceedings of the Philadelphia Academy of Natural Sciences* 80: 483–521.

———. 1936. *Birds of the West Indies*. Philadelphia Academy of Natural Sciences, Philadelphia.

———. 1948. Origin of the bird fauna of the West Indies. *Wilson Bulletin* 60:209–229.

———. 1993. *Birds of the West Indies*. 5th ed. Houghton Mifflin, Boston.

Bond, J., and R. Meyer de Schauensee. 1942. The birds of Bolivia. *Proceedings of the Philadelphia Academy of Natural Sciences* 94:307–391.

———. 1943. The birds of Bolivia. *Proceeding of the Philadelphia Academy of Natural Sciences* 95:167–221.

Bowes, A. L. 1969. The quetzal, fabulous bird of the Maya. *National Geographic* 135:140–150.

Bowes, A. L., and D. G. Allen. 1969. Biology and conservation of the quetzal. *Biological Conservation* 1:297–306.

Brodkorb, P. 1937. Some birds of the Amazonian islands of Caviana and Marajo. *Occasional Papers, Museum of Zoology, University of Michigan* 349:1–7.

———. 1942. A new subspecies of *Trogon citreolus*. *Proceedings of the Biological Society of Washington* 55:183–184.

———. 1971. Catalogue of fossil birds. Part 4, Columbiformes through Piciformes. *Bulletin of the Florida State Museum* 15:163–266.

Brooks, T. M., S. L. Pimm, V. Kapos, and C. Ravilious. 1999. Threat from deforestation to montane and lowland birds and mammals in insular South East Asia. *Journal of Animal Ecology* 68:1061–1078.

Brosset, A. 1983. Parades et chants collectifs chez les couroucou du genre *Apaloderma*. *Alauda* 51:1–10.

Brown, L. H. 1975. Breeding of Stuhlmann's starling and Narina's trogon. *Bulletin of the East African Natural History Society,* pp. 44–45.

Buckley, P. A., M. S. Foster, E. S. Morton, R. S. Ridgely, and F. G. Buckley, eds. 1985. *Neotropical Ornithology*. Ornithological Monographs, no. 36. American Ornithologists' Union, Washington, D.C.

Burton, P.J.K. 1973. Non-passerine bird weights from Panama and Colombia, with notes on "soft-part" colours. *Bulletin of the British Ornithologists' Club* 95: 82–86.

Carriker, M. A., Jr. 1910. An annotated list of the birds of Costa Rica, including Cocos Island. *Annals of the Carnegie Museum* 6:314–915.

Carriker, M. A., Jr., and R. Meyer de Schauensee. 1937. An annotated list of two collections of Guatemalan birds in the Academy of Natural Sciences of Philadelphia. *Proceedings of the Philadelphia Academy of Natural Sciences* 87:411–455.

Chapin, J. P. 1939. The birds of the Belgian Congo. Part 2. *Bulletin of the American Museum of Natural History* 65: 1–632.

Chapman, F. M. 1892. Notes on birds and mammals observed near Trinidad, Cuba, with remarks on the origin of the West Indian bird life. *Bulletin of the American Museum of Natural History* 4:279–330.

———. 1914. Diagnoses of apparently new Colombian birds. *Bulletin of the American Museum of Natural History* 33:603–637.

———. 1917. The distribution of bird life in Colombia. *Bulletin of the American Museum of Natural History* 36: 1–729.

———. 1921. The distribution of bird life in the Urubamba Valley of Peru. *Bulletin of the U.S. National Museum.* 117:1–138.

———. 1926. The distribution of bird life in Ecuador. *Bulletin of the American Museum of Natural History* 55: 1–784.

———. 1931. The upper zonal bird life of Mts. Roraima and Duida. *Bulletin of the American Museum of Natural History* 63:1–135.

———. 1939. The upper zonal bird life of Mt. Auyan-tepui, Venezuela. *American Museum Novitates*, no. 1051:1–15.

Chasen, F. M. 1939. *The birds of the Malay Peninsula*. Vol. 3. H. F. and G. Witherby, London.

Cheng Tso-hsin. 1987. *A synopsis of the avifauna of China.* Paul Parey, Berlin, and Science Press, Beijing.

Chubb, C. 1910. On the birds of Paraguay. *Ibis,* ser. 9, 4:263–285.

———. 1916–1921. *The birds of British Guiana.* 2 vols. Bernard Quaritch, London.

Clark, H. L. 1918. Notes on the anatomy of the Cuban trogon. *Auk* 35:286–289.

Collar, N. G., M. J. Crosby, and A. J. Stattersfield. 1994. *Birds to watch 2. The world list of threatened birds.* BirdLife International, Cambridge, U.K., and Smithsonian Institution Press, Washington., D.C.

Collar, N. G., M. J. Crosby, A. J. Stattersfield, L. P. Gonzaga, N. Krabbe, A. M. Nieto, L. G. Naranjo, T. A. Parker III, and D. C. Wege. 1992. *Threatened birds of the Americas.* Smithsonian Institution Press, Washington., D.C., and International Council for Bird Preservation, Cambridge, U.K.

Cory, C. B. 1919. Catalogue of the birds of the Americas and the adjacent islands. Part 2, no. 2. *Field Museum of Natural History, Zoology Series* 13 (2): 1–607.

Cottam, C., and P. Knappen. 1939. Food of some uncommon North American birds. *Auk* 56:138–169.

Cully, J. F., Jr. 1986. Mobbing behavior by a pair of elegant trogons. *Condor* 88:103–104.

Cunningham-van Someren, G. R. 1973. Calling and flight display assembly of male trogons. *Bulletin of the East African Natural History Society,* pp. 81–83.

Darlington, P. J., Jr. 1931. Notes on the birds of Rio Frio (near Santa Marta), Magdalena. *Bulletin of the Museum of Comparative Zoology* (Harvard University) 71: 349–421.

Davis, L. I. 1972. *A field guide to the birds of Mexico and Central America.* University of Texas Press, Austin.

Davis, W. B. 1945. Notes on Veracruzan birds. *Auk* 62: 272–286.

Deignan, H. G. 1941. New birds from the Indo-Chinese subregion. *Auk* 58:396–398.

———. 1945. The birds of northern Thailand. *Bulletin of the U.S. National Museum* 186:1–616.

Delacour, J. 1947. *Birds of Malaysia.* Macmillan, New York.

Delacour, J., and P. Jabouille. 1931. *Les oiseaux de l'Indochine Française.* 4 vols. Exposition Colonial International, Paris.

Delacour, J., and E. Mayr. 1946. *Birds of the Philippines.* Macmillan, New York.

de la Pena, M. R. 1987. *Nidos y Huevos de Aves Argentina.* Edicion del autor, Santa Fe, Argentina.

de Schauensee, R. Meyer. 1948–1952. The birds of the Republic of Colombia. *Caldasia,* vol. 5 (22–26).

———. 1964. *The birds of Colombia and adjacent areas of South and Central America.* Livingston Publications, Narberth, Penn.

———. 1966. *The species of birds of South America and their distribution.* Livingston Publications, Narberth, Penn.

———. 1970. *A guide to the birds of South America.* Livingston Publications, Narberth, Penn.

———. 1984. *The birds of China.* Smithsonian Institution Press, Washington, D.C.

de Schauensee, R. Meyer, and W. H. Phelps Jr. 1978. *A guide to the birds of Venezuela.* Princeton University Press, Princeton, N.J.

Dick, J. A., W. B. McGillivray, and D. J. Brooks. 1984. A list of birds and their weights from Saul, French Guiana. *Wilson Bulletin* 96:347–365.

Dickey, D. R., and A. J. van Rossem. 1938. The birds of El Salvador. *Field Museum of Natural History, Zoology Series* 23:1–609.

Dickinson, E. C., R. S. Kennedy, and K. C. Parkes. 1991. The birds of the Philippines: An annotated checklist. *British Ornithologists' Union Checklist* 12:1–507.

Dod, A. S. 1978. *Aves de la Republica Dominicana.* Museo Nacional de Historia Natural, Santo Domingo, Dominican Republic.

Dorst, J. 1951. Contribution à l'étude du plumage des Trogonidés. *Bulletin du Muséum Naturelle Histoire* (Paris) (2) 22: 693–699.

Dowsett, R. J. 1970. A collection of birds from Nyika Plateau, Zambia. *Bulletin of the British Ornithologists' Club* 90:49–53.

Dowsett-Lemaire, F. 1983. Ecological and territorial requirements of montane forest birds on the Nyika Plateau, south-central Africa. *Gerfaut* 73:345–378.

Dunning, J. B., Jr. 1993. *CRC handbook of avian body masses.* CRC Press, Boca Raton, Fla.

du Pont, J. E. 1971. *Philippine Birds.* Monograph Series, no. 2. Delaware Museum of Natural History, Greenville, Del.

Durrer, H., and W. Villiger. 1966. Schillerfarben der Trogoniden. Eine Elektronmikroskopische Untersuchung. *Journal für Ornithologie* 107:1–26. (Contains an English summary.)

Eckelbery, D. R. 1963. Some remarks on trogons with special reference to the mountain trogon. *Living Bird* 2:4–6.

Eguarte, L. E., and C. Martinez del Rio. 1985. Feeding habits of the citreoline trogon in a tropical deciduous forest during the dry season. *Auk* 102:872–874.

Eisenmann, E. 1959. The correct specific name of the quetzal, *Pharomachrus mocinno.* *Auk* 76:108.

Elliott, B. G. 1983. Nocturnal moonlight calling by elegant trogon in Arizona. *Western Birds* 14:53.

Espinosa de los Monteros, S.J.E. 1997. Molecular phylogeny of Trogoniformes. Ph.D. diss., City University of New York, New York.

———. 1998. Phylogenetic relationships among the trogons. *Auk* 115:937–954.

Faaborg, J. 1985. Ecological constraints on West Indian bird distributions. In *Neotropical Ornithology.* Ornithological Monographs, no. 36, edited by P. Buckley et al., 621–653. American Ornithologists' Union, Washington, D.C.

Feduccia, A. 1975. Morphology of the bony stapes (columella) in the Passeriformes and related groups: Evolutionary implications. *Miscellaneous Publications, Museum of Natural History, University of Kansas* 63:1–34.

———. 1996. *The origin and evolution of birds.* Yale University Press, New Haven, Conn.

ffrench, R. 1991. *A guide to the birds of Trinidad and Tobago.* 2d ed. Cornell University Press, Ithaca, N.Y.

Fjeldså, J., and N. Krabbe. 1990. *Birds of the High Andes.* Apollo Books, Svendborg, and Museum of Zoology, University of Copenhagen, Sweden.

Fleming, R. L., R. L. Fleming Jr., and L. S. Bangdel. 1975. *Birds of Nepal, with reference to Kashmir and Sikkim.* Published by the authors, Katmandu, Nepal.

Fogden, M.P.L. 1972. The seasonality and population dynamics of equatorial forest birds in Sarawak. *Ibis* 114: 307–342.

Food and Agriculture Organization (FAO). 1984. *FAO production yearbook 1983.* Vol. 37, FAO statistics series no. 55. FAO, Rome.

———. 1993. Forest resources assessment 1990. FAO Forestry Paper 112:1–61.

Friedmann, H. 1948. Birds collected by the National Geographic Society's expeditions to northern Brazil and southern Venezuela. *Proceedings of the U.S. National Museum* 97:373–570.

Friedmann, H., L. Griscom, and R. T. Moore. 1950. Distributional check-list of the birds of Mexico. Part 1. *Pacific Coast Avifauna* 29:1–201.

Friedman, H., and F. D. Smith Jr. 1950. A contribution to the ornithology of northeastern Venezuela. *Proceedings of the U.S. National Museum* 100:411–538.

———. 1955. A further contribution to the ornithology of northeastern Venezuela. *Proceedings of the U.S. National Museum* 104:463–524.

Fry, C. H. 1970. Ecological distribution of birds in northeastern Matto Grosso State, Brazil. *Anais de Academia Brasileira de Ciencias* 42:275–318.

Fry, C. H., S. Keith, and E. K. Urban. 1988. *The birds of Africa.* Vol. 3. Academic Press, New York.

Gilliard, E. T. 1941. The birds of Mt. Auyan-tepui, Venezuela. *Bulletin of the American Museum of Natural History* 77:439–508.

Glenister, A. G. 1951. *The birds of the Malay Peninsula, Singapore, and Penang.* Oxford University Press, London.

Gore, M. E., Jr. 1968. A check-list of the birds of Sabah, Borneo. *Ibis* 110:165–196.

Gould, J. 1835–1838. *A monograph of the Trogonidae or family of Trogons.* 1st ed. Lithographs by John and Elizabeth Gould. H. Sotheran, London.

———. 1858–1875. *A monograph of the Trogonidae or family of Trogons.* 2d ed. Lithographs by John Gould, H. C. Richter, and W. Hart. H. Sotheran, London.

Griscom, L. 1932. The distribution of bird life in Guatemala. *Bulletin of the American Museum of Natural History* 64:1–439.

———. 1935. The ornithology of the Republic of Panama. *Bulletin of the Museum of Comparative Zoology* (Harvard) 78:261–382.

Griscom, L., and J. C. Greenway. 1937. Critical notes on new Neotropical birds. *Bulletin of the Museum of Comparative Zoology* (Harvard University) 81:417–437.

———. 1941. Birds of lower Amazonia. *Bulletin of the Museum of Comparative Zoology* (Harvard University) 88:83–344.

Gross, A. O. 1930. Rainbows on wings. *Nature Magazine* 15:249–250.

Gyldenstolpe, N. 1916. Zoological results of the Swedish zoological expeditions to Siam, 1911–1912 and 1914–1915. Part 4, Birds. *Kungliga Svenska Vetenskapsakademiens Handlingar* (Stockholm) 2:1–160.

Haffer, J. 1974. Avian speciation in tropical South America, with a systematic survey of the toucans (Ramphastidae) and jacamars (Galbulidae). Publications of the Nuttall Ornithological Club, no. 14. Nuttall Ornithological Club, Cambridge, Mass.

———. 1975. Avifauna of northwestern Colombia, South America. *Bonner Zoologische Monographien* 7:1–182.

Hall, L. S. 1996. Habitat selection by the elegant trogon at multiple scales. Ph.D. diss., University of Arizona, Tucson.

Hall, L. S., and J. O. Karubian. 1996. Breeding behavior of elegant trogons in southeastern Arizona. *Auk* 113: 143–150.

Hall, L. S., and R. W. Mannan. Forthcoming. An evaluation of habitat quality for the elegant trogon at multiple scales. *Journal of Wildlife Management.*

Hanson, D. A. 1982. Distribution of the quetzal in Honduras. *Auk* 99:385.

Harcus, J. L. 1976. Presumed anti-predator behaviour of the Narina trogon. *Ostrich* 47:129–130.

Hardy, J. W., G. Reynard, and B. B. Coffey. 1995. *Voices of the New World cuckoos and trogons.* Ara Records, Gainesville, Fla.

Harrisson, T. H., and C. H. Hartley. 1934. Descriptions of new birds from Borneo. *Bulletin of the British Ornithologists' Club* 54:148–160.

Hartman, F. A. 1961. Locomotor mechanisms of birds. *Smithsonian Miscellaneous Collections* 142:1–91.

Haverschmidt, F. 1948. Bird weights from Surinam. *Wilson Bulletin* 60:230–239.

———. 1952. More bird weights from Surinam. *Wilson Bulletin* 64:234–241.

———. 1968. *Birds of Surinam.* Oliver and Boyd, Edinburgh, U.K.

Hayes, F. L. 1993. Status, distribution, and biogeography of the birds of Paraguay. Ph.D. diss., Loma Linda University, Loma Linda, Calif.

Hellebrekers, W.P.J. 1942. Revision of the Penard zoological collection from Surinam. *Zoologische Mededelingen* 24:240–275.

Hellmayr, C. E. 1929. A contribution to the ornithology of northeastern Brazil. *Field Museum of Natural History, Zoology Series* 12 (18): 235–501.

Henry, G. M. 1971. *A guide to the birds of Ceylon.* 2d ed. Oxford University Press, London.

Herklots, G.A.C. 1961. *The birds of Trinidad and Tobago.* Collins, London.

Hernandez, M. 1998. The quetzal and its conservation in the Mexican southeast. *Wilson Bulletin* 110:559.

Heske, F. 1973. *Weltforstatlas.* Paul Parey, Berlin.

Hilty, S. L. 1985. Distributional changes in the Colombian avifauna: A preliminary blue list. In *Neotropical ornithology.* Ornithological Monographs, no. 36, edited by P. Buckley et al., 1,000–1,012. American Ornithologists' Union, Washington, D.C.

Hilty, S. L., and W. L. Brown. 1986. *A guide to the birds of Colombia.* Princeton University Press, Princeton, N.J.

Holmes, D. A., and S. Nash. 1989. *The birds of Java and Bali.* Oxford University Press, Singapore.

———. 1990. *The birds of Sumatra and Kalimantan.* Oxford University Press, Singapore.

Holt, E. G. 1928. An ornithological survey of the Sierra do Itatiaya, Brazil. *Bulletin of the American Museum of Natural History* 57:251–326.

Howe, H. F. 1981. Dispersal of Neotropical nutmeg (*Virola sebifera*) by birds. *Auk* 98:88–96.

Howell, S.N.G., and S. Webb. 1995. *A guide to the birds of Mexico and northern Central America.* Oxford University Press, New York.

Howell, T. R. 1957. Birds of a second-growth rain forest area of Nicaragua. *Condor* 59:7–111.

Inglis, C. M. 1939. The Burmese red-headed trogon *Harpactes erythrocephala erythrocephala* Gould. *Journal of Darjeeling Natural History Society* 14:1–5.

Inskipp, C., and T. Inskipp. 1985. *A guide to the birds of Nepal.* Smithsonian Institution Press, Washington, D.C.

Jenkins, D. V., and G. S. de Silva. 1985. An annotated check-list of the birds of the Mount Kinabalu National Park, Sabah, Malaysia. In *Kinabalu, summit of Borneo.* Sabah Society Monograph, edited by E. R. Dingley, 347–402. Sabah Society, Kota Kinabalu.

Jobling, J. A. 1991. *A dictionary of scientific bird names.* Oxford University Press, Oxford, U.K.

Johansson, U. J. 1998. The early evolution of the Trogonidae. In N. J. Adams and R. H. Slotow, eds. *Proceedings 22d International Ornithological Congress, Durban.* Published in *Ostrich* 69:402.

Johnsgard, P. A. 1997. *The hummingbirds of North America.* 2d ed. Smithsonian Institution Press, Washington, D.C.

———. 1999. *The pheasants of the world: Biology and natural history.* 2d ed. Smithsonian Institution Press, Washington, D.C.

Kalina, J., and J. Baranga. 1991. First East African nest record for the bar-tailed trogon *Apaloderma vittatum Scopus* 15:54–55.

Karr, J. R. 1971. Structure of avian communities in selected Panama and Illinois habitats. *Ecological Monographs* 41: 207–233.

Kern, J. A. 1968. Quest for the quetzal. Audubon 70 (4): 28–39.

———. 1975. Quest for the quetzal. In *The pleasure of birds,* edited by L. Line, 146–153. J. B. Lippincott, Philadelphia.

King, B. F., and E. C. Dickinson. 1975. *A field guide to the birds of Southeast Asia.* Houghton Mifflin, Boston.

Kinnear, N. B. 1927. On new races of *Pyrotrogon* and *Picumnus. Bulletin of the British Ornithologists' Club* 47:111–113.

Kunzmann, M. R., L. S. Hall, and R. R. Johnson. 1998. Elegant trogon (*Trogon elegans*). In *The Birds of North America,* edited by A. Poole, P. Stettenheim, and F. Gill, no. 357. The Birds of North America, Philadelphia.

LaBastille, A. 1973. Establishment of a quetzal cloudforest reserve in highland Guatemala. *Biological Conservation* 9:141–154.

———. 1974. Use of artificial nest-boxes by quetzals in Guatemala. *Biological Conservation* 6:64–65.

———. 1975. Jewel in the clouds. *Wildlife* (London) 17: 556–559.

———. 1976. A question of quetzals. *Animal Kingdom* 79 (4): 18–24.

———. 1983. *Pharomachrus mocinno* (quetzal). In *Costa Rican Natural History*, edited by D. Jansen, 599–600. University of Chicago Press, Chicago.

———. 1985. Resplendent quetzal. *Living Bird Quarterly* 5 (1): 26–31.

LaBastille, A., D. G. Allen, and L. W. Durrell. 1972. Behavior and feather structure of the quetzal. *Auk* 89:339–348.

Lambourne, M. 1992. *Birds of the world*. Rizzoli, New York. (Includes reproductions of eleven 1st-edition and nineteen 2d-edition trogon plates by John Gould, with new text.)

Land, H. 1962. A collection of birds from the Sierra de las Minas, Guatemala. *Wilson Bulletin* 74:267–283.

———. 1963. A collection of birds from the Caribbean lowland of Guatemala. *Condor* 65:49–65.

———. 1970. *Birds of Guatemala*. Livingston Publications, Wynnewood, Penn.

Lawrence, W. F., and R. O. Bierregaard, eds. 1997. *Tropical forest remnants: Ecology, management, and conservation of fragmented communities*. University of Chicago Press, Chicago.

Louette, M. 1987. Additions and corrections to the avifauna of Zaire. *Bulletin of the British Ornithologists' Club* 107:137–143.

Maberly, C.T.A. 1939. Glimpse of the Narina trogon. *Ostrich* 10:36–42.

———. 1951. Some field notes on the Narina trogon. *Bokmakierie* 3:78–79.

MacKinnon, J., and K. Phillipps. 1993. *The birds of Borneo, Sumatra, Java, and Bali*. Oxford University Press, Oxford, U.K.

Mackworth-Praed, C. W., and C.H.B. Grant. 1952–1955. *Birds of eastern and north eastern Africa*. 2 vols. Longman, Green, London.

———. 1957–1960. *Birds of eastern and north eastern Africa*. 2 vols. Longman, London.

———. 1962–1963. *Birds of the southern third of Africa*. 2 vols. Longman, London.

———. 1970–1973. *Birds of west central and western Africa*. 2 vols. Longman, London.

Marshall, J. T., Jr. 1957. Birds of pine-oak woodland in southern Arizona and adjacent Mexico. *Pacific Coast Avifauna* 32:1–125.

Martin, P. S., C. R. Robins, and W. B. Head. 1954. Birds and biogeography of the Sierra de Tamaulipas, an isolated pine-oak habitat. *Wilson Bulletin* 66:38–57.

Maslow, J. E. 1986. *Bird of life, bird of death: A naturalist's journey through a land of political turmoil*. Simon and Schuster, N. Y.

Maurer, D. R., and R. J. Raikow. 1981. Appendicular myology, phylogeny, and classification of the avian order Coraciiformes (including Trogoniformes). *Annals of the Carnegie Museum* 50:417–434.

Mayr, E., and W. Phelps Jr. 1967. The origin of the bird fauna of the south Venezuelan highlands. *Bulletin of the American Museum of Natural History* 136:271–327.

Mayr, G. 1999. A new trogon from the Middle Oligocene of Céreste, France. *Auk* 116:427–434.

McClure, H. E. 1964. Avian bionomics in Malaya. Part 1, The avifauna above 5,000 feet altitude at Mount Brichang, Pahang. *Bird-Banding* 35:141–183.

Medway, Lord, and D. R. Wells. 1976. *The birds of the Malay Peninsula*. Vol. 5. H. F. and G. Witherby, London.

Miller, A. H. 1963. Seasonal activity and ecology of the avifauna of an American equatorial cloud forest. *University of California Publications in Zoology* 66:1–73.

Miller, W. de W. 1918. Relative length of the intestinal ceca in trogons. *Auk* 35:480.

Mindell, D. P., M. D. Sorenson, C. J. Huddleston, H. C. Miranda Jr., A. Knight, S. J. Sawchuk, and T. Yuri. 1997. In *Avian molecular evolution and systematics*, edited by D. Mindell, 213–224. Academic Press, San Diego, Calif.

Mitchell, M. M. 1957. *Observations on birds of southeastern Brazil*. University of Toronto Press, Toronto.

Moermond, T. C., and J. S. Denslow. 1985. Neotropical avian frugivores: Patterns of behavior, morphology, and nutrition, with consequences for food selection. In *Neotropical Ornithology*. Ornithological Monographs, no. 36, edited by P. Buckley et al., 865–897. American Ornithologists' Union, Washington, D.C.

Monroe, B. L., Jr. 1968. *A distributional survey of the birds of Honduras*. Ornithological Monographs, no. 7. American Ornithologists' Union, Washington, D.C.

Moreau, R. E., and W. E. Moreau. 1939. Observations on some West African birds. *Ibis*, ser. 14, 3:296–324.

Mountfort, G. 1988. *Rare birds of the world*. Stephen Greene Press, Lexington, Mass., and Collins, London.

Mourer-Chauviré, C. 1980. The Archaetrogonidae of the Eocene and Oligocene phosphorites of Quercy (France). *Contributions in Science, Los Angeles County Museum*, no. 330:17–31.

Naumburg, E.M.B. 1930. The birds of Matto Grosso, Brazil. *Bulletin of the American Museum of Natural History* 60:1–432.

Nicholson, I. 1967. *Mexican and Central American mythology.* Paul Hamlyn, London.

Oberholser, H. C. 1974. *The bird life of Texas.* Vol. 1. University of Texas Press, Austin.

Ogilvie-Grant, W. R. 1892. Catalogue of Trogones in the British Museum. In *Catalogue of the Piciariae in the collection of the British Museum,* edited by R. B. Sharpe, 17: 429–497. British Museum (Natural History), London.

Olrog, C. C. 1959. *Las aves Argentinas: Una Guia de Campos.* Institute "Miguel Lillo," Tucoman, Argentina.

Olson, S. L. 1975. Oligocene fossils bearing on the origins of the Todidae and Motmotidae (Aves: Coraciiformes). *Smithsonian Contributions to Paleobiology* 27:111–119.

———. 1985. The fossil record of birds. In *Avian biology,* edited by D. S. Farner and J. R. King, 8:79–252. Academic Press, N.Y.

Oriki, Y., and E. O. Willis. 1983. A study of the breeding birds of the Belim area, Brazil. Part 3. *Ciencia e Cultura* 35:1,320–1,324.

Parkes, K. C. 1970. A revision of the Philippine trogon (*Harpactes ardens*). *Natural History Bulletin of the Siam Society* 23:345–352.

Paynter, R. A. 1955. The ornithogeography of the Yucatan Peninsula. *Bulletin of the Peabody Museum of Natural History* (Yale University) 9:1–347.

Peters, J. L. 1929. The identity of *Trogon fulgidus. Auk* 46: 115–116.

———. 1945. *Check-list of birds of the world.* Vol. 5. Harvard University Press, Cambridge, Mass.

Peterson, R. T., and E. L. Chalif. 1973. *A field guide to Mexican Birds.* Houghton Mifflin, Boston, Mass.

Phelps, W. H., and W. H. Phelps Jr. 1958. Lista de las aves de Venezuela con su distribucion. Part 1, No Passeriformes. *Boletin de la Societad Venezolana. Ciencias Natural* 19 (90): 1–317.

Phillips, A. R., J. T. Marshall, and G. Monson. 1964. *Birds of Arizona.* University of Arizona Press, Tucson.

Pinto, O. 1950. Da classificaçao e nomenclatura dos surucuas Brasileiros (Trogonidae). *Papeis Avulsos Deptamento Zoologico* (São Paulo, Brazil) 9:89–136.

Powell, G.V.N., and R. Bork. 1995. Implications of infratropical migration on reserve design: A case study using *Pharomachrus mocinno. Conservation Biology* 92:354–362.

Prigogine, A. 1953. Contribution à l'études de la faune ornithologique de la région à l'ouest du lac Edourd. *Annales du Musée Congo Scientifique Zoologie* 8 (24): 1–117.

Rabor, D. S. 1977. *Philippine birds and mammals.* University of Philippines Press, Quezon City.

Raffaele, H., J. Wiley, K. Allen, and J. Raffaele. 1998. *A guide to the birds of the West Indies.* Christopher Helm, London.

Rand, A., and D. S. Rabor. 1952. Two new birds from the Philippine Islands. *Field Museum of Natural History, Miscellaneous Publications,* no. 107: 1–5.

———. 1960. Birds of the Philippine Islands: Siquijor, Mount Malindang, Bohol, and Samar. *Fieldiana: Zoology* 35:221–441.

Remsen, J. V., Jr., M. A. Hyde, and A. Chapman. 1993. The diets of Neotropical trogons, motmots, barbets, and toucans. *Condor* 95:178–192.

Richmond, C. W. 1893. Notes on a collection of birds from eastern Nicaragua and the Rio Frio, with a description of a supposed new trogon. *Proceedings of the U.S. National Museum* 16:479–532.

Ridgely, R. 1989. *A guide to the birds of Panama.* 2d ed. Princeton University Press, Princeton, N.J.

Ridgely, R. S., and J. A. Gwynne Jr. 1989. *A guide to the birds of Panama.* 2d ed. Princeton University Press, Princeton, N.J.

Ridgway, R. 1911. The birds of North and Middle America. Part 5. *Bulletin* of the *U.S. National Museum* 50: 1–859.

Riley, J. H. 1938. Birds from Siam and the Malay Peninsula at the USNM collected by H. M. Smith and W. L. Abbott. *Bulletin of the U.S. National Museum* 172:1–581.

Robbins, M. B., T. A. Parker, and S. E. Allen. 1985. The avifauna of Cerro Pirre, Darién, eastern Panama. In *Neotropical ornithology.* Ornithological Monographs, no. 36, edited by P. Buckley et al. 198–232. American Ornithologists' Union, Washington, D.C.

Robins, G. R., and W. B. Head. 1951. Bird notes from La Joya de Salsa, Tamaulipas. *Wilson Bulletin* 63:263–270.

Robinson, H. C. 1915. On birds collected by M. C. Boden Kloss, on the coast and islands of southeastern Siam. *Ibis,* ser. 10, 3:718–761.

———. 1927. *The birds of the Malay Peninsula.* Vol. 1. H. F. and G. Witherby, London.

———. 1928. *The birds of the Malay Peninsula.* Vol. 2. H. F. and G. Witherby, London.

Robinson, H. C., and F. M. Chasen. 1936. *The birds of the Malay Peninsula.* Vol. 3. H. F. and G. Witherby, London.

Robinson, H. C., and C. B. Kloss. 1919. On birds from South Annam and Cochin China. Part 1. *Ibis,* ser. 11, 1:392–452.

———. 1945. A distributional survey of the birds of Sonora, Mexico. *Occasional Papers, Museum of Zoology, Louisiana State University* 21:1–379.

Rowley, J. S. 1966. Breeding records of birds from the Sierra Madre del Sur, Oaxaca, Mexico. *Proceedings of the Western Foundation of Vertebrate Zoology* 1:107–204.

———. 1984. Breeding records of land birds in Oaxaca, Mexico. *Proceedings of the Western Foundation of Vertebrate Zoology* 2:73–224.

Ruschi, A. 1979. *Aves du Brasil.* Editoria Rios, São Paulo.

Russell, S. M. 1964. *A distributional study of the birds of British Honduras.* Ornithological Monographs, no. 1. American Ornithologists' Union, Washington, D.C.

Russell, S. M., and G. Monson. 1998. *The birds of Sonora.* University of Arizona Press, Tucson.

Rutgers, A. 1969. *Birds of Asia. Illustrations from the lithographs of John Gould.* Methuen, London. (Includes reproductions of three trogon plates from Gould 1858–1875, with new text.)

———. 1972. *Birds of South America. Illustrations from the lithographs of John Gould.* Methuen, London. (Includes reproductions of twenty-five trogon plates from Gould 1858–1875, with new text.)

Santana, C. E., and B. G. Milligan. 1984. Behavior of toucanets, bellbirds, and quetzals feeding on lauraceous fruits. *Biotropica* 16:15–24.

Santana C. E., T. C. Moermond, and J. S. Denslow. 1986 (actual pub. date 1988). Fruit selection in the collared aracari (*Pteroglossus torquatus*) and the slaty-tailed trogon (*Trogon massena*): Two birds with contrasting foraging modes. *Brenesia* 25–26:279–295.

Schäfer, E., and W. H. Phelps Jr. 1954. Las aves del Parque Nacional "Henri Pittier" (Rancho Grande) y sus funciones ecologicas. *Boletin de la Societad Venezolana Ciencias Natural* 16 (83): 3–167.

Schaldach, W. J., Jr. 1963. The avifauna of Colima and adjacent Jalisco, Mexico. *Proceedings of the Western Foundation of Vertebrate Zoology* 1:1–100.

Schaldach, W. J., Jr., B. P. Escalante, and K. Winker. 1997. Further notes on the avifauna of Oaxaca. *Anales del Instituto de Biologia Universitad Nacional Autonomia de Mexico* 68:91–136.

Schönwetter, M. 1966. *Handbuch der Oologie.* Vol. 1, no. 11. Akademic-Verlag, Berlin.

Schneider, A. 1938. Die Vogelbilder zur *Historia Naturalis Brasiliae* des George Marcgrave. *Journal für Ornithologie* 88:74–104.

Sclater, W. L., and R. E. Moreau. 1932. Taxonomic and field notes on some birds of north-eastern Tanganika Territory. *Ibis,* ser. 13, 2:656–683.

Sharpe, R. B. 1888. Further descriptions of new species of birds discovered by Mr. John Whitehead on the mountain of Kina Balu, north Borneo. *Ibis,* ser. 12, 6:383–396.

Shelton, L. N.d. Unpublished report. Houston Zoological Gardens, Houston, Texas.

Short, L. L. 1975. A zoogeographic analysis of the South American chaco avifauna. *Bulletin of the American Museum of Natural History* 154:163–362.

Sibley, C. G., and J. E. Alquist. 1973. A comparative study of the egg-white proteins of non-passerine birds. *Bulletin of the Peabody Museum of Natural History* (Yale University) 39:1–276.

———. 1990. *Phylogeny and classification of birds.* Yale University Press, New Haven, Conn.

Sibley, C. G., and B. L. Monroe Jr. 1990. *Distribution and taxonomy of the birds of the world.* Yale University Press, New Haven, Conn.

———. 1993. *A Supplement to distribution and taxonomy of the birds of the world.* Yale University Press, New Haven, Conn.

Sick, H. 1993. *Birds in Brazil.* Princeton University Press, Princeton, N.J.

Sinclair, I. 1993. *Illustrated guide to the birds of southern Africa.* New Holland Publications, London.

Skinner, E. H. 1901. Two years with Mexican birds. Part 2, In the haunts of the trogon. *Condor* 3:77–78.

Skutch, A. F. 1944. Life history of the quetzal. *Condor* 46:213–235.

———. 1945. The magnificent quetzal. *Nature Magazine* 38:299–302, 330–331.

———. 1948. Life history of the citreoline trogon. *Condor* 50:137–147.

———. 1953. The elusive Massena trogon. *Animal Kingdom* 56:167–172.

———. 1956. Life history of the collared trogon. *Auk* 73:354–366.

———. 1959a. Life history of the black-throated trogon. *Wilson Bulletin* 71:5–18.

———. 1959b. Trogons and wasps' nests. *Nature Magazine* 52:465–468, 500.

———. 1962. Life history of the white-tailed trogon. *Ibis* 104:301–313.

———. 1969. *Life histories of Central American birds.* Publications of the Nuttall Ornithological Club, no 3. Nuttall Ornithological Club, Cambridge, Mass.

————. 1971. *A naturalist in Costa Rica.* University of Florida Press, Gainesville.

————. 1972. *Studies of tropical American birds.* Publications of the Nuttall Ornithological Club, no. 10. Nuttall Ornithological Club, Cambridge, Mass. (Includes slaty-tailed and violaceous trogon accounts.)

————. 1976. *Parent birds and their young.* University of Texas Press, Austin.

————. 1981. *New studies of tropical American birds.* Publications of the Nuttall Ornithological Club, no. 19. Nuttall Ornithological Club, Cambridge, Mass. (Includes violaceous trogon account.)

————. 1982. Resplendent myth. *Audubon* 84 (5): 74–85.

————. 1983. *The birds of tropical America.* University of Texas Press, Austin. (Includes accounts of resplendent quetzal and Baird's, black-headed, mountain, collared, orange-bellied, and black-throated trogons.)

————. 1985. Clutch size, nesting success, and predation on nests of Neotropical birds, reviewed. In *Neotropical Ornithology.* Ornithological Monographs, no 36, ed. P. Buckley et al., 575–594. American Ornithologists' Union, Washington, D.C.

————. 1987. *A naturalist among tropical splendor.* University of Iowa Press, Iowa City.

————. 1999. *Trogons, laughing falcons, and other neotropical birds.* Texas A&M University Press, College Station. (Includes description of violaceous and slaty-tailed trogon nesting behavior.)

Slud, P. 1960. The birds of Finca "La Selva," Costa Rica: A tropical wet forest locality. *Bulletin of the American Museum of Natural History* 121:49–148.

————. 1964. The birds of Costa Rica: Distribution and ecology. *Bulletin of the American Museum of Natural History* 128:1–430.

————. 1976. Geographic and climatic relationships of avifaunas with special reference to comparative distribution in the Neotropics. *Smithsonian Contributions to Zoology* 212:1–149.

Smithe, F. B. 1966. *The birds of Tikal.* Natural History Press, Garden City, N.Y.

Smythes, B. E. 1953. *The birds of Burma.* 2d ed. Oliver and Boyd, Edinburgh, U.K.

————. 1968. *The birds of Borneo.* 2d ed. Oliver and Boyd, Edinburgh, U.K.

————. 1981. *The birds of Borneo.* 3d ed. Malayan Nature Society, Kuala Lumpur.

Snow, B., and D. Snow. 1988. *Birds and berries: A study of ecological interaction.* T. and A. D. Poyser, Carlton, U.K.

Snow, D., and D. K. Read. 1978. *An atlas of speciation in African non-passerine birds.* Publications of the British Museum (Natural History), no. 787. British Museum, London.

Snyder, D. E. 1966. *The birds of Guyana.* Peabody Museum, Salem, Mass.

Stager, K. E. 1957. The birds of the Tres Marías Islands, Mexico. *Auk* 74:413–432.

Stiles, F. G., and A. F. Skutch. 1989. *The birds of Costa Rica.* Cornell University Press, Ithaca, N.Y.

Stone, W. 1932. The birds of Honduras, with special reference to a collection made in 1930 by John T. Emlen Jr., and C. Brooke Worth. *Proceedings of the Philadelphia Academy of Natural Sciences* 84:291–342.

Storer, R. W. 1961. Two collections of birds from Campeche, Mexico. *University of Michigan Museum of Zoology, Occasional Papers,* no. 621:1–20.

Stotz, D. F., J. W. Fitzpatrick, T. A. Parker, and D. K. Mosovitz. 1996. *Neotropical birds: Ecology and conservation.* University of Chicago Press, Chicago.

Strauch, J. G. 1977. Further bird weights from Panama. *Bulletin of the British Ornithologists' Club* 97:61–65.

Sutton, G. M. 1951. *Mexican birds: First impressions.* University of Oklahoma Press, Norman.

————. 1972. *At a bend in a Mexican river.* Paul Eriksson, New York.

Sutton, G. M., and T. D. Burleigh. 1940. The birds of Tamazunchale, San Luis Potosí. *Wilson Bulletin* 52:221–233.

Taylor, R. C. 1978–1983. Unpublished periodic research reports on the elegant trogon. Files of the Coronado National Forest, Tucson, Ariz.

————. 1994. *Trogons of the Arizona borderlands.* Treasure Chest Publications, Tucson, Ariz.

Thompson, M. C. 1966. Birds from North Borneo. *University of Kansas Publications, Museum of Natural History* 17:377–433.

Thurber, W. A., J. F. Serrano, A. Sermeno, and M. Benitz. 1987. Status of uncommon and previously unreported birds of El Salvador. *Proceedings of the Western Foundation of Vertebrate Zoology* 3:109–293.

Todd, W.E.C. 1943. Critical remarks on the trogons. *Proceedings of the Biological Society of Washington* 56:3–15.

Todd, W.E.C., and M. A. Carriker Jr. 1922. Birds of the Santa Marta region of Colombia: A study in altitudinal variation. *Annals of the Carnegie Museum* 14:3–611.

Traylor, M. A. 1958. Birds of northeastern Peru. *Fieldiana: Zoology* 35:87–141.

van Marle, J. G., and K. H. Voous. 1988. *The birds of Sumatra: An annotated check-list.* B.O.U. Check-list no. 10. British Ornithologists' Union, London.

van Rossem, A. J. 1937. Critical remarks on Middle American birds. *Bulletin of the Museum of Comparative Zoology* (Harvard University) 77:424–490.

Viana, V. M., A.A.J. Tabanez, and J.L.F. Batista. 1997. Dynamics and restoration of forest fragments in the Brazilian Atlantic moist forest. In *Tropical forest remnants: Ecology, management, and conservation of fragmented communities,* edited by W. F. Lawrence and R. O. Bierregaard Jr., 351–365. University of Chicago Press, Chicago.

Webster, J. D. 1984. Richardson's Mexican collection: Birds from Zacatecas and adjoining states. *Condor* 86:204–207.

Wege, D. C., and A. J. Long. 1995. *Key areas for threatened birds in the neotropics.* Birdlife Conservation Series, no. 5. BirdLife International, Cambridge, and Smithsonian Institution Press, Washington., D.C.

Weidenfield, D. A., T. S. Schulenberg, and M. B. Robbins. 1985. Birds of a tropical deciduous forest in extreme northwestern Peru. In *Neotropical Ornithology.* Ornithological Monographs, no. 36, edited by P. Buckley et al., 305–315. American Ornithologists' Union, Washington, D.C.

Wells, D. R. 1998. *The birds of the Thai-Malaysian peninsula.* Vol. 1. Academic Press, London.

Weske, J. S. 1972. The distribution of the avifauna in the Apurimac Valley of Peru with respect to environmental gradients, habitat, and related species. Ph.D. diss., University of Oklahoma, Norman.

Wetmore, A. 1967. Further systematic notes on the avifauna of Panama. *Proceedings of the Biological Society of Washington* 80:229–242.

———. 1968. *The Birds of the Republic of Panamá.* Part 2. Smithsonian Institution Press, Washington, D.C.

Wetmore, A., and B. H. Swales. 1931. The birds of Haiti and the Dominican Republic. *Bulletin of the U.S. National Museum* 155:1–483.

Wheatley, N. 1995. *Where to find birds in South America.* Princeton University Press, Princeton, N.J.

———. 1996a. *Where to find birds in Asia.* Princeton University Press, Princeton, N.J.

———. 1996b. *Where to find birds in Africa.* Princeton University Press, Princeton, N.J.

Wheelwright, N. T. 1983. Fruits and the ecology of resplendent quetzals. *Auk* 100:286–301.

———. 1991. How long do fruit-eating birds stay in the plants where they feed? *Biotropica* 23:29–40.

Wheelwright, N. T., W. A. Haber, K. G, Murray, and C. Guindon. 1984. Tropical fruit-eating birds: A survey of a Costa Rican lower montane forest. *Biotropica* 16: 173–192.

Wildash, P. 1968. *Birds of South Vietnam.* Chas. Tuttle, Rutland, Vt.

Wille, C. 1993. The shrinking kingdom of the quetzal. *Living Bird Quarterly* 12 (3): 8–13.

Williams, S. O. 1992. The summer season, New Mexico. *American Birds* 46:1,162–1,165.

Willis, C. 1993. The shrinking kingdom of the quetzal. *Living Bird Quarterly* 12 (3):8–13.

Willis, E. O., and E. Eisenmann. 1979. A revised list of birds of Barro Colorado Island, Panama. *Smithsonian Miscellaneous Contributions to Zoology,* no. 291: 1–31.

Willis, E. O., D. Wechsler, and S. Kistler. 1982 (actual pub. date 1983). *Galbula albirostris* (Aves, Galbulidae), *Trogon rufus* (Trogonidae), and *Electron platyrhynchum* (Motmotidae) as army ant followers. *Revista Brasiliera Biologia* 42:761–766.

Winter, S. 1998. The elusive quetzal. *National Geographic* 164 (6): 35–45.

Wong, M. 1986. Trophic organization of understory birds in a Malaysian dipterocarp forest. *Auk* 103:100–116.

Zimmer, J. T. 1930. *Birds of the Marshall Field Peruvian expedition, 1922–23.* Field Museum of Natural History, Zoological Series, vol. 17, no. 7: 233–480.

———. 1949. Studies of Peruvian birds. No. 53, The family Trogonidae. *American Museum Novitates,* no. 1380:1–56.

Zimmer, J. T., and W. H. Phelps Jr. 1946. Twenty-three new subspecies of birds from Venezuela and Brazil. *American Museum Novitates,* no. 1312:1–23.

Zimmerman, D. A. 1978. Eared trogon–immigrant or visitor? *American Birds* 32:135–139.

Zimmerman, D. A., D. A. Turner, and D. J. Pearson. 1996. *Birds of Kenya and Northern Tanzania.* Princeton University Press, Princeton, N.J.

Zolta, A. R. 1939. Los Trogoniformes Argentina. *El Hornero* 7:125–139.

INDEX TO SCIENTIFIC AND VERNACULAR NAMES OF TROGONS AND QUETZALS

This index is limited to the scientific epithets (subspecies, species, and higher taxonomic categories) and English vernacular names used in this volume for the Trogoniformes. Alternate or obsolete vernacular names are cross-listed only to English species' names used here, and subspecies are indexed to the principal species account. **Page numbers in boldface** refer to the principal account of each species, genus, or higher taxon. *Page numbers in italics* point to any maps or figures that are outside the principal account of the taxon. "Pl." indicates the number of a color plate. Neither the front matter nor the appendix has been indexed here.

aequatoriale, Apaloderma, 18, **36–38**
African trogons, **27–42**
amazonicus, T. rufus, 139
ambiguus, T. elegans, 120
annamensis, H. erythrocephalus, 189
antisianus, Pharomachrus, 5, 17, **57–59**
Apaloderma, 8, **30–42,** *160*
Apaloderminae, **27–42**
Apalodermini, **27–42**
ardens, Harpactes, 17, 160, 172, **173–175**
Asian trogons, **156–194**
assimilis, T. personatus, 135
aurantius, T. surrucura, 144
aurantiventris, Trogon, 6, 17, **132–134**
auriceps, Pharomachrus, 5, 17, 59, **63–67**
australis, T. massena, 85

Bahia trogon. *See* Violaceous trogon
bairdii, Trogon, 6, 17, **100–103**
Baird's trogon, Pl.17, *16, 22, 95,* **100–103,** *113*
Bang's trogon. *See* Golden-headed quetzal
Bare-cheeked trogon, Pl.3, *22, 33, 35,* **36–38**
Bare-cheeked trogons, **30–42**
Bar-tailed trogon, Pl.4, **40–42**
Beautiful train-bearer. *See* Crested quetzal
behni, T. curucui, 147
Behn's trogon. *See* Blue-crowned trogon
Black-faced trogon. *See* Masked trogon
Black-headed citreoline trogon. *See* Black-headed trogon

Black-headed trogon, Pl.20, *11, 16, 22,* 109, 110, **111–114,** 152
Black-tailed trogon, Pl.14, *14, 16, 22,* **89–92,** *97, 98, 129*
Black-throated trogon, Pl.26, *16* 106, *129, 138,* **139–143,** *153*
Black-winged trogon. *See* White-tailed trogon
Blasius's trogon. *See* Diard's trogon
Blue-billed trogon. *See* Blue-tailed trogon
Blue-crowned trogon, Pl.28, *16, 146,* **147–149**
Blue-tailed trogon, Pl.29, **89–92,** *186. See also* White-eyed trogon
Bolivian trogon. *See* Blue-crowned trogon
bolivianus, T. curucui, 147
Booted trogon. *See* Violaceous trogon
braccatus, T. violaceus, 150
brachyurum, A. narina, 31
Brazilian trogon. *See* Surucua trogon

caligatus, T. violaceus, 150
Cameroon bar-tailed trogon. *See* Bar-tailed trogon
camerunensis, H. vittatus, 40
canescens, T. elegans, 120
casteneus, T. collaris, 127
Central Indian trogon. *See* Malabar trogon
Ceylon trogon. *See* Malabar trogon
Chapman's trogon. *See* Slaty-tailed trogon
chaseni, H. erythrocephalus, 189
chionurus, T. viridis, 104
chrysochloros, T. rufus, 139

ABOUT THE AUTHOR

Paul A. Johnsgard is author of more than fifty books, including *The Hummingbirds of North America* (1997) and *Grassland Grouse and Their Conservation* (2002), both also available from Smithsonian Institution Scholarly Press.

COVER CREDITS

FRONT COVER: Resplendant quetzal, male and female. Acrylic painting by John O'Neill.

BACK COVER: Blue-crowned trogon, males and female. Hand-colored lithograph by J. Gould, H. C. Richter, and W. Hart. Courtesy of Kenneth Spencer Research Library.

01 14